# Agroforestry
## Principles and Practices

### AP Dwivedi
MSc (Ag.), AIFC (Hons.)

## Oxford & IBH Publishing Co. Pvt. Ltd.
New Delhi
( *A Unit of* CBS Publishers & Distributors Pvt Ltd )

CBSPD

# CBS Publishers & Distributors Pvt Ltd
New Delhi • Bengaluru • Chennai • Kochi • Kolkata • Mumbai
Hyderabad • Jharkhand • Nagpur • Patna • Pune • Uttarakhand

**Agroforestry**
**Principles and Practices**

ISBN-13: 978-81-204-0703-9
ISBN-10: 81-204-0703-2

## OXFORD & IBH
New Delhi
*( A Unit of CBS Publishers & Distributors Pvt Ltd )*

Published by **Satish Kumar Jain** and produced by **Varun Jain** for

**CBS Publishers & Distributors** Pvt Ltd

4819/XI Prahlad Street, 24 Ansari Road, Daryaganj, New Delhi 110 002, India.
Ph: 011-23266838, 23289259    Website: www.cbspd.com
e-mail: delhi@cbspd.com

*Corporate Office:* 204 FIE, Industrial Area, Patparganj, Delhi 110 092
Ph: 011-4934 4934    Fax: 011-4934 4935
e-mail: publishing@cbspd.com; publicity@cbspd.com

*Branches*

- **Bengaluru:** Seema House 2975, 17th Cross, KR Road, Banasankari 2nd Stage, Bengaluru 560 070, Karnataka, India
  Ph: +91-80-26771678/79    Fax: +91-80-26771680    e-mail: bangalore@cbspd.com
- **Chennai:** 18/8B, Subbarayan Street, Shenoy Nagar, Chennai 600 030, Tamil Nadu, India
  Ph: +91-44-42032115, 26681266    e-mail: chennai@cbspd.com
- **Kochi:** 42/1325, 1326, Power House Road, Opp KSEB, Power House, Ernakulum Kochi 682 018, Kerala, India
  Ph: +91-484-4059061-65,67    Fax: +91-484-4059065    e-mail: kochi@cbspd.com
- **Kolkata:** 147, Hind Ceramics Compound, 1st Floor, Nilgunj Road, Belghoria, Kolkata-700056, West Bengal, India
  Ph: +033-25633055, 033-25633056    e-mail: kolkata@cbspd.com
- **Lucknow:** Basement, Khushnuma Complex, 7 Meerabai Marg (Behind Jawahar Bhawan), Lucknow-226001, UP India
  Ph: +0522-4000032    e-mail: tiwari.lucknow@cbspd.com
- **Mumbai:** PWD Shed, Gala no 25/26, Ramchandra Bhatt Marg, Next to JJ Hospital Gate no. 2, Opp. Union Bank of India, Noorbaug,
  Mumbai-400009, Maharashtra, India
  Ph: 022-66661880/89    e-mail: mumbai@cbspd.com

*Representatives*

| | | | | | |
|---|---|---|---|---|---|
| • Hyderabad | 0-9885175004 | • Jharkhand | 0-9811541605 | • Nagpur | 0-8692091830 |
| • Patna | 0-9334159340 | • Pune | 0-9664372571 | • Uttarakhand | 0-9716462459 |

*Printed at* Chaman Enterprises, Daryaganj, New Delhi, India

# Preface

Man's desire to live in peaceful co-existence in a community gave birth to settled agriculture. Increasing population necessitated bringing more and more land under the plough. Besides the horizontal, vertical expansion of agriculture continued to meet the increasing needs of the increasing population. Further, the pressure of the burgeoning population has forced man to seek unconventional methods of agriculture and utilise land to the maximum possible extent. Thus, in this quest of increasing productivity, the multi-tier system 'agroforestry' has assumed wider recognition. Although agroforestry is an old practice, the world over, the term and the science of agroforestry are comparatively of recent origin. Some of the agroforestry systems in which both agricultural and forestry crops are combined were previously included under farm forestry and range management practices. Shifting and taungya cultivations are old agroforestry systems which were in vogue and are practised in different forms even today.

The recent agroforestry systems involve a combination of crops and trees and sometimes animals in the land use in such a way so as to obtain maximum production of food, fodder, wood and other products and provide greater financial return. The combinations and crop types vary from region to region. It has been realised that monocultures involving either agricultural crops or forestry and forage crops do not give the maximum production. Agroforestry systems involving a mixture of agricultural crops, pasture and trees provide maximum production as they are able to tap solar radiation and soil more efficiently. A large number of combinations of trees and agricultural crops have already been developed through the experience of farmers and researchers. It is therefore, necessary to document all such information on combined production systems involving crops of agriculture, forestry, pasture, sericulture, apiculture, lac culture, etc. This publication is an effort in that direction. All existing agroforestry systems as practised in India have been discussed, including the crop composition and crop interaction. Several other factors, such as the criteria for the selection of suitable tree species, the interaction of trees with agricultural crops, their effect on the total yield and income of the farmers, management considerations, economics and ecological aspects, etc., which are relevant to the agroforesters, have also been discussed.

Tree planting by farmers in an organised manner is comparatively of recent development. With the rapid acceptance of agroforestry practises by farmers, the collection and collation of all available information on different types of tree species, their nature, rate of growth, their interaction with annual crops, yield and produce obtainable from them and marketing facilities available are of great significance. The silvicultural and general information of about 150 tree species commonly grown in different regions of India, given in the book will be useful to practising agroforesters and other interested in agroforestry.

Several colleagues of mine have helped me in the preparation of this book; I would like to thank all of them. I am especially grateful to Dr. R.V. Singh (Former Director General, ICFRE) and Dr. D.N. Tewari, Director General, ICFRE for encouraging me to write this book. I am thankful to my friends and colleagues Dr. R.M. Singhal, Dr. K.K. Sharma and Dr. G.N. Gupta for helping me in several ways. My thanks are also due to Tripti Chaudhari, Neelam Verma, Scaria, V.T., Sarita Mutha and Vinod Kumar Sahani who gave me their willing support and helped in proof reading. Lastly, I record my thanks to my wife Sumi, my daughters, Shuchi and Ruchi and son, Sameer who always provided me the moral support. I needed to carry through this venture.

Jodhpur                                                                 A.P. DWIVEDI
June 28, 1992

# Contents

# List of Plates

# CHAPTER 1
# Background

## Land

Air, water and land are basic resources for biological systems. Among these, the land is the most important basic resource for several biological production systems. The land and its soil profile support the plants and other living organisms. The land provides a medium and stores nutrients and water for the growth of plants and animals. Production of food grains, vegetables, fruits, firewood, spices, fodder, timber and other crops largely depends on the land area, type of soil, availability of water, technology and several other physical and socio-economic factors. The soil, which is the uppermost layer of earth crust is important for plant growth and production of several kinds of goods, e.g., agricultural, horticultural, forestry, etc.

Of the total land area of 13,300 million ha in the world, India has only 329 million ha, i.e., approximately 2.4 per cent of the world's area. India has, however, a large human and livestock population. According to the 1991 census, the human population in India is 845 million while that of livestock (1982 census) 416 million, which constitute about 15 per cent and 16 per cent of the world's human and livestock population respectively. Looking to the large human and livestock population, India is very poorly placed as far as land resource is concerned.

### LAND PROBLEMS

Though the total land area of the country is 329 million ha, all of it is not productive. There are several factors which limit the productivity of the land. Some of the important factors include: water stress, physiography, soil erosion, land degradation, floods, etc.

### (a) Water stress

For good growth of crops and plants, a sufficient supply of moisture is necessary. Most of the moisture is obtained from rain, dew, snow, etc. This moisture is retained by the soil as soil moisture or ground water and can be available to plants. About 70 per cent of the total land area in the country faces a varying degree of water stress. The area subject to wind erosion is about 32.0 million ha. It includes 18.8 million ha desert and about 6.5 million ha coastal sandy lands (NCA, 1976a). The desert area is spread over the states of Rajasthan (16.68 million ha), Gujarat (0.70 million ha) and Haryana (1.4 million ha). Coastal sandy lands fall in Orissa, West Bengal, Gujarat, Kerala, Karnataka, Maharashtra and Tamil Nadu. Nearly 2/3rd of the country's cropped area is rainfed. Failure of the monsoon, therefore, adversely affects the productivity of these lands. These lands are primarily used for production of coarse foodgrains, pulses, oilseeds, etc.

A large area of the country is prone to drought. Drought is a general term which implies a deficiency of precipitation of sufficient magnitude so as to affect adversely the agricultural production and, in turn, the economy. Drought causes water stress, which adversely affects the plants and crop yields. It also depletes the surface water and leads to drying of reservoirs, streams, lakes, rivers, etc. During the last one hundred years, the ten worst drought years have been recorded in which more than 50 per cent of the area of the country was affected. Such droughts occurred during 1899, 1901, 1918, 1920, 1941, 1951, 1965, 1966, 1979 and 1987. While widespread severe droughts occur once in a period of about 10 years, the monsoon fails repeatedly in arid and semi-arid areas causing drought conditions in these areas (Table 1).

**Table 1: Periodicity of drought in different parts of India (Anon, 1973)**

| | Area/Region | Periodicity of drought |
|---|---|---|
| 1. | Assam | Very rare, once in 50 years |
| 2. | West Bengal, Madhya Pradesh, Konkan, Andhra Pradesh, Karnataka, Maharashtra, Orissa, Bihar & Kerala | Once in 5 years |
| 3. | Eastern UP, South interior Karnataka, Vidarbha | Once in 4 years |
| 4. | Eastern Rajasthan, Gujarat, West UP, Rayalaseema, Tamil Nadu, Kashmir | Once in 3 years |
| 5. | Western Rajasthan | Once in 2.5 years |

The average productivity of rainfed arable land is low. The monsoon is not certain. There is always some problem in the annual rainfall either in its quantity or in its distribution. Drought causes reduction in agricultural production. It also reduces the availability of fodder. This brings down the national productivity level and also neutralises the efforts to improve the agriculture production per unit area (Anon., 1982; Anon., 1984).

**Plate 1.** Village wastelands in Uttar Pradesh

**Plate 2.** Eroded hill slopes

**Plate 3.     Floods cause severe losses**

**Plate 4.     Firewood shortage is a serious problem in rural India**

Plate 5.     Cowdung made as cakes for fuel

Plate 6.     Degraded forest areas

Plate 7.    Good forest areas provide environmental protection and supply forest products

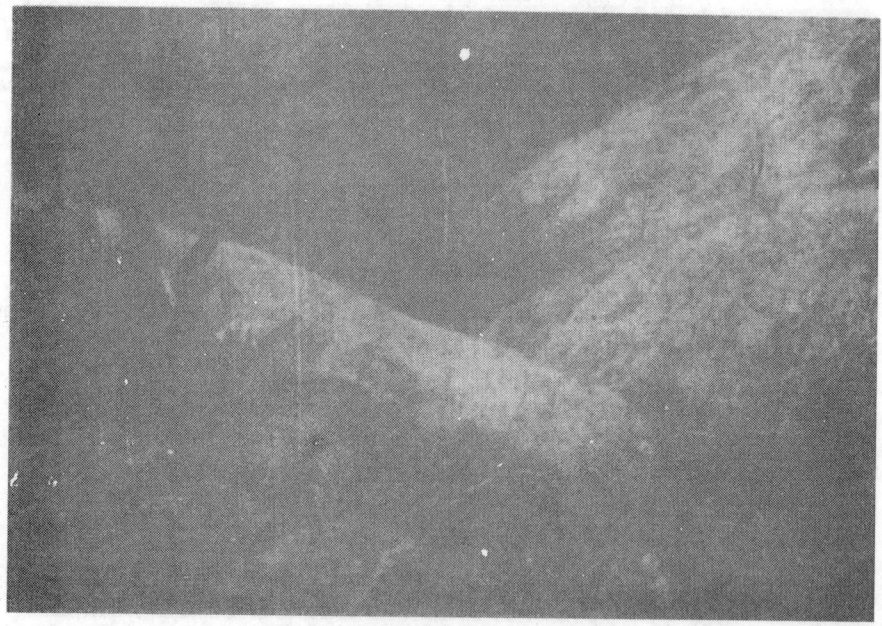

Plate 8.    Shifting cultivation in Mizoram

Plate 9.    Eucalyptus on field bunds

Plate 10.    Agrisilviculture system consisting of wheat + mango + poplar

**Plate 11.** Poplar ETP are being planted in crop lands

Plate 12.    Bund planting of *Syzygium cumini*

Plate 13.    *Grevillea robusta* on field bunds

Plate 14.    Bamboo, a useful tree in home gardens

Plate 15.    Heavy incidence of grazing in forests and grazing lands

Plate 16.    Goat grazing is detrimental to trees and shrubs

Plate 17.    Protection leads to abundant grass production for cut and carry

Plate 19. Fodder tress are severely lopped (*Grewia optiva*)

Plate 18. Camels are severe browsers

**Plate 20.** Heavy lopping causes damage to the forest *(Ailanthus excelsa)*

**Plate 21.** A medium dense windbreak provides protection to agricultural crops

Plate 22. Plantation of subabul on village roads

Plate 23. Plantation along canals

Plate 24.　Village woodlots of subabul, babul, sissoo, etc.

Plate 25.　Saline-alkaline areas

**Plate 26.    In waterlogged area mounds are prepared for planting**

**Plate 27.    No adverse effect of trees on crops is visible when they are young**

**Plate 28.** Eucalyptus trees cause adverse effect on agricultural crops in their vicinity.

**Plate 29.** Adverse effect of Eucalyptus is less when sufficient water is available

*(b) Physiography*

Productivity of the land is also controlled by the physiography, which emerges over long periods and is manifested in the form of mountains, plateaus and plains. The physiography controls altitude, soil depth, slope, drainage and various other soil parameters which are important for influencing the productivity. The water flow in rivers originating from mountains considerably influences the land productivity in the plains. Usually, mountains with high altitudes are associated with steep slopes, low soil depth, adverse climatic factors, etc., and therefore, have low productivity. In India, about 93.06 million ha of 329 million ha of geographical area, is mountainous. The distribution of these areas is given in Table 2 (Anon., 1988).

**Table 2: Distribution of montane areas in India**

| Sl. No. | Region | Area in million ha |
|---------|--------|-------------------|
| 1. | Himalayan region | 51.43 |
| 2. | Vindhyan region | 9.27 |
| 3. | Satpura ranges | 6.60 |
| 4. | Western ghats | 7.74 |
| 5. | Eastern ghats | 18.02 |
| | Total | 93.06 |

The productivity in these areas is comparatively low compared to the Indo-gangetic plain. The Satpura and Vindhyan hill ranges consist of hard rocks and the soils derived from them are low in fertility. Therefore, the productivity of these areas is also low. Similarly, the Eastern and Western ghats are montane areas with poor soil conditions. The upper reaches of the Himalayas above 3500 m support no significant vegetation.

*(c) Soil erosion problems*

About 150 million ha land area is subject to the problem of a varying degree of soil erosion (Das, 1981; Singh *et al.*, 1981; Anon., 1982). The total loss of surface soil is estimated to be about 6000 million tonnes with major plant nutrients of NPK varying from 5.37 to 8.4 million tonnes (Singh *et al.*, 1981; Anon., 1988). Wind erosion is common in an area of 32.0 million ha lying mainly in Rajasthan, Gujarat and Haryana. Various forms of water erosion, e.g., sheet, rill and gully, are prevalent in all land classes, particularly in agricultural lands of high rainfall areas. Various types of lands which are affected by soil erosion and their extent are given in Table 3 (Singh, *et al.*, 1981).

Due to soil erosion, there is a gradual loss in soil productivity. In highly eroded soils, tillage operations become difficult as there is formation of gullies of various dimensions. Soil conservation works have been taken up but

**Table 3: Area affected by soil erosion (million ha)**

| Land use | Total area | Problem area |
|---|---|---|
| **A. *Cultivable land*** | | |
| 1. Rainfed-non-paddy | 82.1 | 82.1 |
| 2. Current fallows | 14.3 | 3.5 |
| 3. Fallows other than current fallows | 9.6 | 4.8 |
| 4. Permanent pastures and other grazing grounds | 12.5 | 4.8 |
| 5. Miscellaneous tree crops & groves | 3.9 | 0.8 |
| 6. Culturable wastelands | 17.1 | 8.6 |
| Total of A | 139.5 | 104.6 |
| **B. *Forest lands*** | | |
| 1. Reserved forests | 39.0 | 3.9 |
| 2. Protected forests | 23.2 | 9.3 |
| 3. Unclassed forests | 12.6 | 6.3 |
| Total of B | 74.8 | 19.5 |
| **C. *Area not available for cultivation*** | | |
| 1. Area under non-agricultural use | 17.5 | 3.5 |
| 2. Barren and unculturable wastelands | 21.9 | 4.4 |
| Total of C | 39.4 | 7.9 |
| Total of A + B + C | 253.7 | 132.0 |

progress is slow. The state-wise area subject to soil erosion, land degradation and the area so far treated are given in Table 4.

*(d) Wastelands or degraded lands*

Wastelands refer to degraded lands which can be brought under vegetative cover with reasonable effort and which are currently underutilised and land which is deteriorating for lack of appropriate water and soil management or because of natural causes (NWDB, 1987). These lands can result from inherent/imposed disabilities such as location, environment, chemical and physical properties of the soil or financial and management constraints. These lands usually suffer from one or more problems such as: (i) low nutrient status, (ii) top soil completely removed due to soil erosion, (iii) difficult land surface due to formation of ravines, gullies, slides, etc., (iv) lack of moisture due to aridity, (v) development of toxicity in the soil, particularly in the root zone, (vi) poor physical conditions and (vii) waterlogging due to impeded drainage conditions.

NCA (1976) took a diagnostic view of wastelands and estimated that the area of such lands in the country is 175 million ha. While making the estimates, NCA included all such lands which were in need of attention and treatment. More recently, attempts were made to identify different kinds of wastelands. Various types of wastelands have been listed which include: (i) gullied and ravined land, (ii) upland with or without scrub, (iii) waterlogged or marshy areas, (iv) lands affected by salinity/alkalinity, both inland

**Table 4: State-wise estimated area subject to soil erosion, land degradation and area so far treated (Anon., 1986)**

(Area in thousand ha)

| State/Union Territories | Problem area | | Total | Area treated up to 1984–85 |
|---|---|---|---|---|
| | Soil erosion | Land degradation | | |
| Andhra Pradesh | 11,502 | 729 | 12,231 | 772 |
| Assam | 2,217 | 782 | 2,999 | 158 |
| Bihar | 4,260 | 2,292 | 6,552 | 902 |
| Gujarat | 9,946 | 2,640 | 12,586 | 2,283 |
| Haryana | 1,591 | 2,571 | 4,162 | 436 |
| Himachal Pradesh | 1,914 | — | 1,914 | 240 |
| Jammu & Kashmir | 883 | 10 | 893 | 187 |
| Karnataka | 10,989 | 414 | 11,403 | 3,108 |
| Kerala | 1,757 | 178 | 1,935 | 195 |
| Madhya Pradesh | 19,610 | 1,107 | 20,717 | 3,825 |
| Maharashtra | 19,181 | 665 | 19,846 | 9,550 |
| Manipur | 374 | 360 | 734 | 88 |
| Meghalaya | 837 | 265 | 1,102 | 91 |
| Nagaland | 405 | 77 | 482 | 81 |
| Orissa | 4,578 | 3,225 | 7,803 | 619 |
| Punjab | 1,007 | 2,223 | 3,230 | 583 |
| Rajasthan | 19,902 | 17,492 | 37,392 | 1,553 |
| Sikkim | 303 | — | 303 | 69 |
| Tamil Nadu | 3,640 | 182 | 3,822 | 1,081 |
| Tripura | 167 | 112 | 279 | 92 |
| Uttar Pradesh | 7,110 | 6,005 | 13,115 | 2,778 |
| West Bengal | 1,033 | 3,270 | 4,303 | 310 |
| Union Territories | 3,414 | 400 | 3,814 | — |
| Total | 1,26,620 | 44,999 | 1,71,619 | 29,383 |
| | | 1,465* | 1,465* | |
| | | 4,64,604 | 1,73,084 | |

*Coastal sandy area not reported by state.

and coastal, (v) shifting cultivation area, (vi) desertic and coastal sands and (vii) wastelands due to mining and industries (NWDB, 1987). The area under these wastelands is not precisely known. However, some of the estimates indicate that the area under such lands may be about 58 million ha as per details given in Table 5 (Singh *et al.*, 1981; Das, 1981; Anon., 1986).

The above list of wastelands does not include the area under uplands, with or without scrub, and wastelands due to mines and industries. If we take into consideration the wastelands under agriculture, forests and pastures, the total area may be approximately 100 million ha (Singh *et al.*, 1981, Anon., 1984a). The productivity of these lands is very poor. Some of these lands could be made productive with proper inputs and adopting proper management systems.

### Table 5:  Area under wastelands

| Category of wastelands | Area in million ha |
|---|---|
| 1.  Gullied and ravines | 4.0 |
| 2.  Alkali soils | 7.9 |
| 3.  Saline soils | 7.9 |
| 4.  Coastal sandy area | 6.5 |
| 5.  Shifting cultivation | 4.4 |
| 6.  Riverine land and torrents | 2.9 |
| 7.  Deserts | 18.8 |
| 8.  Waterlogged area | 6.0 |
| Total | 58.4 |

### (e) Floods

Several areas, particularly those located along big rivers such as the Ganga, the Yamuna, the Brahmaputra, the Godavari, the Mahanadi, the Krishna, the Narmada and their tributaries are flooded, which causes serious loss to crops, houses and other properties. Some of the river systems bring repeated floods along their banks. The total area affected by repeated floods in the country is about 34 million ha (Anon., 1978). Of this, 24 million ha lie in Uttar Pradesh, Bihar, West Bengal, Haryana and Punjab and the remaining 10 million ha lie in Assam, Orissa, Andhra Pradesh, Tamil Nadu and Madhya Pradesh. The main reasons for floods are: (i) shape of watershed, (ii) continuous high-intensity rainfall, (iii) high rate of run-off, (iv) siltation of river beds, (v) change of the course of rivers and (vi) lack of drainage facilities. The area affected and the loss due to floods in some of the past years are given in Taole 6 (Anon., 1978; Anon., 1980).

### Table 6:  Area affected and losses due to floods

| Year | Area affected (million ha) | Losses to crops and other properties (million Rs.) |
|---|---|---|
| 1971 | 13.25 | 9,319 |
| 1973 | 13.72 | 5,619 |
| 1975 | 6.15 | 4,711 |
| 1977 | 16.09 | 12,315 |
| 1978 | 10.41 | 14,369 |
| 1980 | 15.45 | 8,350 |

Of a flood affected area of 34 million ha, about 12.75 ha have been provided with reasonable protection. The Rashtriya Barh Ayog (National Commission on Floods) has estimated the total flood-prone area in the country to be 40 million ha, which also includes damage to protected areas (Anon., 1980). On the basis of data for the last few years, the average annual flood affected area is 9.0 million ha (Anon., 1988). The floods are not restricted to humid areas, but have also become frequent and devastating in arid and semi-arid areas of Rajasthan and Gujarat.

**Present Land Use**

Of the total geographic area of 329 million ha, land use statistics are available for 304.17 million ha. The present land use statistics and their progress over the last 30 years are indicated in Table 7 (Anon., 1988).

**Table 7: Land utilisation in India**

| Land utilisation classes | Area in million ha | | | |
|---|---|---|---|---|
| | 1950–51 | 1960–61 | 1970–71 | 1980–81 |
| 1. *Forests | 40.4 | 60.2 | 63.9 | 67.42 |
| 2. Land not available for cultivation (a+b) | 47.6 | 50.7 | 44.6 | 39.62 |
| (a) Land put to non-agricultural uses | 9.4 | 14.8 | 16.5 | 19.45 |
| (b) Barren and uncultivable land | 38.2 | 35.9 | 28.1 | 20.17 |
| 3. Other uncultivable lands excluding fallow land (a+b+c) | 49.3 | 37.2 | 35.1 | 32.23 |
| (a) Permanent pastures | 6.7 | 14.1 | 13.3 | 12.01 |
| (b) Land under miscellaneous tree crops and groves | 19.7 | 4.5 | 4.3 | 3.49 |
| (c) Cultivable waste | 22.9 | 18.6 | 17.5 | 16.73 |
| 4. Fallow land (a+b) | 28.1 | 21.6 | 19.4 | 24.63 |
| (a) Fallow other than current fallows | 17.4 | 10.5 | 8.8 | 9.82 |
| (b) Current fallows | 10.7 | 11.1 | 10.6 | 14.81 |
| 5. Net area sown (6–7) | 118.7 | 135.6 | 140.8 | 140.27 |
| 6. Total cropped area | 131.9 | 156.4 | 165.8 | 173.32 |
| 7. Area sown more than once | 13.2 | 20.8 | 25.0 | 33.05 |
| 8. *Reported area for land utilisation* | *284.1* | *305.3* | *303.8* | *304.17* |
| (i) Net irrigated area | 20.8 | 24.6 | 31.1 | 38.81 |
| (ii) Gross irrigated area | 22.6 | 27.9 | 38.2 | 49.58 |

*According to area statistics of forest departments, the area under forest lands is 74.89 million ha.

In the table given above, the reported area stands for the area for which data on land use classification are available either on the basis of land records/village papers or on the basis of *ad hoc* estimates. Area under *forests* includes all lands classed as forests under any legal enactment dealing with forest, whether state-owned or private and whether wooded or maintained as potential forest land. The area *not available for cultivation* covers all lands occupied by buildings, roads, railways or water, e.g., rivers, and other lands put to uses other than agriculture. This category of land comprises almost 40 million ha. The area under *grazing lands and pastures* includes all such lands which are primarily used for grazing. It includes village common grazing lands but does not include forest area used for grazing. The area under *miscellaneous tree crops and groves* includes all cultivable land which is not covered in the net area sown, but is put to some agricultural use. Lands under *Casuarina, Eucalyptus,* Bamboo, thatching grass, other tree groves for fuel, fodder, etc., which are not included under orchards, fall under this category of land use. *Culturable wastes* mean all lands available for cultivation, whether or not taken up for cultivation, but not cultivated during the current year and the last five years in succession. The land may be fallow or covered with bushes and jungle which are not put to use. *Fallow lands other than current fallow* applies to all lands which were taken up for cultivation but have been temporarily out of cultivation for a period of not less than one year and not more than five years. The reasons for keeping such lands fallow may be: (i) poverty of cultivators, (ii) inadequate supply of water, (iii) adverse climate, (iv) silting of canal and rivers, etc. *Current fallows* comprise cropped areas which are kept fallow during the current year. *Net area sown* represents the area sown with crops and orchards, counting the area sown more than once in the same year only once.

It can be seen that over the past three decades, the net area sown has increased by 22 million ha. If the fallow lands, which are temporarily out of cultivation, are also taken into consideration, the area under agriculture is 164.90 million ha.

It would appear that almost all such areas which could be cultivated have been taken up for agriculture and perhaps no more additional land will be available for agricultural purposes. The area under tree crops and groves, though remaining unchanged for the last 20 years, is likely to increase in view of increased prices of wood and scarcity of fuel and fodder. The land put to non-agricultural purposes has been on the increase. During 1950–51, only 9.4 million ha land were under this category which rose to 19.45 million ha in 1980–81. Human settlements alone account for 17.8 million ha of land (Anon., 1988). Due to increase in population, this will continue to increase and encroach upon the land under agricultural and other uses.

**Present Production**

*(i) Foodgrains*: Agriculture is an important sector in the country's economy. It provides a livelihood for about 70 per cent of the labour force, contributes nearly 35 per cent of the net national product and accounts for a sizable share of the total value of exports. Per capita net availability of foodgrains was up to a level of 478 gms/day in 1986 as compared to 395 gms/day in the fifties. The foodgrain production has been showing an increasing trend due to use of improved seeds, fertilisers, increase in irrigation and other inputs. Table 8 shows some of the key indicators of agricultural progress

(Anon., 1988a). It would appear from this table that agricultural production has been increasing continuously. There has been a proportionate increase in the use of fertilisers, improved seeds, availability of credits, etc.

The production of foodgrain is such that the country is not only able to meet domestic requirements but is also able to export some rice and wheat to other countries. During 1986, about 250,000 tonnes of wheat were exported to Nepal, Vietnam and some other countries. However, there is a shortage of oilseeds and pulses.

*(ii) Fodder:* Production of sufficient fodder to meet the demand of a high livestock population is one of the goals of land use. At present, the important sources of fodder are agricultural crop residues, grass and grazing, agricultural green fodders, tree leaf fodders, etc. Among agricultural residues, the important ones are straw of cereals, such as wheat, paddy, barley, maize, millets, pulses, etc. Grass and grazing also constitute important sources of fodder in India. Several kinds of areas, e.g. forests, pasture and grazing lands, culturable wastelands, fallow lands, lands under miscellaneous tree crops, etc., which constitute about 136.99 million ha provide grazing facility. Agricultural green fodders are grown over an area of 4 per cent of the total cultivated area. Of this, 20 per cent is under irrigated fodder production. The present production of fodder in India is estimated as under (Table 9).

*(iii) Firewood and timber:* Firewood occupies a predominant position as an energy source in rural India. As against the estimated requirement of about 157 million tonnes of fuelwood per annum, the supply is about 40 million tonnes, of which, recorded production is only 15 million tonnes (Anon., 1988b). Because of scarcity of firewood, 73 million tonnes of cow-dung and 41 million tonnes of agricultural residues are burnt as fuel (Anon., 1979; Anon., 1981). The constructional and industrial timber which are important forest produce are also not in adequate supply. The present recorded production of timber is about 12.5 million $m^3$ as against the requirement of 27.5 million $m^3$ (Anon., 1988b). The country is already importing a large quantity of timber logs and paper pulp to meet the domestic needs.

**Population**

India has the second largest human population in the world, only next to China. According to the 1981 census, the total population of the country is 685.17 million which was only 361.13 million during 1950–51 (Table 10).

Of the rural population, 51 per cent represent cultivators, 30 per cent agricultural workers and 19 per cent other workers. Thus, almost 81 per cent population is engaged in agriculture.

India is primarily an agricultural country where more than 80 per cent of the population depend upon agriculture and animal husbandry for their

**Table 8: Some key indicators of agricultural production**

| Item/Unit | 1950–51 | 1960–61 | 1970–71 | 1980–81 | 1981–82 | 1982–83 | 1983–84 | 1984–85 | 1985–86 | 1986–87 |
|---|---|---|---|---|---|---|---|---|---|---|
| 1) FOODGRAINS PRODUCTION (lakh tonnes) | | | | | | | | | | |
| i) Rice | 206 | 346 | 422 | 536 | 532 | 471 | 601 | 583 | 641 | 600 |
| ii) Wheat | 65 | 110 | 238 | 363 | 375 | 428 | 455 | 441 | 469 | 490 |
| iii) Other Cereals | 153 | 237 | 306 | 291 | 311 | 278 | 339 | 312 | 265 | 290 |
| iv) Total Cereals | 424 | 693 | 966 | 1,190 | 1,218 | 1,177 | 1,395 | 1,336 | 1,375 | 1,380 |
| v) Total Pulses | 84 | 127 | 118 | 106 | 115 | 118 | 129 | 119 | 130 | 130 |
| vi) Total Foodgrains | 508 | 820 | 1,084 | 1,296 | 1,333 | 1,295 | 1,524 | 1,455 | 1,505 | 1,510 |
| 2) INPUTS | | | | | | | | | | |
| (a) Seed | | | | | | | | | | |
| i) Production of Breeder seed (thousand qtls) | — | — | — | 5.27 | 3.91 | 17.07 | 30.00 | 29.03 | 23.64 (p) | 24.84 |
| ii) Production of Certified seed (lakh qtls) | — | — | — | 21.86 | 24.18 | 36.61 | 41.26 | 49.97 | 64.84 | 56.51 |
| iii) Distribution of Certified/Quality seed (lakh qtls) | — | — | — | 25.01 | 29.81 | 42.06 | 44.97 | 48.46 | 55.01 (p) | 55.83 |
| (b) Fertiliser consumption (NPK) | | | | | | | | | | |
| i) Total (lakh tonnes) | 0.69 | 2.92 | 21.77 | 55.16 | 60.64 | 63.88 | 77.10 | 82.11 | 87.37 | 92.00 |
| ii) Per hectare (kg) | Negligible | 1.9 | 13.33 | 31.83 | 34.25 | 37.00 | 44.66 | 47.56 | 50.61(p) | N.A. |
| (c) Area under High-yielding varieties (Lakh ha) | — | 18.9 | 153.3 | 430.7 | 464.9 | 474.3 | 537.4 | 541.4 | 554.2 | 540.4 |
| (d) Cooperative credit disbursed (Rs. crores) | 24.23 | 214.35 | 678.79 (1) | 2,126.31 | 2,478.95 (p) | 2,765.88 (p) | 2,905.34 (p) | 2,995.99 (p) | 3,206.06 (p) | N.A. |

[1] Relates to 1966–67; N.A.—Not available; p—Provisional.

Table 9: **Estimated fodder production in India (Anon., 1986a)**

| Type of fodder | Estimated production (million tonnes) | |
|---|---|---|
| | Dry fodder | Green fodder |
| 1. Agricultural crop residue | 236 | — |
| 2. Grass | 205 | — |
| 3. Green fodder | | |
| (a) Cultivated green fodder | — | 208 |
| (b) Top feed including sugar-cane tops | — | 4 |
| (c) Weeds | — | 14 |
| 4. Leaf fodder from trees | — | 24 |
| Total | 441 | 250 |

Table 10: **Population in India (in millions)**

| Year | Rural population | Urban population | Total |
|---|---|---|---|
| 1951 | 298.15 | 62.98 | 361.13 |
| 1961 | 359.77 | 78.24 | 438.01 |
| 1971 | 439.10 | 109.10 | 548.20 |
| 1981 | 525.46 | 159.71 | 685.17 |

livelihood. Livestock is an important component in the rural economy. India has about 1/6 of the world livestock population. As per the census carried out in 1982, the total livestock population was 416 million. The livestock population has also been increasing steadily (Table 11).

Table 11: **Livestock population in India (in millions)**

| Sl. No. | Category of livestock | 1951 | 1961 | 1972 | 1982 |
|---|---|---|---|---|---|
| 1. | Cattle | 155.3 | 175.6 | 178.3 | 190.8 |
| 2. | Buffaloes | 43.4 | 51.2 | 57.4 | 69.0 |
| 3. | Sheep | 38.4 | 40.0 | 40.0 | 48.1 |
| 4. | Goats | 47.1 | 60.9 | 67.5 | 94.7 |
| 5. | Pigs | 4.4 | 5.2 | 6.9 | 9.5 |
| 6. | Horses & ponies | 1.5 | 1.3 | 0.9 | 0.9 |
| 7. | Camels | 0.6 | 0.9 | 1.1 | 1.0 |
| 8. | Other livestock | 1.3 | 1.2 | 1.1 | 1.8 |
| | Total | 292.0 | 336.3 | 353.2 | 415.8 |

## Future Projections

Growth of population is an important determinant of total demand. Several projections of the future population are available (Raghavachari, 1974).

Based on fertility assumptions, the estimated population of India in 1991 and 2001 is presented in Table 12 (Raghavachari, 1974; NCA, 1976a).

**Table 12: Projected population under different fertility assumptions**

| Fertility assumptions* | Population (in millions) | |
| --- | --- | --- |
| | 1991 | 2001 |
| High–2 | 837.6 | 1032.1 |
| High–1 | 817.7 | 996.3 |
| Medium–2 | 801.2 | 945.4 |
| Medium–1 | 786.2 | 924.3 |
| Low–2 | 751.3 | 846.4 |
| Low–1 | 736.5 | 830.6 |

* High, medium and low fertility assumptions correspond to a birth rate of about 50, 35 and 20 per thousand respectively.

The estimates for 1981 were 668.2 million and 677.5 million for high-1 and high-2 growth assumptions respectively, which were very near to the exact population of 685, which indicates that the population of India should be somewhere about 995 million during the year 2001 (NCA, 1976a).

The livestock population has also been continuously increasing. The population of livestock for the years 1990 and 2000 has been projected as under (Table 13).

**Table 13: Projected population of livestock (Anon., 1987a)**

| Category | Population (in millions) | |
| --- | --- | --- |
| | 1990 | 2000 |
| Cattle | 202.0 | 218.0 |
| Buffaloes | 81.0 | 100.0 |
| Sheep | 51.2 | 55.1 |
| Goats | 106.8 | 131.8 |
| Camels and others | 2.9 | 3.0 |
| Total | 443.9 | 507.9 |

The basic requirements of food, fodder, fuel, shelter, clothing, etc., for this large human and livestock population have to be worked out. NCA (1976a) made estimates of some of the requirements of agricultural commodities for the year 2000 as under (Table 14).

The requirement for fodder is also very important. Various estimates of the requirements of feed and fodder for livestock are available. NCA (1976b), on the basis of average rate of feeding worked out the requirements of green fodder, dry fodder and concentrates for the projected livestock population of 2000 A.D. as 590, 373 and 70 million tonnes respectively.

Table 14: Demand for selected agricultural commodities (million tonnes)

| Item | *Projected requirement for 2000 A.D. | |
|---|---|---|
| | Low | High |
| 1. Foodgrains | 205.3 | 225.1 |
| 2. Sugar and gur | 24.0 | 29.9 |
| 3. Oilseeds | 8.0 | 10.0 |
| 4. Cotton (million bales) | 10.4 | 17.2 |

*These are exclusive of export demands.

On the basis of fodder requirement for different categories of livestock, recommended by the National Dairy Research Institute, the total requirement works out to be as under (Table 15).

Table 15: Estimated fodder requirement for 2000 A.D. (Anon., 1987a)

| Item | Fodder requirement (million tonnes) |
|---|---|
| Dry fodder | 949 |
| Green fodder | 1136 |

The above estimates are considerably higher than the estimates made by the National Commission on Agriculture.

Fuelwood and timber are also the basic necessities of life. Fuelwood is important domestic energy source for cooking, heating, etc., in rural India. Timber is important for construction and for various industries as a raw material. The demand for wood and wood products is going to increase due to increase in population and standard of living of the people. The projected demand for timber, fuelwood and bamboo is given in Table 16.

Table 16: Projected demand for forest products (NCA, 1976b; Anon., 1981; Anon., 1988b)

| Item | Projected demand for 2000 A.D. (million m³) |
|---|---|
| 1. Firewood | 300 |
| 2. Industrial timber | 65 |
| 3. Bamboo* | 35 |

*The figures of bamboo are in million tonnes.

There are similar other projections. Vohra (1986) estimates the need of firewood for 2000 at 325 million m³. The Advisory Board of Energy also puts this figure as 300 million m³ (Anon., 1982b). It can, therefore, be concluded that the firewood need is quite large, is not going to be met from the present production levels and concerted efforts are urgently required.

If we look at the demand position for 2000 A.D. and the present production of foodgrains, fodder, fuel and other agricultural and forestry products, the situation is not at all satisfactory. There is no need to be alarmed but there is no room for complacency either. We have to find ways and means to maximise production so that increasing demands are met.

The projected demands of food, fodder, firewood, timber, etc., can be met from the existing land resource of the country but with the following conditions:

  (i) The demand for agricultural commodities can be met from existing arable lands with improved management.
 (ii) The demand of firewood and timber can be met substantially by increasing the productivity of existing forest lands and partially bringing more areas under forest vegetation. Agroforestry practices in arable lands can help substantially in meeting the demand.
(iii) The demand of fodder can also be met by fodder cultivation in arable lands and practising silvipasture and agroforestry systems in grazing lands and wastelands.
 (iv) The production systems have to be so designed that the land is able to produce the maximum on a sustained basis without deterioration of the land.
  (v) The lands classified as 'other uncultivated lands' are likely to be under maximum pressure from competing and conflicting land use demands. Some land use adjustments according to land suitabilities might be possible in lands presently under agricultural and non-agricultural uses.
 (vi) The most productive lands are to be utilised for the production of agricultural crops.

**Land-use Planning**

For meeting the projected demands for 2000 A.D. one way could be the procurement of additional land for agriculture, forestry and fodder production. Some of the wastelands could certainly be reclaimed for agriculture, forestry and range land development. Experience in wasteland development, however, indicates that proper restoration of productivity is a long-term process and substantial gain is not available immediately. Secondly, the process of wasteland development requires heavy investment which is not easily available.

Another way of increasing the production could be to increase the inputs per unit area. There is certainly some scope for increasing the production but there are several limitations. One subject which has not been properly attended to in India is proper land use planning and an efficient production system.

Man's demand for food from the land has increased considerably. The productive capacity of the land is limited. Also, some areas converted for agriculture are not suitable and have caused serious land degradation. In several areas, reduction in quality and quantity of land under agriculture and forestry is being observed. The land use pattern may vary from area to area depending upon the land characteristics and the climate. It has also to be seen that the land has a limited population-supporting capacity, beyond which, there will be degradation and irreversible loss of productivity because of improper and excessive use.

NCA (1976c), regarding land use observes that: (i) A primary concern of the land use policy should be to continuously increase the productive capacity of the land and to prevent its deterioration. Soil conservation should be looked upon as an integral part for the programmes for maximising land use. (ii) Land use should be optimised by putting it to the best use as is consistent with ecology and capability of the land. Good agricultural land should not ordinarily be diverted to other uses. The land use planning should be based on a resources survey and production potential of the land. (iii) Diversification in land use should be encouraged by introducing various cropping patterns including pulses, oilseeds, fodder, etc. (iv) For diversifying production, increasing returns from the land, employment and income and ensuring balanced supply of food and fodder, the farmer should be encouraged to take up mixed farming, dovetailing cultivation with subsidiary occupations such as animal husbandry, poultry, pisciculture, sericulture, apiculture, etc., to provide year-round use of resources and labour. (v) The cropping patterns should be so structured as to suit the rainfall and soil conditions. The main policy regarding crops should be to evolve technologies to secure substantial increase in yields. Considerable emphasis should be given to grow crops and plants resistant to insect pests and diseases, droughts, floods, etc. (vi) The domestic needs of the people for timber, firewood and fodder should be met from forest and social forestry programmes.

The land should be under such a production system which does not deteriorate the land and produces the most. Most annual crops require intensive soil working and addition of manures and fertilisers as there is considerable drain of nutrients. In most of the agricultural lands, there is considerable soil erosion. Therefore, several farming systems aim at providing effective soil cover and providing soil conservation measures. Sometimes the lands are left fallow to regain the productivity. In some farming systems, annual and perennial crops are combined. The perennial crops, e.g., shrubs, trees, etc., help to conserve soil productivity, meet the demand of timber and firewood and bring about ecological restoration. Agroforestry systems hold promise to provide such land use systems. The production of foodgrains, fodder, fuel, etc., can be obtained simultaneously. The total production-mix can be selected depending upon the area, climate, demand, etc.

CHAPTER 2

# Features of Indian Agriculture and Forestry

## Agriculture

Agriculture in India is the most important sector. It has the responsibility of producing enough foodgrains, oilseeds, cotton and jute, sugar, fruits, nuts, tea, coffee, spices, fodder, etc., not only for meeting a large domestic requirement, but also for export. Of 525.26 million rural population, 51 per cent cultivators and 30 per cent agricultural workers depend entirely on agriculture. During 1985–86 agriculture contributed about Rs. 214,500 million, i.e., about 35 per cent in the total net domestic product at a factor cost of Rs. 605,480 million (Anon., 1986). This sector is emerging as an important sector for export as is evident from Table 17 (Anon., 1988d).

**Table 17: Imports and exports of foodgrains**

(Thousand tonnes)

|        | Imports | | | Exports | | |
|--------|------|-------|-------|------|-------------------------------------|-------|
|        | Rice | Wheat | Total | Rice | Wheat and wheat products (Flour) | Total |
| 1981–82 | 78.3  | 2,113.7 | 2,192.0 | 332.95 | —      | 332.95 |
| 1982–83 | —     | 1,952.1 | 1,952.1 | 361.43 | 105.02 | 466.45 |
| 1983–84 | 465.5 | 3,738.8 | 4,204.3 | —      | 23.40  | 23.40  |
| 1984–85 | 380.9 | 689.5   | 1,070.4 | —      | 32.41  | 32.41  |
| 1985–86 | 10.0  | —       | 10.0    | —      | 336.95 | 336.95 |
| 1986–87 | —     | —       | —       | —      | 439.65 | 439.65 |
| 1987–88 | —     | —       | —       | —      | 320.62 | 320.62 |

Some features of Indian agriculture which are important from the point of production, land use, land degradation, cropping pattern, soil management, etc., and have relevance in the study of agroforestry are discussed below.

## 1. PRODUCTION SYSTEM

Except for some areas in the north-east, most agricultural area is under settled agriculture. Most farms in India practise *mixed farming*, i.e., farmers grow agriculture and also rear animals. Sometimes they also adopt piggery, poultry, etc. A large number of small farms practise sustenance agriculture. Agriculture meets only their self needs. Commercial and economic angles are not realised. Pure agriculture, horticulture and dairy farms are common near urban areas. A large number of farmers grow several crops together, i.e., practise *mixed cropping*, particularly in rainfed areas. This ensures some crop being successful in the event of irregular monsoon. It also helps in meeting domestic divergent needs of the family.

The farming practises followed in the country can be grouped into three broad categories; (i) shifting cultivation, (ii) sedentary peasant farming and (iii) capitalist farming. *Shifting cultivation* is practised in small areas in Orissa, Andhra Pradesh, Madhya Pradesh, Assam, Meghalaya, Nagaland, Manipur, Tripura, Mizoram and Arunachal Pradesh in forested areas by the tribals. *Sedentary peasant farming* is practised by the majority of the farmers in the country. This type of farming covers most of the cropped area and accounts for most of the agricultural production. It is characteristic of small holdings and most operations are performed with the help of draught animals. *Capitalist farming* is highly capital intensive and practised on large farms or estates with the help of modern machines. Plantations of tea, coffee, mechanised farms, etc., can be grouped in this category.

## 2. SOIL CHARACTERISTICS

Most agricultural lands in India are deficient in organic matter, nitrogen and phosphorus and comparatively rich in potassium. The alluvial soil in the Indo-Gangetic basin is the richest foodgrain-producing tract in the country. The soils being alluvial suffer from several limitations. The texture is coarse, cohesion among the particles is little and soil erosion is rampant. Almost all ravined areas are located in this region. The alluvial soils differ greatly in texture and structure. They range from sands and loams to silts and clays that are ill-drained and charged with injurious sodium salts. Alluvial soils are suitable for growing almost all kinds of crops. *Black cotton soils* are derived from trap rocks and occur in Madhya Pradesh, Maharashtra, Andhra Pradesh and Gujarat. These soils are clayey in texture and black in colour. They are sticky when wet and form deep cracks when dry. These soils produce good crops of cotton, jowar, arhar, cereals, oilseeds, citrus, etc. *Red and yellow soils* occur in Madhya Pradesh, Orissa, Andhra Pradesh, Tamil Nadu, Karnataka, Bihar and West Bengal. These soils are suitable for cultivation of rice, jowar, ragi, tobacco, millets, etc. *Lateritic soils* occur in high rainfall areas of the eastern and western ghats. These are good for growing rice. *Desert soils* are confined in arid areas of Rajasthan, Gujarat and Haryana. These soils are usually sandy and possess a high concentration of sodium salts. They are poor in nitrogen and organic matter and relatively better in phosphorus. A large area is occupied by shifting sand dunes. Moisture is the main limiting factor for plant growth. Where irrigation is possible, good yields of crops are obtained. *Montane or hill soils* occur in the Himalayas and other hills. These soils are characterised by the presence of a high amount of organic matter and low pH. The montane soils, e.g., brown earths and podsols, usually possess a prominent layer of undecomposed leaf litter. Other soil groups include skeletal soils, peaty and organic soils, etc., which present several limitations for plant growth. The total area distributed under these soil groups is given in Table 18.

**Table 18: Area under board soil groups in India (Anon., 1986).**

| Major soil groups | Area (million ha) | Percentage |
|---|---|---|
| 1. Red and yellow soils | 105.5 | 32.6 |
| 2. Lateritic soils | 11.7 | 3.6 |
| 3. Black cotton soils | 64.4 | 19.9 |
| 4. Mixed red and black soils | 14.8 | 4.6 |
| 5. Alluvial soils | 72.2 | 22.3 |
| 6. Desert soils | 21.9 | 6.7 |
| 7. Montane soils | 28.8 | 8.9 |
| 8. Other soil groups | 4.6 | 1.4 |
| Total | 323.9 | 100.0 |

## 3. CROPPING SYSTEM

A wide variety of crops are grown in the country. These crops can be grouped into; (i) foodgrain crops, (ii) commercial crops or cash crops, (iii) horticulture crops, (iv) plantation crops and (v) fodder crops. Among food crops, the important crops include cereals, e.g., rice, wheat, jowar, bajra, maize, barley, etc.; pulses, e.g., arhar, gram, black gram, peas; and oilseeds, e.g., groundnut, mustard, linseed, rape seed, sesame, etc. The area and production of some of the important food crops are given in Table 19.

Several cash crops, e.g., sugar-cane, cotton, jute, tobacco, etc., are also grown on a large scale in different regions of the country. Sugar-cane is grown in Uttar Pradesh, Bihar, Punjab, Haryana, Maharashtra, Karnataka and Andhra Pradesh. Cotton is grown widely in Gujarat, Maharashtra, Karnataka, Punjab and Haryana. Jute is a common crop of West Bengal, Assam, Bihar and Orissa. Tobacco is commonly grown in Andhra Pradesh and Gujarat. The total area and production of some of the commercial crops are given in Table 20.

Table 19: Area and production of some food crops as of 1985—86 (Anon., 1986)

| Crops | Area (million ha) | Production (million tonnes) | Yield (kg/ha) |
|---|---|---|---|
| 1. Rice | 40.9 | 64.15 | 1,568 |
| 2. Wheat | 23.1 | 46.88 | 2,032 |
| 3. Jowar | 15.78 | 10.12 | 641 |
| 4. Bajra | 10.70 | 3.70 | 345 |
| 5. Maize | 5.88 | 6.89 | 1,172 |
| Total cereals | 103.25 | 137.50 | 1,332 |
| Total pulses | 23.81 | 12.96 | 499 |
| Total foodgrains | 127.08 | 150.46 | 1,184 |
| Total oilseeds | 18.87 | 11.13 | 591 |

Table 20: Area and production of some commercial crops as of 1985—86 (Anon., 1986)

| Crops | Area (million ha) | Production (million tonnes) | Yield (kg/ha) |
|---|---|---|---|
| Sugar-cane | 2.86 | 171.68 | 59,986 |
| Cotton | 7.58 | 8.61 | 193 |
| Jute | 1.14 | 10.95 | 1,717 |
| Tobacco | 0.40 | 0.43 | 1,097 |

Plantation crops include tea, coffee, cocoa, coconut, arecanut, rubber and spices, e.g., pepper, cardamum, chillis, turmeric, ginger, etc. Tea is grown mainly in Assam, West Bengal, Tamil Nadu and Kerala. Coffee is grown in Karnataka, Kerala and Tamil Nadu. Coconut and arecanut are grown in all coastal areas mainly in Kerala, Tamil Nadu, Karnataka, Andhra Pradesh and Assam. Rubber is grown mainly in Kerala and to a small extent in Karnataka and Tamil Nadu. The area and production of some of these crops are given in Table 21.

Table 21: Area and production of crops as of 1985—86 (Anon., 1986)

| Crops | Area (thousand ha) | Production (thousand tonnes) |
|---|---|---|
| Tea | 398 | 657 |
| Coffee | 235 | 209 |
| Coconut | 1,126 | 612 |
| Rubber | 363 | 201 |
| Arecanut | 193 | 166 |
| Black pepper | 121 | 36 |
| Cardamum | 115 | 9 |

Garden crops or horticulture crops include mainly fruits and vegetable crops. Important fruit crops are mango, banana, citrus, apple, guava, grapes, pineapple, peaches, plums, etc. Potato, tomato, onion, cabbage, cauliflower, brinjal, etc., are important vegetable crops. The area and production of some of the important fruit and vegetable crops are given in Table 22.

**Table 22: Area and production of important horticulture crops as of 1986—87 (Anon., 1988d)**

| Crops | Area (thousand ha) | Production (thousand tonnes) |
|---|---|---|
| Mango | 1022.22 | 10,113.33 |
| Banana | 317.60 | 5,808.03 |
| Citrus | 262.32 | 2,530.68 |
| Apple | 176.29 | 1,311.37 |
| Guava | 179.54 | 1,159.79 |
| Pineapple | 95.00 | 144.92 |
| Grapes | 14.12 | 318.02 |
| Other fruits | 960.88 | 3,185.87 |
| Potato* | 848.00 | 10,696.00 |
| Onion* | 268.00 | 2,870.00 |

*Figures are for 1985—86.

Fodder crops include leguminous and non-leguminous crops. Leguminous crops include berseem, shaftal, white clover, cowpea, lucerne, guar, kulthi, etc. Important non-leguminous crops are maize, sorghum, ragi, oats, etc. The area under green fodder is about 5.6 million ha, of which 20 per cent is irrigated. The total production of cultivated fodder is about 208 million tonnes annually.

## 4. SOIL EROSION

About 150 million ha land is subject to soil erosion (Das, 1981; Singh *et. al.*, 1981). Land degradation has set in—in another 25 million ha (Anon., 1982). The total loss of surface soil is estimated to be about 6000 million tonnes with loss of major plant nutrients of NPK varying from 5.37 million tonnes to 8.4 million tonnes (Singh *et. al.*, 1981; Anon., 1988). Of 164 million ha cultivable land, 104.6 million ha is reported to be affected by soil erosion (Singh *et. al.*, 1981).

Agricultural lands which are not properly managed suffer from erosion losses. *Sheet erosion*, which is not quite evident, keeps on taking place every year. *Rill erosion* is common in lands which are not bunded or terraced. Gullies are formed when the rills are not properly attended. Deep gullies and ravines have already covered about 4 million ha area and are scattered in Uttar Pradesh (1.23 million ha), Madhya Pradesh (0.68 million ha), Rajasthan (0.45 million ha), Gujarat (0.4 million ha), Bihar (0.6 million ha), Punjab and West Bengal (0.11 million ha each) and smaller areas in other states.

*Sheet erosion* is common in relatively slopy areas with heavy rainfall. This type of erosion is common in the Himalayan foothills, lower north-eastern parts of the Peninsula in Assam and in the eastern and western ghats. *Rill erosion* is active in Bihar, Uttar Pradesh, Madhya Pradesh and semi-arid parts of the Peninsula in Maharashtra, Karnataka, Andhra Pradesh and Tamil Nadu. The *chos* of Haryana and Punjab have resulted due to gully erosion and deposition of bouldery sediments on a large scale.

*Wind erosion* is active in dry areas. About 32 million ha area is subject to wind erosion which includes desert areas and coastal sands.

## 5. DEPENDENCE ON RAINFED AGRICULTURE

Indian agriculture largely depends on the vagaries of nature. Drought which causes water stress is perhaps the most important single factor which is very important for Indian agriculture. Except for some areas of the north-east and southern coastal region most areas of the country receive precipitation during June—July to September—October leaving a long dry period of about 8 to 9 months. Even in the normal rainfall years, irrigation facilities are necessary for rabi crops over almost the entire country. Approximately 41.77 million ha, i.e., about 27 per cent of the cropped area, is under irrigation (Table 23).

Table 23: **Area irrigated by sources (Anon., 1988d)**

| Sources | Area in thousand ha | | | |
|---|---|---|---|---|
| | 1950—51 | 1970—71 | 1982—83 | 1984—85 |
| Government canals | 7,158 | 11,972 | 14,875 | 15,366 |
| Private canals | 1,137 | 866 | 495 | 495 |
| Tanks | 3,613 | 4,112 | 3,112 | 3,330 |
| Tube-wells | N.A. | 4,461 | 10,684 | 11,265 |
| Other wells | 5,978 | 7,426 | 8,428 | 8,723 |
| Other sources | 2,967 | 2,266 | 2,375 | 2,600 |
| Total | 20,853 | 31,103 | 39,969 | 41,779 |

It is believed that India cannot be assured of success in agriculture unless more than 50 per cent of the cropped area is brought under assured irrigation. This is confirmed by the experience in Punjab, Haryana, Western Uttar Pradesh and Tamil Nadu. During the periods of droughts, even the irrigation sources do not effectively work but the intensity of droughts are considerably reduced. Trees and shrubs are better capable of surviving during a drought than annual cereals.

## 6. SMALL LANDHOLDINGS

The average size of an Indian farm holding is about 2.6 ha which is too small for several agricultural operations and does not yield enough income even for sustenance. Purchase of irrigation water, fertilisers, pesticides, machinery, etc., are beyond the capacity of the average Indian farmer. Nearly three-fourths of the total number of about 8 million holdings found in the country vary in size from 0.19 to 2.00 ha (Table 24).

Most small farms grow cereals for their sustenance. Only those located near towns and cities grow vegetables, fruits, etc. In remote areas and in arid areas where success of the crops depends purely on rainfall, people tend to grow food crops. They are not attracted by other commercial crops or perennial trees where chances for success are comparatively better. Larger farms tend to put their lands under commercial crops, vegetables, fruits, and other trees besides cereals.

Not only are the holdings small in India, they are also fragmented. There are several disadvantages of this fragmentation: (i) There is wastage of labour in movement from one field to another. (ii) The fields are small which make developmental works difficult and costly. (iii) A lot of agricultural land is wasted in boundaries. (iv) Supervision of farm operations becomes difficult and damage from thefts, etc., is more. Though State governments have made enactments for consolidation of holdings, progress has not been satisfactory.

## 7. Declining Soil Fertility and Use of Manures and Fertilisers

Depletion of soil fertility is one of the main reasons for low crop yields in India. The soils under agriculture have been cultivated for hundreds of years without much replenishing. Addition of cow dung and other organic manures, e.g., compost, animal urine, etc., is done only in reduced measure. The practice of leaving the land fallow to regain the soil fertility is being gradually given up in view of increase in population and large demands for foodgrains.

Larger quantities of cow dung and agricultural wastes are being used as fuel because of reduced availability of firewood due to dwindling of forest resources. Chemical fertilisers, though a good source for regaining soil fertility are costly and usually beyond the reach of the ordinary farmer.

However, the consumption of fertilisers has increased substantially during recent years. During 1950–51, the total fertiliser consumption was only 69 thousand tonnes; it increased to 9,200 thousand tonnes in 1985–86 and about the one-fourth of this quantity was met through imports. Per hectare consumption, which was negligible during 1950–51, inreased to 50.6 kg/ha during 1986–87. There is, however, urgent need to augment the supply of organic manures.

## 8. Natural Calamities

Indian agriculture suffers from several natural calamities, e.g., droughts, floods, frosts, cyclones, etc. As discussed earlier, the arid and semi-arid ar-

**Table 24: Distribution of operational holdings by major size groups: 1976–77 and 1980–81 (Anon, 1988d)**

| Size-Group | 1976–77 | | | | 1980–81 | | | |
|---|---|---|---|---|---|---|---|---|
| | Number of operational holdings (in millions) | Percentage of total | Operated area (in million ha) | Percentage of total operated area | Number of operational holdings (in millions) | Percentage of total | Operated area (in million ha) | Percentage of total operated area |
| Marginal (below 1.0 ha) | 44.52 | 54.6 | 17.51 | 10.7 | 50.12 | 56.4 | 19.73 | 12.1 |
| Small (1.0-2.0 ha) | 14.73 | 18.1 | 20.90 | 12.8 | 16.07 | 18.1 | 23.17 | 14.1 |
| Semi-medium (2.0-4.0 ha) | 11.67 | 14.3 | 32.43 | 19.9 | 12.45 | 14.0 | 34.65 | 21.2 |
| Medium (4.0-10.0 ha) | 8.21 | 10.0 | 49.63 | 30.4 | 8.07 | 9.1 | 48.54 | 29.6 |
| Large (10.0 ha & above) | 2.44 | 3.0 | 42.87 | 26.2 | 2.17 | 2.4 | 37.71 | 23.0 |
| Total | 81.57 | 100.0 | 163.34 | 100.0 | 88.88 | 100.0 | 163.80 | 100.0 |

eas are prone to repeated droughts which cause failure of agricultural crops. About a 40 million ha area is prone to floods. On average, about 10 million ha are damaged by floods annually. The northern part of the country is prone to frost damage in the winter. The coastal areas sufffer repeatedly from cyclones in which considerable loss of crops and property takes place. The eastern coast, particularly the coasts of Orissa, West Bengal and Andhra Pradesh, are more susceptible to serious cyclone damages than the western coasts. In hills, particularly in the Himalayas, serious localised damages occur due to landslips, landslides and avalanches. Some of the details regarding area affected by these land-use problems are given in Table 25 (Anon., 1988d).

**Table 25: Estimated areas under land-degradation and land-utilisation problems**
(Area: million hectare unless otherwise stated)

| | | With 1976–77 land-utilisation statistics and reports to 1980–81 | With 1981–82 land-utilisation statistics and reports to 1984–85 |
|---|---|---|---|
| 1. | Geographic area | 329 | 329 |
| 2. | Area subject to water and wind erosion | 150 | 144.43 |
| 3. | Area degraded through special problems | 25 | 29 |
| 4. | Annual average loss of nutrients from land due to soil erosion | 5.37 to 8.4 million tonnes | 5.37 to 8.4 million tonnes |
| 5. | Average annual loss of production for not developing ravines | 3 million tonnes | 3 million tonnes |
| 6. | Average annual rate of encroachment of tablelands by ravines | 8000 ha | 8000 ha |
| 7. | Average area annually subject to damages through shifting cultivation | 1 | 1 |
| 8. | Total flood-prone area | 40 | 40 |
| 9. | Annual average area affected by flood | 9 | 8 |
| 10. | Maximum area affected in the worst year | 17.6 | 18.6 |
| 11. | Annual average cropped area affected by flood | 3.6 | 3.7 |
| 12. | Maximum cropped area affected in any year | 10 | 10 |
| 13. | Total drought-prone area | 260 | 260 |
| 14. | Area affected by cyclone | — | 3.5 |
| 15. | Maximum damage in cyclone at one time | — | Rs. 10000 million |

## 9. ROLE OF ANIMALS AND MECHANISATION

Animals play an important role in Indian agriculture. They render useful service in agriculture. Almost all agricultural operations, e.g., ploughing, sowing, threshing, transport, etc., are done with the help of draught animals.

Animals are regarded as the backbone of Indian agriculture. That is why people in India keep a large number of animals. Until recently, the economic status of a farmer was judged by the number of animals he kept in his farm.

The livestock population in India is the highest in the world. The population increased from 292 million in 1951 to 416 million in 1982. India has about 20 per cent (191 million) of the world's cattle and 50 per cent (69 million) of the world's buffalo population. In addition, there are 95 million goats, 48 million sheep, besides a large number of other animals such as pigs, camels, horses, etc.

The quality of Indian livestock is very poor. Livestock rearing is highly neglected in India. It is often a subsidiary occupation to crop production and does not receive proper attention. Various factors which are responsible for the poor condition of livestock include: poor genetic base, acute shortages of feed, fodder, grazing and unhygienic condition of living. The present availability of green and dry fodders is only 250 million tonnes and 441 million tonnes respectively against actual demand of 932 million tonnes and 780 million tonnes respectively (Anon., 1986a). The supply position of concentrates is also bad. The majority of cattle and buffalo are maintained on agricultural by-products which consist of roughages of low nutritive value. Concentrates are given to the milch and drought animals only if the farmer can afford them. Feeding green fodder as a component of daily ration is not commonly practised. Due to these reasons, yields of animal products are very low. The average annual milk yield of a cow in India is 175 kg as against 3710 kg in Denmark. An Indian sheep on average yields only 0.89 kg of wool per year as against 7 kg in several countries.

In order to achieve proper growth of livestock production which would be commensurate with the needs of the people, it is imperative that adequate arrangements be made for fodder production for balanced nutrition and genetic improvement achieved through selection and cross-breeding. There is also a need to control the rate of growth of low-value and less productive livestock.

Agricultural machinery such as tractors, power-tillers, harvesters, combines, etc., has been gradually increasing. The use is limited mostly to rich and bigger farmers. However, increased indigenous production of this machinery and liberal agricultural credit would help in increased use of these machines. Table 26 shows a gradual increase in production and use of tractors during the last 15 years.

During 1970–71, power-tillers were produced at the rate of 1450 units per year. The indigenous production had picked up to 4225 units by 1984–85 (Anon., 1986). The use of these machines has limited scope due to: (i) very small size of farms, (ii) low income of the farmers, (iii) comparatively high cost of machines and (iv) poor repair and maintenance facilities in several regions.

**Table 26:  Production, import and sale of tractors (Anon., 1986)**

| Year | Indigenous production | Import | Total | Sale |
|------|------------------------|--------|-------|------|
| 1970–71 | 20,104 | 13,300 | 33,404 | 33,000 |
| 1974–75 | 31,129 | 793 | 31,922 | 30,229 |
| 1978–79 | 54,799 | – | 54,799 | 54,134 |
| 1982–83 | 62,054 | – | 62,054 | 63,013 |
| 1984–85 | 85,005 | – | 85,005 | 80,317 |

Though manufacture and popularisation of tractors and tractor-driven implements and mechanisation of agricultural operations have progressed, not much has been done to develop improved bullock-drawn implements and tools required by small farmers. There is a need to concentrate on designing, developing and popularising such implements.

10. AGRICULTURAL LABOUR

The condition of agricultural workers is not satisfactory. They suffer from chronic problems of underemployment and poor wages resulting in poor productivity. The number of cultivators and agricultural workers in India is given in Table 27.

**Table 27:  Classification of workers (Anon., 1986)**

| Class of workers | Workers (in millions) | Percentage of total |
|------------------|------------------------|---------------------|
| Cultivators | 92.5 | 37.8 |
| Agricultural labour | 55.5 | 22.7 |
| Other workers | 96.6 | 39.5 |
| Total workers | 244.6 | 100.0 |

Agricultural operations are seasonal in nature and therefore agricultural workers do not get job opportunities equally during all seasons. Therefore, any additional activity in farming such as tree planting, silvicultural operation, harvesting, etc., will help to provide enough job opportunities.

**Forest**

The word 'Forest' is derived from the Latin 'foris' meaning outside the village boundary or away from inhabited land. Generally, forest refers to an area occupied by different kinds of trees, shrubs, herbs and grasses and maintained as such. Technically, a forest is an area set aside for the production of timber and other forest produce, or maintained under woody vegetation for certain indirect benefits which it provides, e.g., climatic or protective (Anon., 1966). Ecologically, it is defined as a plant community predominantly of trees and other woody vegetation, usually with a closed canopy. Legally, a forest is an area of land proclaimed to be a forest under a forest law. The FAO classified all such lands into forests which bear vegetative associations dominated by trees of any size, exploited or not, capable of producing wood or other forest products, or exerting an influence on the climate or water regime or providing shelter for livestock and wildlife.

From the above definitions it becomes clear that a forest has five constituents, namely: (i) it has a land area, (ii) the land area should be occupied by different vegetation types, essentially by trees or it is proposed to establish essentially tree crops, shrubs, etc., and other forms of vegetation, (iii) the trees should form a closed or a partially closed canopy, (iv) the trees and other forms of vegetation should be managed for obtaining forest produce and/or benefits, (v) it should provide shelter to wildlife, birds and other fauna.

## CLASSIFICATION OF FORESTS

Forests have been classified on the basis of: (i) age, (ii) method of regeneration, (iii) composition, (iv) ownership, (v) object of management and growing stock. On the basis of age, forests are classified into: (a) even-aged forests and (b) uneven-aged forests. Even-aged forests, also called *regular forests*, are those consisting of even-aged woods. Even-aged woods mean trees of approximately the same age. True, even-aged forests can be only man-made forests. In the case of forests which regenerate naturally, some age difference is often allowed. A difference up to 25 per cent of the rotation is usually allowed in cases where a forest is not harvested for 100 years or more. The forest is called uneven-aged or *irregular* when stems vary widely in age.

On the basis of regeneration, forests are identified as (i) *high forests* when regeneration is obtained from seed and (ii) *coppice forests* when regeneration is through the coppice or some vegetative part of the tree. When this regeneration is obtained naturally, the forests are called *natural forests* and when it is obtained artificially the forests are called *man-made forests* or plantations.

On the basis of composition, forests are classified as (i) pure forests and (ii) mixed forests. *Pure forests* are composed almost entirely of one species, usually to the extent of not less than 50 per cent. Mixed forests are defined as forests composed of trees of two or more species intermingled in the same canopy. The species composing the mixture are usually distinguished into: (i) principal, (ii) accessory and (iii) auxiliary species. The *principal species* is the species in the mixture first in order or importance either by frequency, volume or silvicultural value. The *accessory species* is useful but of less value than the principal species which assists in the growth of the latter and influences to a smaller degree the method of the treatment. The *auxiliary species* is a species of inferior quality or size, or relatively little silvicultural value or importance associated with the principal species. These species are also called *secondary species*.

On the basis of management, forests are classified as: (i) protection forests, (ii) production forests and (iii) social forests. *Protection forests* are those which are managed primarily for ameliorating climate, checking soil erosion and floods, conserving soil and water, regulating stream flow, increasing water yields, and exerting any other beneficial influences. *Production forests* are those which are managed primarily for their *produce*. *Social forests* are also, in fact, production forests where produce is to be utilised by neighbouring society.

On the basis of ownership, forests are classed into: (i) government forests, (ii) private forests and (iii) forests owned by corporations, panchayats, societies and other bodies. On the basis of legal status, government forests are further classified as: (i) reserved forests, (ii) protected forests and (iii) village forests. A *reserved forest* is an area with complete protection constituted according to the provisions of the Indian Forest Act or other forest laws. A *protected forest* is an area subject to a limited degree of protection constituted under the provisions of Chapter IV of the Indian Forest Act, 1927. A *village forest* is a state forest assigned to a village community under the provisions of the Indian Forest Act.

On the basis of growing stock, a forest may be classified as: (a) normal forest and (b) abnormal forest. A *normal forest* is an ideal forest with regard to growing stock, age-class distribution and increment and from which the annual or periodic removal of produce is equal to the increment and can be continued indefinitely without endangering future yields. Such forests serve a standard and are rarely found in nature. An *abnormal forest* is one which is not normal, i.e., growing stock, age-class distribution of stems, increment, etc., are either in excess or more usually in deficit than the normal forest.

## FORESTRY

Forestry is defined as the theory and practice of all that constitutes the creation, conservation and scientific management of forests and the utilisation of their resources (Anon., 1966). It includes all thinking and all actions which pertain to the creation, conservation, protection and management of forests, including harvesting, marketing and utilisation of all forest products and services. It includes not only management of existing forests but also the creation of new forests.

Present forestry practices have been classified in different ways. The National Forest Policy of 1952 recommended that on the basis of function all forest lands could be classified as: (i) protection forests, (ii) national forests, (iii) village forests and (iv) tree lands. The National Commission on Agriculture (1976) proposed that forests be classified as: (i) protection forests, (ii) production forests and (iii) social forests. Production forests were proposed to be divided into: (i) mixed quality forests, (ii) valuable forests and (iii) inaccessible forests. The practice of managing forests for their protective function is called protection forestry. In protection forestry, the object is to protect the site due to instability of terrain, nature of soil, geological formations, etc. Such areas where manipulation of the forest cover is not desirable may be classed as protection forests. The forests located on higher hill slopes, in national parks and sanctuaries, preservation plots, biosphere or nature reserves, and wilderness area may be included under protection forests. The practice of forestry with a view to conserving soil and water, increasing water yields, reducing flood and droughts, amelioration of climatic conditions, etc., is called protection forestry. The ever-increasing demand for timber, fuelwood and various other forest products has to be met from production forests and the practice of forestry with the object of producing the maximum quantity of timber, fuelwood and other forest produce is called *production forestry*. Production forestry may be further divided into: (i) commercial forestry and (ii) industrial forestry. *Commercial forestry* aims to get the maximum production of timber, fuelwood and other forest products as a business enterprise. *Industrial forestry* aims at producing raw material required for industry. In production forestry, there is a greater concern for the production and economic returns. In industrial forestry, the object is to meet the requirements of raw materials of various industries.

*Social forestry* is the practice of forestry which aims at meeting the requirements of rural and urban populations. The object of social forestry is to meet the basic needs of rural community aiming at bettering the conditions of living through (i) meeting the fuelwood, fodder and small timber requirements, (ii) protecting agricultural fields against wind, (iii) meeting recreational needs and (iv) maximising production and increasing farm returns. Various forms of social forestry have been recognised. The practice of forestry on lands outside the conventional forest area for the benefit of the local population has been called *community forestry*. Community forestry seeks the involvement of the community in the management of such forests. *Farm forestry* is defined as the practice of forestry in all its aspect on farms or village lands generally integrated with other farm operations. The National Commission on Agriculture (1976) observed that what is actually practised might possibly be better termed as *extension forestry*, which includes the activity of raising trees on farm lands, village wastelands and community forest areas and on lands along the sides of roads, canal banks and railway lines. More recently there has been emphasis on dynamic land-use planning and efforts are being made to maximise production on farm lands under agroforestry. *Agroforestry* has been defined as a sustainable land-management system which increases the yield of the land, combines the production of crops and forest plants and/or animals simultaneously or sequentially on the same unit of land and applies management practices that are compatible with the cultural practices of the local population (King and Chandler, 1978).

More recently there has been considerable demand for *recreational forestry*, which is defined as the practice of forestry with the object of developing and maintaining forests of high scenic value. Recreational forests are being developed near towns and cities. The areas are being planted with flowering trees, shrubs and creepers to provide forest atmosphere near towns and cities.

Forestry has also been classified on the basis of intensity of management into: (i) intensive and (ii) extensive forestry. *Intensive forestry* refers to the practice of forestry wherein the objective is to maximise production per unit area by applying suitable silviculture and management techniques. When, due to some constraints, maximum production from the unit area is not obtained, the practice is usually referred to as *extensive forestry*. In intensive forestry, the site potential is utilised to the maximum possible extent while in extensive forestry the site is not effectively utilised.

EXTENT OF FORESTS IN INDIA

Of the total geographic area of 329.0 million hectare approximately 75.2 hectare area is classified as forest. About 95 per cent of the forests of the country are owned by the government and the remaining 4 per cent and 1 per cent are owned by corporate bodies and private individuals respectively. Government forests are classified as reserved, protected and unclassed forests. The reserved forests are declared by the government under the provisions of the Indian Forest Act, 1927. In these forests, very little or no rights are allowed. In protected forests local people are allowed certain rights. The extent of various categories of forest is given in Table 28 (Anon., 1987a).

Indian forests are greatly varied in their composition and species distribution. Distribution of forest area on the basis of composition, ownership and exploitation is indicated in Table 29.

Most forests of India belong to tropical climate. The forest areas under subtropical and temperate climates are limited to the higher ranges of the North and South. The subalpine and alpine forests are located in the higher hill ranges of the Himalayas where such climates are prevalent due to higher altitudes (Table 30).

ACTUAL FOREST COVER

Of 75.2 million ha recorded forest area in the country, the area under a good density forest is only 36 million ha (NRSA, 1984; Anon., 1988b). Approximately 28 million ha forest area is classified as open and scrub land and the remaining 11 million ha under other categories (Anon., 1988b). It can be seen in Table 31 that more than half of the forest area is degraded

**Table 28: Distribution of various categories of forests (Area in Km²)**

| State/U.T | Geographic area | Reserved | Protected | Unclassed | Total Forest Area |
|---|---|---|---|---|---|
| Andhra Pradesh | 275,100 | 49,921 | 12,343 | 1,507 | 63,771 |
| Arunachal Pradesh | 83,700 | 13,653 | 8 | 37,879 | 51,540 |
| Assam | 78,400 | 17,277 | 3,373 | 10,058 | 30,708 |
| Bihar | 173,900 | 5,051 | 24,169 | 7 | 29,227 |
| Goa, Daman & Diu | 3,800 | 42 | – | 1,208 | 1,250 |
| Gujarat | 196,000 | 13,490 | 1,020 | 4,810 | 19,320 |
| Haryana | 44,200 | 229 | 1,109 | 361 | 1,699 |
| Himachal Pradesh | 55,700 | 1,825 | 17,196 | 2,304 | 21,325 |
| Jammu & Kashmir | 222,200 | 20,182 | – | – | 20,182 |
| Karnataka | 191,800 | 28,611 | 3,931 | 6,103 | 38,645 |
| Kerala | 38,900 | 11,036 | – | 182 | 11,218 |
| Madhya Pradesh | 443,400 | 80,996 | 69,082 | 5,336 | 155,414 |
| Maharashtra | 307,700 | 42,823 | 15,366 | 5,969 | 64,158 |
| Manipur | 22,300 | 1,377 | 4,171 | 9,606 | 15,154 |
| Meghalaya | 22,400 | 978 | 12 | 7,524 | 8,514 |
| Mizoram | 21,100 | 9,048 | 1,647 | 5,240 | 15,935 |
| Nagaland | 16,600 | 86 | 507 | 8,032 | 8,625 |
| Orissa | 155,700 | 26,108 | 33,427 | 20 | 59,555 |
| Punjab | 50,400 | 43 | 1,104 | 1,676 | 2,823 |
| Rajasthan | 342,200 | 13,970 | 14,170 | 3,150 | 31,290 |
| Sikkim | 7,100 | 2,650 | – | – | 2,650 |
| Tamil Nadu | 130,100 | 18,375 | 3,390 | 614 | 22,379 |
| Tripura | 10,500 | 3,571 | 291 | 2,436 | 6,298* |
| Uttar Pradesh | 294,400 | 34,461 | 884 | 5,412* | 51,337 |
| West Bengal | 88,800 | 7,054 | 3,772 | 1,053 | 11,879 |
| A & N Islands | 8,300 | 3,059 | 4,112 | – | 7,171 |
| Dadar Nagar Haveli | 100 | 203 | 3 | – | 206 |
| All India** | 32,87,330 | 4,06,119 | 2,15,087 | 1,31,067* | 7,52,273 |

*10,580 Sq. km. Unclassed Forest area outside forest department.

**Union territories of Chandigarh, Delhi, Pondicherry and Lakhsha Dweep have no forest.

where the density of trees is less than 40%. Degraded forest areas neither provide ecological protection nor any other produce. The area of degraded forest has been continuously increasing.

The details of forest cover in each state are given in Table 32 (Anon., 1988b).

Due to degradation of forest cover over a large area, various adverse effects are evident. It has been continuously reducing the life-supporting capacity of the earth and adversely affecting environment. Frequent floods, drought, severe soil erosion, reduced water yields, decreasing productivity and reduced harvests from the forests are some of the consequential adverse effects (Fig. 1).

**Table 29: Distribution of forest area (million ha)**

| Description | Area as of 1985–86 | Percentage of total |
|---|---|---|
| A. Ownership | | |
| (i) State owned | 71.0 | 94.7 |
| (ii) Corporate bodies | 3.1 | 4.1 |
| (iii) Private | 0.9 | 1.2 |
| Total | 75.0 | 100.0 |
| B. Composition | | |
| (i) Broad-leaved forests | 68.0 | 90.7 |
| (ii) Coniferous forests | 7.0 | 9.3 |
| Total | 75.0 | 100.0 |
| C. Exploitation | | |
| (i) Exploitable forests (forests in use) | 40.0 | 53.3 |
| (ii) Potentially exploitable forests | 16.0 | 21.3 |
| (iii) Others | 19.0 | 25.4 |
| Total | 75.0 | 100.0 |

In 1987, of the recorded forest area of 75.2 million ha, the total forest cover was only 64.2 million ha. Of this, only 35.77 million ha are of adequate forest density. The effective forest cover is therefore limited to 10.8 per cent of the geographic area. For maintaining ecological balance, the country needs about 33 per cent of the land area under adequate forest cover.

ROLE OF FORESTS

Forests are known as the world's air-conditioner and the earth's blanket. Without forests, this world would be an inhospitable place to live in. Forests are important for maintaining ecological balance and preserving the life-supporting system of the earth. They are essential for food production, health and other aspects of human survival and sustainable development. Of all the eco-systems, forests are the largest, most complex and self-perpetuating. The maintenance of forests is vital for all sections of society regardless of their stage of development. Forests perform various functions. Some of these functions are: (i) productive, (ii) protective, (iii) ameliorative, (iv) recreational and (v) developmental. Some of these functions are as under:

*(A) Productive functions of the forests*

(1) Forests are valuable natural resources. The goods provided by forests are of immense importance. Wood is a major forest produce and is used extensively for various purposes. In India, most of the wood produced is

**Table 30: Distribution of forest types (Anon., 1985)**

| Sl. No. | Forest Types | Area (million ha) | Area of occurrence |
|---|---|---|---|
| 1. | *Tropical* | | |
| | (a) Evergreen and semi-evergreen | 6.5 | Western side of Western Ghats, upper Assam and Andamans (rainfall over 2500 mm). |
| | (b) Moist deciduous | 22.4 | Foothills of Himalayas, east side of Western Ghats, Chhota Nagpur and Khasi hills (Rainfall 1500 mm to 2500 mm) |
| | (c) Littoral and swamp | 0.7 | Along the coast |
| | (d) Dry deciduous | 29.7 | Almost entire Indian Peninsula with rainfall of 750–1500 mm. |
| | (e) Thorn | 5.2 | Rajasthan and adjoining areas with 250 to 750 mm rainfall |
| | (f) Dry evergreen | 0.1 | Karnatic coast with no or little summer rain |
| 2. | *Subtropical* | | |
| | (a) Broad-leaved hill forest | 0.3 | Lower Himalayas |
| | (b) Pine forest | 3.7 | Central & Western Himalayas with 1000 mm to 2000 mm rainfall |
| | (c) Dry evergreen | 0.2 | Western Himalayas with lower rainfall. |
| 3. | *Temperate* | | |
| | (a) Wet temperate | 1.6 | Eastern Himalayas between 1800–2000 m elevation and tops of southern hills. |
| | (b) Dry temperate | 0.2 | Inner ranges of Himalayas with low rainfall |
| | (c) Moist temperate | 2.7 | Central and western Himalayas betwen 1500–3000 m elevation. |
| 4. | *Subalpine and alpine forest* | 1.8 | In Himalayas above 3000 m elevation |
| | Total | 75.1 | |

used for construction of houses, agricultural implements, bridges, sleepers, etc. In India, about 12 million m³ of timber is produced from our forests. Many species, e.g., teak, sal, deodar, sissoo, babul, chir, haldu, axlewood, rosewood, dipterocarp, etc., yield valuable timber.

(2) Wood is a universal fuel. For thousands of years, until the advent of coal, oil, gas, electricity, etc., wood constituted man's chief source of fuel. Even today more than half of the total world consumption of wood is for fuel. Wood remains the major source of domestic fuel in India. Approximately 175 million m³ of wood is used as fuel in the country, most of which comes from our forests (NCA, 1976).

**Table 31: Area under different categories of forest cover (Anon., 1988b)**

| Category | Area (million ha) | % age of total geographic area |
|---|---|---|
| A. *Forest* | | |
| Dense forest (Density more than 40 per cent) | 35.77 | 10.88 |
| Open forest (Density 10–40 per cent) | 27.66 | 8.41 |
| Mangrove forest | 0.40 | 0.12 |
| Coffee plantation | 0.37 | 0.11 |
| Total forest | 64.20 | 19.52 |
| B. Scrub area | 7.68 | 2.34 |
| C. Uninterpreted area (under clouds, shadows) | 1.15 | 0.35 |
| D. Non-forest (also includes tea gardens) | 255.85 | 77.79 |
| | 328.88 | 100.00 |

(3) Forests provide raw material to a large number of industries, e.g., paper and pulp, plyboard and other boards, saw-mills, furniture, packing cases, matches, toys, etc.

(4) A large number of non-wood products are also obtained from forests. These are commonly called minor forest products (M.F.P.) not because they are of minor significance, but because they are harvested in smaller quantities. Some of the important minor forest products are as under:

*(i) Fibres and flosses:* Fibres are obtained from bast tissues of certain woody plants which are used for making ropes. Flosses are obtained from semal (*Bombax ceiba*) and kapok (*Ceiba pentandra*).

*(ii) Grasses and bamboos:* A large variety of grasses are found in the forests. About 30 per cent of the 416 million livestock population graze in the forests. Among valuable grasses, e.g., sabai (*Eulaliopsis binata*) is harvested annually to the tune of 80,000 tonnes. About 6.5 million tonnes of bamboo are harvested from our forests every year. Approximately 30 per cent of the bamboo is used for housing, 25 per cent for rural/agricultural works, 25 per cent for paper pulp and the remainder for packaging and other uses.

*(iii) Essential oils:* India produced about 1,500 tonnes of essential oils during 1980 which was utilised in making soaps, detergents and chemicals. Many species, e.g., *Eucalyptus, Bursera, Cymbopogon, Santalum album*, etc., produced these oils.

*(iv) Oilseeds:* Many tree species, e.g., *Madhuca indica, Pongamia pinnata, Shorea robusta, Azadirachta indica, Schliechera oleosa, Vateria indica*, etc.,

Table 32: **Classification of Tree-covered Area into Density Classes**

(Area in km²)

| State/U.T. | Dense Forest (Crown density above 40%) | Open Forest (Crown density 10–40%) | Mangroves | Total Tree-covered area |
|---|---|---|---|---|
| Andhra Pradesh | 28,580 | 21,119 | 495 | 50,194 |
| Arunachal Pradesh | 51,096 | 9,404 | — | 60,500 |
| Assam | 18,415 | 7,971 | — | 26,386 |
| Bihar | 13,490 | 15,258 | — | 28,748 |
| Goa, Daman & Diu | 763 | 522 | — | 1,285 |
| Gujarat | 7,850 | 5,293 | 427 | 13,570 |
| Haryana | 43 | 601 | — | 644 |
| Himachal Pradesh | 9,908 | 2,974 | — | 12,882 |
| Jammu & Kashmir | 12,978 | 7,902 | — | 20,880 |
| Karnataka | 16,394 | 15,870 | — | 32,264 |
| Kerala | 8,569 | 1,833 | — | 10,402 |
| Madhya Pradesh | 72,174 | 55,575 | — | 127,749 |
| Maharashtra | 27,244 | 20,032 | 140 | 47,416 |
| Manipur | 4,670 | 13,009 | — | 17,679 |
| Meghalaya | 5,749 | 10,762 | — | 16,511 |
| Mizoram | 2,938 | 16,154 | — | 19,092 |
| Nagaland | 6,379 | 7,972 | — | 14,351 |
| Orissa | 28,573 | 24,391 | 199 | 53,163 |
| Punjab | 96 | 670 | — | 766 |
| Rajasthan | 3,048 | 9,430 | — | 12,478 |
| Sikkim | 1,867 | 972 | — | 2,839 |
| Tamil Nadu | 10,866 | 7,491 | 23 | 18,380 |
| Tripura | 340 | 5,403 | — | 5,743 |
| Uttar Pradesh | 18,876 | 12,567 | — | 31,443 |
| West Bengal | 3,512 | 3,223 | 2,076 | 8,811 |
| A & N Islands | 6,807 | 110 | 686 | 7,603 |
| Chandigarh | — | 2 | — | 2 |
| Dadar Nagar Haveli | 187 | 50 | — | 237 |
| Delhi | — | 15 | — | 15 |
| Laksha Dweep | — | — | — | — |
| Pondicherry | — | 8 | — | 8 |
| All India | 3,61,412 | 2,76,583 | 4,046 | 6,42,041 |

produce oil-bearing seeds which are commercially important. Some of these oils can be made fit for human consumption. Presently these seeds are used in the soap industry. Tribals use these oils for various purposes. There is a potential production of about 1 million tonnes of oil every year from forest tree seeds.

*(v) Tans and dyes:* A variety of vegetable tanning materials is produced in the forests. Important vegetable tanning materials are the myrobalan nuts, bark of wattle (*Acacia mearnsii, A. decurrens, A. nilotica*) and *Cassia auriculata*, etc. Other tanning materials include leaves of *Emblica officinalis*

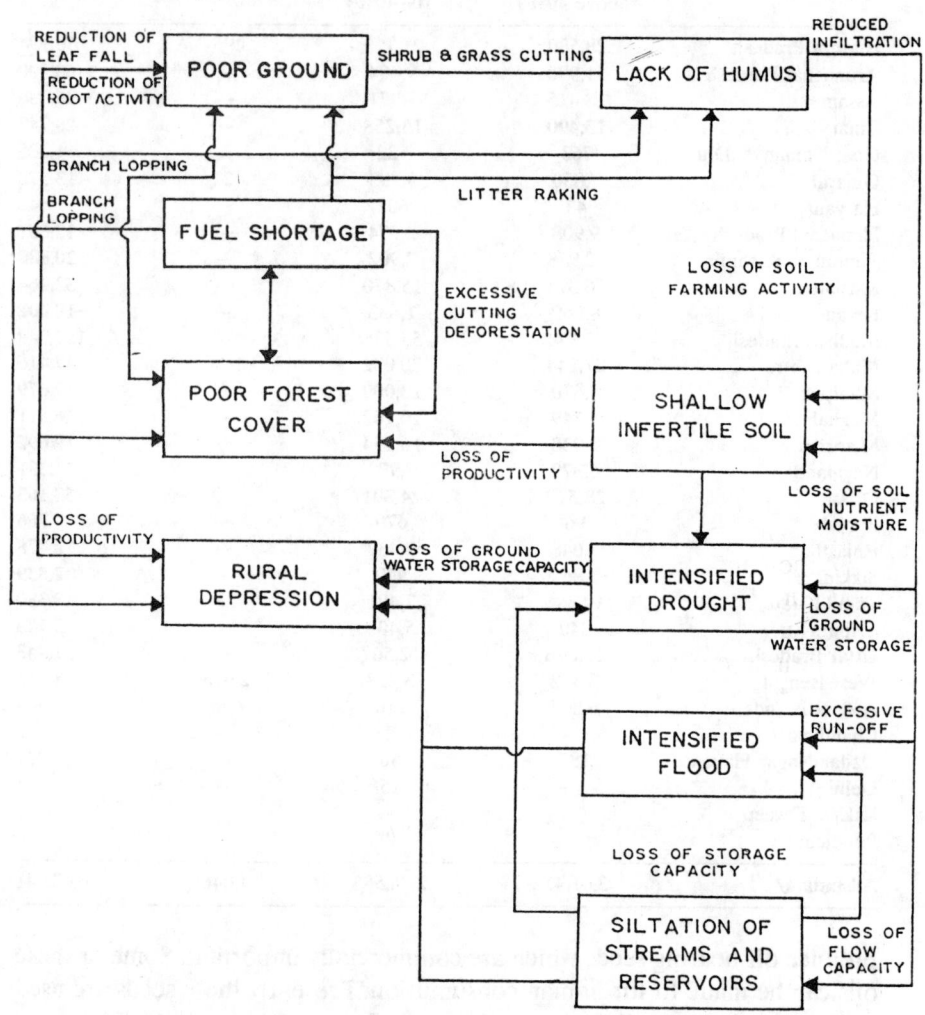

**Fig. 1:  Consequences of forest destruction**

and *Anogeissus latifolia*, bark of *Cleistenthus collinus*, fruits of *Ziziphus xylopara*, bark of *Cassia fistula*, *Terminalia alata*, *T. arjuna*, etc. Katha and Kutch are obtained from *Acacia catechu* trees.

(*vi*) *Gums and resins:* Gums and resins are exuded by trees as a result of injury to the bark or wood. Gums are collected from several tree species, viz., *Sterculia urens*, *Anogeissus latifolia*, *Lannea coromandelica*, *Acacia nilotica*, *Cochlospermum religiosum*, *Pterocarpus marsupium*, *Butea monosperma* etc. Resin is obtained from *Pinus roxburghii*.

(*vii*) *Drugs, spices and insecticides:* Of the total of about 2000 items of drugs mentioned in the Indian Materia Medica, over 1800 are of vegetable origin. A large number of these are obtained from forests. All parts of the plants, e.g., roots, shoots, leaves, fruits, seeds, bark, etc., are used for drugs. Important species yielding drugs include *Rauvolfia serpentina*, *Ephedra* sp., *Hemidesmus indicus, Swertia chirata, Dioscorea* spp., Podophyllus spp., *Atropa* spp., *Datura innoxia*, etc. A variety of spices, such as Kalajira (seeds of *Carum carvi*), dalchini (bark of *Cinnamomum zeylanicum*), cardemom (dried capsules of *Ellattaria cardomomum*), tejapatta (leaves of *Cinnamomum tamala*), etc., are obtained from forests. *Pyrethrum* and *Neem* are used as insecticides.

(*viii*) *Tendu and other leaves:* Tendu leaves are used to produce bidis and therefore are also called bidi leaves. The annual collection of tendu (*Diospyros melanoxylon*) leaves is about 70,000 tonnes in the country. In Madhya Pradesh alone, about 45 per cent of this quantity is produced. Leaves of trees such as *Bauhinia* spp., *Butea* spp., etc., are used for making plates, dona, etc.

(*ix*) *Edible products:* Fruits, flowers, seeds, tubers, etc., of several forest species are eaten. Fruits and seeds of *Anacardium occidentale, Tamarindus indica, Syzygium cumini, Emblica officinalis, Buchanania lanzan*, etc., flowers of *Madhuca indica*, green pods of *Moringa oliefera*, new shoots of bamboos, etc., are edible. There are several other species which yield edible products.

(*x*) *Lac and other products:* Lac is a resinous secretion of insects which feed on forest trees, particularly on *Butea monosperma*. Similarly, the silkworm is raised on *Terminalia alata* or *Morus alba* plantations. Honey is another product which is obtained from forests.

(*xi*) *Fodder and grazing:* Forests provide fodder leaves and grazing facility to rural animals. About 30 per cent of 416 million livestock population depend upon forest grazing and leaf fodder supply. The leaf fodder of several tree species is almost as nutritious as agricultural fodder crops. Good fodder-yielding trees include: *Ailanthus excelsa, Moringa oliefera, Sesbania* spp., *Morus alba, Albizia lebbeck, Leucaena leucocephala, Pongamia pinnata, Hardwickia binata*, etc.

*(B) Protective and ameliorative functions of forests*

The protective and ameliorative functions of forests are of great significance to mankind. Some of the important ones are discussed below.

(i) Forests play a significant role in maintaining the $CO_2$ balance in the atmosphere. Without sufficient forest cover, all the $CO_2$ released in the atmosphere will not be utilised, resulting in a higher per cent of $CO_2$ in the atmosphere. This, according to scientists, will result in warming of the world temperature, disturbance in the climate, melting of polar ice caps, increase in sea levels, etc. The $CO_2$ per cent in the atmosphere has already reached 0.042 per cent against the normal of 0.030 per cent. If this increases continuously, higher temperature and other disturbances on the earth may bring unimaginable miseries to mankind.

(ii) Forests increase local precipitation by about 5 to 10 per cent due to their orographic and microclimatic effect and create conditions favourable for the condensation of clouds.

(iii) Forests reduce temperature and increase humidity. The temperature in forests is 3° to 8°C less than in adjoining open areas. Reduced temperature makes life comfortable. It also reduces evaporation losses. The effect of forests on temperature is not limited to forested areas; it extends far beyond the boundaries of the forests.

(iv) Forests maintain the productivity of the soil through adding a large quantity of organic matter and recycling of nutrients. The leaves of trees are used as manure. Supply of firewood from forests releases dung for use, as manure.

(v) Tree crowns reduce the violence of rain and check splash erosion. Forests increase the infiltration and water-holding capacities of the soil, resulting in much lower surface run-off. This in turn results in checking of soil erosion.

(vi) Forests check floods. Forests intercept 15 to 30 per cent of the total rainfall. Most floods are caused due to siltation of river channels caused due to erosion.

(vii) Forests conserve both soil and water. Forests prolong the water cycle from its inception to the final disposal as run-off into streams and ocean. The longer the water is retained in the land, the greater is its usefulness in nurturing crops and trees, and in maintaining a regular supply of water in streams throughout the year. Forests increase subsurface run-off which is much slower than surface run-off and the subsurface run-off does not cause erosion.

(viii) Forests and trees reduce wind velocity considerably. Reduction of wind velocity causes considerable reduction in wind erosion, checks shifting of sand dunes, and halts the process of desertification.

(ix) Forests by reducing erosion check the siltation of irrigation and hydel reservoirs. Rapid siltation of various reservoirs in the country is the result of deforestation in the catchment areas of these reservoirs.

(x) Forests are the storehouse of genetic diversity. Several unknown plants may be the potential for medicines and food. Deforestation eliminates several species of plants and animals. In India, about 5000 species of plants, 180 species of mammals, 180 species of birds and several thousand species of insects are already on the verge of extinction due to loss of forests.

(xi) Forests protect us from physical, chemical and noise pollution. Dust and other particulates and gaseous pollutants cause serious health problems. Forests protect us from these pollutants.

(xii) Forests and trees provide a shelterbelt and windbreak effect which is beneficial to agricultural crops, particularly in arid and semi-arid areas, and increase agricultural production.

*(C) Recreational and educational function of forests*

(i) Forests provide recreational facilities to the people. A large variety of trees and shrubs, animals and birds attract a large number of people towards them. National parks and sanctuaries rich in flora and fauna are visited by a large number of people every year.

(ii) Forests provide an experimental area and laboratory for college and university students. Forests provide sites for ecological studies.

(iii) Forests have a natural healing effect for a number of diseases. Most of the sanitoria are found in a forested locality.

*(D) Developmental functions*

(i) Forests provide employment to a large number of people. Almost all forestry activities are labour intensive and provide considerable employment in primary and secondary sectors.

(ii) Forests and various forest activities help tribals to improve their socioeconomic condition through collection, processing and marketing of various forest products and by providing gainful employment. Forestry is an important activity in a poverty alleviation programme.

(iii) Forests provide a good sum as revenue to the government which is used for various developmental works. During 1985, forests provided revenue worth Rs. 14,500 million.

Over the years, forests in the country have suffered serious depletion. During the British period, the forests were largely exploited for meeting the demands of railway and navy. Large forest areas were cleared for agriculture because the collection of revenue from agricultural land was more and easier. The pace of forest destruction set in during the British period continued even after independence. Though the forest policy of 1952 realised the importance of forests and recommended that one-third of the geographic area be kept under tree cover, this objective could not be achieved. Due to increase in human and livestock population, great pressures were built which adversely affected the forests of the country. Extremes of climate, repeated droughts, serious floods, rampant soil erosion, serious ecological imbalances, particularly in ecologically fragile zones, scarcity of firewood, timber and bamboo, etc., are some of the results of deforestation.

The tree cover which was 22 per cent during 1952 reduced to almost half, i.e., 11 per cent, during 1987. The degraded forest areas need to be revegetated and more areas to be brought under tree cover. If trees are combined with agricultural crops, the tree cover could easily be increased. The forest policy announced in 1988 recommended among other things that: (i) existing forests should be fully protected and their productivity should be improved; (ii) forest cover should be rapidly increased on hilly slopes, in the catchment areas of rivers, lakes, reservoirs and in the semi-arid and arid areas of the country; (iii) the country should have a minimum of 33 per cent of the total area under forest cover. The montane regions should have two-thirds of the area under forest cover to prevent erosion and land degradation; (iv) massive need-based afforestation particularly for fuelwood and fodder be taken up; (v) the diversion of forest land for non-forest purposes should be after most careful examination and (vi) the removals and concessions should relate to the carrying capacity of the forest. It also recommended that good agricultural land should not be diverted for forestry.

# CHAPTER 3

# Agroforestry—Definition, Scope and Advantages

## Historical

Man (*Homo sapiens*) appeared on this earth about one million years ago (*Eolithic age*). In the Eolithic age, primitive man lived in dense forests in caves and trees and depended entirely on plant parts. In the Palaeolithic age (3 lakh years ago), he took to hunting of wild birds and animals and covered himself with their skins. In the Neolithic age (15,000 years ago), he used fire for cooking. About 10,000 years ago, man discovered that seeds of certain grasses could be eaten. To obtain these seeds, he cleared forests and grew them. Subsequently he cleared larger tracts by felling trees and burning them. It was found that the productivity went down after sometime and therefore, other areas were cleared and burnt to grow those seeds. Thus the practice of *shifting cultivation* began. Shifting cultivation gave way to *regular cultivation* in due course of time.

Trees and forests were an integral part of the Indian culture. This is amply supported by ancient scriptures and historical records. The best of Indian culture was born in the forests. The Aryan civilisation was cradled in our forests and our *rishis* who evolved the Hindu philosophy, lived in forests in complete harmony with nature. The *ashrams* were the centres which harmonised agriculture and pasture with trees, animals and birds. Lord Buddha sought the shelter of the *bodhi* tree to become the Buddha. Many trees are worshipped even today. We have records of *kalpabriksh* which could satisfy all human needs.

It was widely believed that destruction of forests and cutting of trees created famine conditions, as has been said in one of the Puranas written about 1000 B.C.:

क्रियते यत्र विच्छेदः सपुष्प फलानि नस्तरो ।
अनावृष्टि भयंधोरं तस्मिन् देशे प्राजायैन ।।

Planting and maintaining trees were considered noble acts. They were considered better acts than having sons, as has been said in Agnipuran (1000 B.C.):

दश कूप समावापी दशवापी समाहदः ।
दश हद समपुत्रौ दशपुत्र समातरुः ।।

(Ten wells are equal to one tank, ten tanks are equal to one lake, ten lakes are equal to one son and ten sons are equal to one tree for giving *punya*). In the Puranas it is said अपुत्रस्य पुत्रत्वं पाद्या इह कुर्वते ।, which means for those who do not have sons, trees are like sons for them. According to the Puranas, planting of trees not only gives leaves, fruits, shade, roots, bark and wood, but trees also help in achieving salvation for parents.

पत्र पुष्प फलच्छाया मूल वल्कल दारुभिः ।
परेषामुप कुर्वन्ति तारयन्ति पिता महान ।।

In fact, so much has been said about trees in our ancient literature that planting trees was being done by individuals on their own along with agricultural crops. *Krishishukti* (about 700 B.C.), written by *Maharshi Kashyap*, classifies lands into several categories and identifies areas which are suitable for planting trees, all wet and dry lands and areas around houses, wells, tanks are specifically identified for tree planting.

However, during recent periods due to intensive agriculture, tree planting was subjugated to agriculture because firewood, timber, bamboo, tree, fodder, etc., were available to people from nearby forests. More recently however, the forest area has receded and resources have shrunk considerably. People are no longer able to meet their requirements for firewood, fodder, timber, bamboo, etc., from the forests. The prices of these commodities have, therefore, increased substantially. Forest-based industries have been facing problems in supply of raw material. Many farmers quite recently started planting trees on their farm lands to meet these shortages. In several areas, farmers have now taken up plantation of the tree species required for firewood, small timber, paper and pulp, match boxes, and other uses.

## Definition

Agroforestry is not a new system or concept. The practice is very old, as explained already, but the term is definitely new. People raised together trees, crops and animals traditionally on the same farm. This practice of mixed farming developed over centuries for meeting most of the requirements of a family. The crops provided foodgrains for livelihood. They also gave fodder. The trees gave wood for construction of houses. They also yielded firewood.

The animals provided milk and meat. They also pulled the plough and the carts. The farming system yielded almost everything which people needed.

Agroforestry means practice of agriculture and forestry on the same piece of land. Bene *et al.*, (1977) defined agroforestry as a sustainable management system for land that increases overall production, combines agricultural crops, tree crops and forest plants and/or animals simultaneously or sequentially and applies management practices that are compatible with the cultural patterns of a local population. This definition was modified by King and Chandler (1978) as: "Agroforestry is sustainable land management system which increases the overall yield of the land, combines the production of crops (including tree crops) and forest plants and/or animals simultaneously or sequentially, on the same unit of land and applies management practices that are compatible with the cultural practices of the local population."

Nair (1979) defines agroforestry as a land use system that integrates trees, crops and animals in a way that is scientifically sound, ecologically desirable, practically feasible, and socially acceptable to the farmers.

In Australia, integrated agriculture, grazing and forestry systems have been called 'Agroforestry'. This could be of two types; either grazing and/or cultivation in natural forests or woodlands. For example, range management in the United States of America or grazing and/or cultivation in combination with trees established by man (Bartle, 1977). Thomas (1977) observes that in Australia agroforestry includes: (a) combination of agriculture and forestry in a physical and functional relationship, for example, grazing within forest plantations, (b) diversification of forest activities into agriculture, or those of farmers into forestry even if no physical interaction is involved and (c) development of mixed forest/farm landscape in the interest of community as a whole, possibly through rural planning, for example, in order to protect water resources, control erosion, or reclaim salt-affected lands, etc.

In New Zealand, agroforestry is referred to as forest farming. Various combinations of trees and pastures are being tried by individual farmers, forest companies and land development agencies and it is defined as any situation where trees and grazed pastures are grown together as an integrated management system; the prime objective is to increase long-term net profit per hectare. McQueen (1977) observes that within this definition a number of situations would occur where agroforestry is attractive including: (a) a future and alternative source of income to farmers, (b) a source of intermediate income to the forest companies, (c) a means of erosion control, (d) an aid to tree establishment on steep or broken country, (e) an economic use of land currently marginally profitable, (f) an intensified land use and (g) shelter for animals and crops.

Cannel (cited by Nair) defines agroforestry as a land use system (a) in which woody perennials, and herbaceous crops are grown together in mixtures, zonally and/or sequentially, with or without animals, and (b) which provides greater benefits for land use than agriculture or forestry alone, including one or more of the following: sustained soil fertility, soil conservation, increased yield, diminished risk of crop failure, ease of management, pest and disease control and/or greater fulfilment of the socioeconomic needs of the local population (Nair, 1989).

Agroforestry is a collective name for land use systems and technologies where woody perennials (trees, shrubs, palms, bamboos, etc.) are deliberately used on the same piece of land management units as agricultural crops and/or animals in some form of spatial arrangement or temporal sequence. In agroforestry systems, there are both ecological and economical interactions between the different components (Lundgren and Raintree, 1982).

In India, the word agroforestry is often used as a synonym for farm forestry. *Farm forestry* is defined as the practice of forestry in all its aspects on farms or village lands generally integrated with other farm operations (NCA, 1976). Farm forestry includes forestry on farm lands in the form of raising rows of trees on bunds or boundaries of the fields and individual trees on private agricultural land as well as the creation of windbreaks and shelterbelts, which are protective vegetative screens created round a farm or an orchard.

Agroforestry practices vary considerably from place to place, region to region and country to country. Quite naturally, the word conveys a different meaning at different places. In India, agroforestry is a sustainable land use system which integrates growing of agricultural crops and forest trees together on the same piece of land for maximum production of food, fodder and wood, and other products in such a way that the system is economically and ecologically desirable and acceptable to the local population.

In simple terms, *agriculture* is a land use system when agricultural crops are grown for the production of foodgrains, fodder, etc. Forestry, on the other hand, is land use system where forest trees, shrubs, etc., are grown for the production of wood, fodder and other benefits. Agroforestry is a hybrid of both land use systems where the object is to obtain yield of grains, fodder, wood and other benefits. That is why during the process of the evolution of the term 'agroforestry', some workers held that from a linguistic point of view proper usage of the term would be 'agriforestry' and not agroforestry (Stewart, 1981). But since the word agroforestry is well established, it would be confusing to introduce another word. Vergara (1985) observes that agroforestry is a land use system combining agriculture and tree crops of varying longevity (ranging from annual through biannual and perennial plants) arranged either temporarily (crop rotation) or spatially (intercropping) to maximise and sustain aggregated yields. At times livestock and fish may be added as components. Under Indian conditions, agroforestry has two main components e.g., (i) growing of agricultural crops including fodder and (ii) forestry, i.e., growing of trees and other woody perennials. The number of livestock in India is large and regulated grazing is not usually practised. Only to a limited extent are some areas managed as pastures. But individual farms being small, animals usually are not maintained as a component of agroforestry as a form of land use. Generally the by-products of agriculture, tree fodder, grasses, etc., are used for feeding the animals.

The system of agroforestry involves interaction between various components. The interaction of one component with another need not be very intimate. The intimacy of mixture, width and extent of zones, blocks, strips and rows, etc., are also quite relevant questions. King (1978) suggests that agroforestry might be considered to be practised whenever trees and agricultural crops are grown in mixture, provided that the combined widths of the rows of agricultural crops do not exceed the heights of the forest trees at maturity and provided further that the combined widths of rows of forest trees do not exceed the height of the tree crops at maturity or at some selected rotations. Such suggestions assume that trees and crops can have an interaction only in close proximity. The practice of crop geometry in raising trees and agricultural crops differs so widely from place to place that it would not be appropriate to suggest the mixture type, width of rows/strips, etc. Shelter-belts and wind-breaks and trees grown in rows exert an influence in areas up to 25 to 30 times of their height. Also, the present level of research in interactions of trees and agricultural crops is not sufficient to delimit the area of such possible interactions.

It can be seen from the definitions that agroforestry has the following components (Nair, 1989):

(i) It is deliberate growing of woody perennials on the same unit land as agricultural crops and/or animals either in some form of spatial mixture or in sequence.

(ii) There must be significant interaction (positive/negative) between the woody and non-woody components of the system either ecological and/or economical.

(iii) This is a production system which tends to harmonise the production of various components and also maximises the total production from a given unit of land.

(iv) The production and use is sustainable and makes use of modern technologies and traditional local experience and is compatible with the social and cultural life of the local population.

(v) It is a long-term land management system and the cycle of agroforestry system is always more than one year.

(vi) Agroforestry is a more complex form of land management both ecologically and economically than other agricultural or forestry systems.

**Advantages of Agroforestry**

Agroforestry systems have several advantages. Some of these advantages are as under:

MAXIMISE PRODUCTION

In tropical conditions, tree crops and forests have greater advantages. In tropical conditions, temperature and humidity levels are high and the cultivation of annuals fails to conserve the optimum solar radiation throughout the year in contrast to trees and forests. Under agriculture, repeated ploughing of top soil leads to excessive leaching of nutrients during the monsoon. Soil fertility is reduced by accelerated decomposition of organic matter and denitrifying processes. Loss of soil and nutrients due to erosion in agricultural monoculture is tremendous.

Trees, on the other hand, are capable of utilising available solar radiation throughout the year and, therefore, total productivity under the forestry land use system is likely to be more than agriculture. Forestry systems maintain soil fertility through recycling of nutrients and prevent soil erosion and loss of nutrients through leaching and run-off. Many leguminous tree species fix nitrogen from the atmosphere and return much more in leaf-fall than they take from the soil. Leaves of the trees could be used as a green manure and help the farmer to increase the soil fertility. Agroforestry systems are therefore helpful in maintaining the soil productivity at optimum levels over a long period of time. Combining agricultural crops with trees helps in increasing the productivity of the land.

Higher yields of crops have been observed in forest-influenced soils than in ordinary soils (Chaturvedi, 1981; Sanghal, 1983; Verinumbe, 1987). In the terai area of Uttar Pradesh, taungya cultivators harvested higher yields of crops such as maize, wheat, pulses, etc., in leased-out areas without fertilizer than in agricultural lands. Verinumbe (1987) reported higher yields of dry matter, particularly maize and sorghum, in forest-influenced soils in Sahel (Africa) than in ordinary field soils. Higher crop yields were obtained in the soils under *neem* followed by *Prosopis* and *Eucalyptus* than in ordinary fields. In the semi-arid areas of Africa, 30–40 per cent higher yields of maize have been found with *Leucaena leucocephala*, *Gliricidia sepium* and *Cassia siamea* intercropped as hedgerow and prunings used as mulch and green manure (ICRAF, 1988). The International Institute of Tropical Agriculture (IITA) has developed a system of alley cropping in which *Leucaena* is intercropped with corn, yams, and rice. In the growing season of agricultural crops, the trees are cut and pruned so that they do not shade the nearby crops. The pruned leaves are used as mulch and the larger branches as fuel and poles. In the dry season the trees are allowed to regrow. On infertile sandy soils, alley cropping has given corn yields of more than 3–5 tonnes per hectare in the second season (Anon., 1984b).

Approximately 20 per cent higher yields of grains and wood have been reported in agroforestry areas of Haryana and western Uttar Pradesh than from pure agriculture (Dwivedi and Sharma, 1989). In Haryana and western Uttar Pradesh, farmers have planted *Eucalyptus* hybrid and *Populus deltoides* in rows on field boundaries. Observations taken from different areas indicate that the total yield of agricultural crops and wood was more than simple agriculture without trees. In these areas, however, it was seen that growth and yields of crops near the edges of tree rows were poorer. However, the loss in growth and yields of crops was more than compensated when the wood produced was also taken into consideration. Experiments conducted at IGFRI, Jhansi indicate that the total yield of fodder is more when fodder grasses are grown with fodder trees than pure fodder grass cultivation (Deb Roy, 1990). *Leucaena leucocephala* intercropped with agricultural crops and fodder grasses increases the total yield of foodgrain, fodder and fuel (Tiwari, 1970; Anon., 1984b; Pathak, 1989).

The reasons for higher production under agroforestry systems may include: (i) greater efficiency of perennial crops for photosynthesis, (ii) tapping nutrients and water from deeper layers by perennial crops, (iii) creating better environmental conditions for the growth of annual crops, etc. However, under several conditions, intense competition for light, moisture and nutrients has been observed. The annual crops suffer under such conditions. If suitable trees and agricultural crops are selected, production can be substantially increased.

## SUPPLEMENT FOOD AND FODDER

Several trees, shrubs, herbs and climbers yield a substantial quantity of food materials which are used by the rural poor and particularly by tribals. About 213 species of large and small trees, 17 species of palms, 128 species of shrubs, 116 species of herbs, 4 species of ferns and 15 species of fungi are known to yield edible food materials (Solanki, 1981). Important trees and shrubs yielding food are given in Table 33 (Solanki, 1981; Biswas and Bhuyan, 1983; Singh, 1988).

Among these species, perhaps the most important species having a great promise is *Madhuca indica* (mahua tree). It yields an edible flower, fruit and seeds. The flowers are rich in sugar (65 to 70 per cent), calcium (14 per cent), protein (4.4 per cent) and vitamin C and vitamin B complex (Anon., 1962). These flowers are regarded good as a coolant, tonic and demulcent. In view of their high sugar content and absence of any toxic effects, the flowers of mahua are traditionally eaten by tribals and other rural population. The flowers are dried and cooked for eating. They are also powdered and made into *chapatis*, either pure or mixed with wheat flour.

**Table 33:  Important forest species yielding food**

| Species | Part used as food |
|---|---|
| A) Trees | |
| *Aegle marmelos* | Pulp of ripened fruits |
| *Anacardium occidentale* | Fruit; seeds very nutritive |
| *Annona squamosa* | Fruit |
| *Artocarpus communis* | Fruit |
| *A. heterophyllus* | Ripened fruit eaten as fruit; unripened used as vegetable |
| *Bauhinia malabarica* | Tender shoots used as vegetable |
| *B. purpurea* | Flower buds used as vegetable |
| *B. variegata* | Flower buds used as vegetable |
| *Buchanania lanzan* | Seeds edible, very nutritive, yield edible oil |
| *Cordia dichotoma* | Fruit used as vegetable or pickled |
| *Dillenia indica* | Ripened fruit eaten as fruit; unripened used as vegetable |
| *D. pentagyna* | Flower bud and young fruit |
| *Dendrocalamus strictus* | Young shoots cooked for vegetable or pickled |
| *Emblica officinalis* | Fruit eaten or pickled |
| *Ficus hispida* | Green fruit used as vegetable |
| *F. palmata* | Fruit |
| *F. recemosa* | Ripened fruit eaten as fruit; unripened cooked as vegetables |
| *Madhuca indica* | Flowers and ripened fruit eaten; seeds yield edible oil |
| *Mangifera indica* | Fruit |
| *Moringa oleifera* | Flower and fruit used as vegetable |
| *Morus alba* | Fruit |
| *Phoenix sylvestris* | Fruit |
| *Syzygium cumini* | Fruit |
| *Tamarindus indica* | Fruit |
| B) Shrubs | |
| *Asparagus filicinus* | Tuberous roots and tender shoots eaten as vegetable |
| *Capparis decidua* | Fruit and flower bud |
| *Carissa carandis* | Fruit |
| *Salvadora oleoides* | Fruit |
| *Ziziphus* spp. | Fruit |

Mahua trees are widely distributed in the forests of Gujarat, Maharashtra, Madhya Pradesh, Andhra Pradesh, Uttar Pradesh, Orissa, Bihar and West Bengal. The total yield of mahua flowers in the country is reckoned to be of the order of about 1 million tonnes (Subhash Chand and Mahapatra, 1983). There is tremendous scope for increasing this production by raising mahua in agroforestry. The species can be raised either pure or in mixture on bunds in rows or blocks. It can be grown in a variety of edaphic conditions. It can also be grown on wastelands such as saline-alkaline soil, ravined and gullied areas, lateritic, rocky and other areas.

A single tree of mahua can yield about 50 to 320 kg of dried mahua flower (Anon., 1962). One hectare can easily support about 100 trees of medium size. If mahua is introduced into social forestry and agroforestry plantations in its natural zone, the area under this species could easily be increased to about 15 to 20 million hectares. This area with about 100 trees per hectare would be capable of producing about 50 million tonnes of mahua flowers annually, taking only 50 kg mahua flower production per tree (Dwivedi, 1989a). This would be a huge supplement to our food production with almost no investment. The seeds of mahua yield edible oil, which would be helpful in meeting our oil shortages. The only thing to be done is to encourage plantation of mahua in our afforestation programmes, particularly in drought-prone areas and in agroforestry plantations.

Sufficient supply of fodder is necessary to increase milk production and to maintain productive cattle. Trees of several species yield palatable and nutritious fodder. If fodder trees are planted in large areas, they could serve as a fodder bank and be lopped during scarcity periods. During drought years, one of the serious problems is fodder for livestock. Production of tree fodder is perhaps the only solution for meeting the fodder shortages during drought years. Several tree species are known to yield palatable and nutritious green leaf fodder (Sen 'et. al., 1978; Singh, 1982; Gulati et.al., 1982). Some of these species are *Acacia nilotica, Aegle marmelos, Ailanthus excelsa, Bauhinia variegata, Celtis australis, Dalbergia sissoo, Dendrocalamus strictus, Grewia optiva, Hardwickia binata, Leucaena leucocephala, Moringa oleifera, Morus alba, Ziziphus* spp., etc.

Trees are able to produce more fodder per unit area than agricultural fodder crops. They do not die due to drought. Fodder yields from some trees are indicated in Table 34 (Singh, 1982).

**Table 34: Fodder yields from some trees**

| Species | Fodder yield per tree per annum (kg) |
| --- | --- |
| 1. *Acacia nilotica* | 50–120 |
| 2. *Ailanthus excelsa* | 100–200 |
| 3. *Bauhinia variegata* | 30–50 |
| 4. *Grewia optiva* | 25–50 |
| 5. *Prosopis cineraria* | 25–50 |

Fodder trees planted on a large scale could serve as a *fodder bank.* Fodder grasses could be introduced to further increase fodder production.

The leaf fodder production from trees offer another advantage as fuelwood can be obtained as by-product. This would help to meet the energy requirements of the rural population. The trees and shrubs have a great advantage in dry areas over agricultural crops as they can withstand drought because of their capability of tapping the lower layers of soil for water and nutrients. Trees and shrubs are capable of growing in degraded lands as well.

## MEET DIVERSE NEEDS OF THE PEOPLE

As already explained, in 2000 A.D. we will need about 225 million tonnes of foodgrains, 29.5 million tonnes of sugar and gur, 10.0 million tonnes of oilseeds, 24 million bales of cotton, 16.7 million bales of jute, 1136 million tonnes of green fodder, 949 million tonnes of dry fodder, 300 million $m^3$ of firewood, 130 million $m^3$ of industrial and commercial wood and 3.5 million tonnes of bamboo besides a large quantity of tea, coffee, rubber, cashewnuts, spices, etc. Our land resources are limited and, therefore, the production systems have to be efficient enough to produce enough of these commodities on a sustained basis so that our requirements are met. Trees and annual crop associations have been found to increase the total production. Windbreaks and shelterbelts are known to have beneficial effects on agricultural production throughout the world (Caborn, 1957; Frank, 1976). Increase in agricultural production due to windbreaks and shelterbelts in India is also reported (Bhimaya et al., 1958; Kaul, 1959; Rao, 1980). Windbreaks and shelterbelts also increase agricultural yield in cooler regions by protecting the crops from snow and frost. Increase in total production under agroforestry has been observed in different agroclimatic areas. The total production under an agroforestry system is almost always better than in agriculture. Similarly, fodder production is always more under the silvipasture system. If suitable crop combinations are selected, mixed cropping always gives more production. Several examples are available in support of this concept for various edapho-climatic conditions (Verinumbe, 1983; Budelman, 1988; Verma, 1988).

Agroforestry aims to maximise production of biomass of trees and agricultural crops. Experience in Punjab, Haryana, Uttar pradesh, Gujarat and some parts of the southern states indicates that a tree and agricultural crop production system is more productive and is capable of meeting almost all the demands of timber and firewood. Dwivedi and Sharma (1989) worked out, on the basis of actual measurements, that irrigated agricultural land can support as much as 2.5 $m^3$/ha/annum; the unirrigated can support about half of the quantity only as boundary plantation. The total production of wood expected from agroforestry alone is about 350 million $m^3$, which is enough to meet the foreseeable demand of wood in the near future (Dwivedi and Sharma, 1989).

The total production and value of fuel, fodder and small timber in degraded lands are reported to be many times more than the coarse grains usually produced on them (Chaturvedi, 1981; Gupta and Mohan, 1982; Shyam Sunder, 1983; Shanmugasundaram, 1983). In some of the eastern states and coastal areas, fish culture in paddy fields may increase the total production and income. In coffee plantations, a good growth of trees and spices (black pepper and elaichi) etc., are reported. In some areas, mango, cashew-nuts, jamun, etc., are also grown in agricultural lands increasing fruit production. We have trees yielding a variety of products, e.g., food, fruit, fuelwood, timber, fodder, fibre, flosses, oils, medicines, gums, resin, tannins, dyes, spices, etc. Trees yield almost everything which man needs. Therefore, agroforestry systems tend to satisfy all human needs.

IMPROVE SOIL

Agroforestry systems tend to protect soil from several adverse effects. Most agroforestry systems constitute sustainable land use and help to improve soils in a number of ways. Some of these beneficial effects are evidenced in a number of experiments carried out in different parts of the world (Nair, 1987; Young, 1989). Some of the beneficial effects of agroforestry systems for which enough experimental evidence is available include: (i) reduction of loss of soil as well as nutrients through reduction of run-off; (ii) addition of carbon and its transformations through leaf, twig, bark falls, etc.; (iii) nitrogen enrichment by fixation of nitrogen by nitrogen-fixing trees, shrubs, etc.; (iv) improvement of physical conditions of soil such as water-holding capacity, permeability, drainage, etc.; (v) release and recycling of nutrients by affecting biochemical nutrient cycling; (vi) more microbial associations and addition of more root biomass; (vii) moderating effect on extreme conditions of soil acidity and alkalinity; (viii) creating more favourable microclimate by windbreak and shelterbelt effects and (ix) lowering effect on the water-table in areas where the water-table is high. Some other beneficial effects for which sufficient research evidence is yet to be confirmed include: (i) increased precipitation and addition of nutrients; (ii) uptake of nutrients from deeper layers of the soil and deposition on the surface via leaf litter; (iii) withholding of nutrients which are usually lost by leaching and (iv) augmenting of soil water availability in arid and semi-arid conditions by reducing surface evaporation.

It should not be concluded, however, that the above advantages of agroforestry systems are available under all conditions. Considerable drain of nutrients has been reported by short-rotation, fast-growing species, e.g., *Eucalyptus, Casuarina* and others (Cornforth, 1970; Lundgren, 1978; Negi and Sharma, 1984; Singh, 1984). Some tree species, particularly those with a faster rate of growth, strongly compete for moisture and create moisture stress conditions for annual crops, particularly in the zone of competition (Dwivedi, 1987). Some of the tree species create adverse chemical and biological effects leading to acidification, allelopathy, etc. (Del Moral and

Muller, 1970; Rao and Reddy, 1984). The shading effect of trees causes an effect on soil temperature and moisture which sometimes adversely affects the growth of agricultural crops.

Most of the beneficial effects of trees on soils are primarily due to their roots, crown, litter-fall and nitrogen-fixing ability; the effects depend on the density of the trees. Tree crowns protect the soil from the impact of raindrops and create specific microclimate (Pradhan, 1973; George, 1978; Ghosh *et al.*, 1980). The leaf litter from trees absorbs a large quantity of water, which means less quantity of water is available for surface run-off. Soil erosion from vegetated watersheds is always less in comparison to cultivated lands (Sinha, 1975; Anon., 1975). The rate of infiltration of soil water is 3 to 5 times more in forested areas in comparison to agricultural lands (Ghosh, 1974).

The nitrogen-fixing ability of different trees and shrubs being grown in agroforestry has not been thoroughly investigated. The nitrogen fixation of 50 kg to 500 kg per ha is reported in some tree species (Vergara, 1982; Anon., 1984b; John, 1988). The lands under pure agriculture are always deficient in soil organic matter. The leaf litter after decomposition forms humus, releases nutrients and improves various soil properties. The nutrient content of leaf litter may be 150 kg to 300 kg nitrogen, 10 to 20 kg phosphorus, 75 to 150 kg potassium and 100 to 300 kg of calcium per ha per annum which may mean that nutrient recycling can substantially reduce the fertilizer needs (Negi and Sharma, 1984; Singh, 1984; Anon., 1989). Roots in several trees develop a mycorrhizal association which helps in nutrition and water absorption. It also helps against infection with diseases. The organic matter increases the biochemical activities in the soil.

The protective role of windbreaks and shelterbelts on soil and vegetation in hot arid and colder regions has been very well demonstrated (Caborn, 1957; Bhimaya *et al.*, 1958; Rao, 1980; Shankarnarayan *et al.*, 1987). Control of sand dunes and their reclamation has been obtained by raising windbreaks and afforesting them with resistant species. Afforestation of these areas leads to gradual improvement of soil conditions. Trees tend to lower the water-table and, therefore they can be effective in drying waterlogged and areas with a high water-table.

Forests are reported to increase 5 to 15 per cent annual precipitation (Voelcker, 1897; Martin, 1944; Ranganathan, 1949; Nicholson, 1960; Rakhamnov, 1966; Warren, 1974). It is reported that annual nutrient supply through rain-water in temperate regions is about 1 to 10 kg K, 3 to 19 kg Ca, 4 to 11 kg Mg, 0.2 to 0.6 kg P, and 0.8 to 4.9 kg N (Ovington, 1965; 1968). The increase in precipitation will result in a higher quantity of nutrient supply from which agricultural crops can benefit. Trees with long taproots recover nutrients lost to the subsoil through leaching and infiltration and recycle them to the surface in the form of litter-fall. The litter after decomposition releases nutrients back to the surface soil. The nutrients lost to the rivers and lakes through leaching and surface run-off cannot be recovered but are replaced when the trees absorb minerals just released by newly weathered parent rocks in the lower soil strata and pump them to the surface in the form of litter-fall (Vergara, 1982). The trees cast shade on the ground which reduces evaporation. This reduces the irrigation requirement for agricultural crops by 30 per cent in some cases (Sharma, 1988).

UTILISE WASTELANDS AND DEGRADED LANDS

It has already been discussed in Chapter 1 that we have more than a 100 million ha area under different kinds of wastelands. These wastelands and degraded lands are not being effectively utilised under agricultural land use because of several reasons. However, these lands can be gainfully utilised for the cultivation of trees. Once the area is vegetated, ecological restoration processes start which lead to improvement in soil conditions. Once the soil conditions improve, these lands could be utilised for agriculture also. Experience of the past confirms that trees have been successfully used to reclaim and improve various kinds of wastelands and degraded lands. Almost all types of wastelands have successfully been planted. A number of species have been found capable of growing in saline-alkaline, waterlogged, ravined, montane hill slopes, landslides, mine overburden, eroded areas, etc. The important considerations in afforestation of wastelands and degraded lands are: (i) selection of suitable species, (ii) selection of appropriate technology, (iii) sufficient inputs and (iv) effective organisation.

Though some of the species which are capable of growing on wastelands and degraded lands are fairly identified, there is a need to scruitinise different species and provenances for specific situations. The technology for soil working, planting, maintenance, etc., needs to be further developed so that it can become more cost effective (NWDB, 1989a). The afforestation of degraded lands and wastelands requires comparatively more inputs which need to be provided. The areas under private ownership are likely to be managed by the individual owners but for areas under community and government ownerships, suitable micro-level organisation for managing such lands has been found necessary (Verma, 1988). Microplanning of all operations helps in achieving the desired results (NWDB, 1987a).

There are several examples that afforestation of wastelands with suitable species of trees and grasses has led to improvement in soil conditions and good financial returns. It has been found that the afforestation programme of these lands needs support from the government as initial investments are usually higher than ordinary site conditions. Research evidence indicates that several categories of land such as arid areas and montane areas can be better managed with greater production and sustainability by tree-based land use systems (Gupta and Mohan, 1982; Martins and Nautiyal, 1987).

PROVIDE EMPLOYMENT OPPORTUNITIES

Unemployment is the country's main problem. About 7 million people enter the job market every year in India (Ranganathan, 1979). The industrial sector provides about half a million jobs and about 6.5 million are left unemployed, totalling more than 80 million at present, three-quarters of which live in rural areas. In spite of various programmes of rural employment, there are serious problems of rural unemployment and underemployment in the country.

Agroforestry systems increase employment opportunities. Agricultural labour which is underemployed finds alternative labour opportunities in forestry works. Most of the forestry activities are labour intensive and considerable employment opportunities are generated from agroforestry works. Plantations, including nursery operations, generate employment of about 200 to 500 man-days/ha (NCA, 1976; Parasnis and Gogate, 1979; Pant, 1984; Dwivedi, 1986). Subsequent care and maintenance also provide employment to the tune of 50 to 75 man-days per ha/annum. Harvesting of trees is another major activity which provides about 10 to 15 man-days/m$^3$ of wood harvested. Besides generation of employment opportunities at primary level, about 10–20 times more employment is generated at the secondary and tertiary levels, e.g., in wood-based industries such as sawmilling, furniture, sport goods, pulp and paper, plywood and panel products, match-splints, bamboo and cane furniture, agricultural implements, doors, windows and other constructional materials, photoframes, handles, packing cases, musical instruments, etc.

Several activities such as production of nursery stock and development and production of tools of forestry operations could be taken up to boost labour opportunities. Effort should be to supplement agriculture with forestry activities. Complete replacement of agriculture with forestry might adversely affect labour opportunities. Therefore, effort has to be made to properly plan both activities so that more labour opportunities are generated.

INCREASE FARM INCOME

Agroforestry gives more income to the farmer per unit of land than pure agriculture of forestry. Several studies in different parts of the world suggest that agroforestry is more profitable to farmers than agriculture or forestry (Chaturvedi, 1981; Lahiri, 1983; Pillai, 1983; Mathur and Sharma, 1983; Mathur et al., 1984; Chandra, 1986; Patel, 1988).

In fact, agroforestry provides the farmer with a large number of alternatives of agricultural, forestry and horticultural crops. The farmer can select such combinations as are suitable and most profitable to him. In India, during the last two decades the prices of agricultural commodities have, in general, increased by almost three or four times but in the case of forestry products, e.g., timber, firewood and charcoal, etc., the prices have increased by fifteen to twenty times (Dwivedi, 1985). This change in price structure has made tree growing a highly profitable and competitive land use in some areas. During the last 15 years or so, farmers have resorted to planting of trees on farm bunds, along canals, paths, around houses/wells, etc., and harvested rich dividends. In some areas even good agricultural land has been diverted for tree farming.

Trees grown with agricultural crops either horizontally or vertically do compete for water, nutrients, light and space to a certain extent. Crop yields have been found to decrease under several conditions near the vicinity of trees. Nevertheless, the total annual returns to the farmer were found to be more under agroforestry in comparison to pure agriculture. In order to obtain maximum net returns per unit area, suitable tree, crop and fodder combinations need to be identified for each agroclimatic region.

Growing of trees and fodder crops (including fodder trees) is more economical, particularly in marginal lands. Observations taken in hot arid and semi-arid areas of Rajasthan indicate that marginal lands are incapable of sustaining stable and dynamic cultivation of agricultural crops. Silvipastural use consisting of growing of trees such as *Prosopis, Albizia, Ziziphus, Acacia,* etc., may provide many times more returns per unit of land than agriculture under such conditions (Gupta and Mohan, 1982). *Eucalyptus* in agroforestry has been found to be more profitable than pure agriculture in Haryana (Pillai, 1983; Hooda, 1983). *Populus deltoides* increases the farm return by 50 per cent in terai regions of Uttar Pradesh (Chaturvedi, 1981; Mathur *et al.*, 1984).

Patel (1988) found that with high density plantation with adequate water and fertilisers in agroforestry, a farmer can earn as much as Rs. 37,500 to 60,000 per ha per year through a 3 to 6 year tree cutting cycle in comparison to an annual return per ha of only Rs. 8700 from banana, Rs. 6500 from sugar-cane and Rs. 5700 from potato crops.

Some of these results may be true for specific areas where timber and fuelwood demands and prices are very high, particularly near cities, but these results indicate the trend. Some of the farmers have therefore started converting good agricultural lands for tree cultivation. The Government of India, therefore, in a recently announced forest policy declared that good agricultural land will not be allowed to be converted for pure forestry (Anon., 1988c).

MINIMISE ADVERSE EFFECT OF CLIMATIC FACTORS

A large part of the country is periodically affected by drought and floods. According to available records, the area affected by moderate, large, severe and disastrous drought is 76.2, 47.0, 11.68 and 19.7 million ha respectively.

These areas, however, overlap. The area affected by periodical floods is 40.0 million ha.

During droughts and floods, the agricultural crops are either destroyed or their production is considerably reduced. Trees and shrubs are deep rooted and have the capacity to survive even under severe drought conditions. Several species of trees survive waterlogging and inundation. Several trees and shrubs may provide food, fodder and other products during the most critical periods. During such difficult periods, trees provide insurance for survival. If a farmer cuts a few trees and sells the wood in the market, he will have ready money for sustenance during such difficult periods. Agroforestry, therefore, is the most appropriate land use system for areas which periodically suffer due to drought and floods.

During severe drought years, there are serious problems of food, fodder and water. Foodgrains are easily transported due to improvement in communication and transport modes and the problem is mitigated to a considerable extent. Fodder is bulky in nature and transport over long distances is not feasible. Serious fodder problems were recently faced in some parts of the country during 1979, 1985 and 1987 when droughts occurred. Fodder trees for such periods are very useful. Similarly during floods, when agricultural crops and grass fodders are destroyed, tree fodder is useful.

AID INDUSTRIAL GROWTH

Agroforestry systems are able to sustain both kinds of industries, i.e., those based on agricultural and forestry raw materials. Due to dwindling of forest resources in the past few decades in the country, serious problems are anticipated for forest-based industries. Some of the existing industries are facing acute raw material shortage and are threatened with closure. Some of them have been able to manage for the time being, with imported raw material.

Agroforestry systems are capable of meeting the demands of raw materials of several agricultural and forest-based industries. Some of the industries, e.g., paper and pulp mills, sport goods, furniture, sawmills, etc. located in Haryana, Punjab, Uttar Pradesh, etc., are meeting their total raw material requirements from agroforestry produce. *Eucalyptus* hybrid grown by farmers on a large scale is now being used for a variety of purposes. *Poplar* has been widely cultivated in the terai area of Uttar Pradesh and Haryana and is being used by several industries, e.g., match splints, plywood, packing cases, etc. Some industries have started buying back guarantee schemes for planting specific tree species required by the industry. A similar system, which yielded good results, was initiated by the Paper Industries Corporation of the Philippines (PICOP) in the seventies which aimed at the development of agriculture and tree plantation, firstly to ensure a constant supply of raw material for its paper and pulp mills and secondly to improve the socio-economic conditions of the farmers (FAO, 1978).

The Government of India, in its revised forest policy, declared in 1988, in paragraph 4.9 recommended that: (i) Industry should establish a relationship with individuals who can grow raw materials and should provide inputs including credit, technical advice and harvesting the transport services. (ii) Forest-based industry should not only provide employment to local people on priority but also involve them fully in raising trees and raw material. (iii) Farmers, particularly, small and marginal farmers should be encouraged to grow on marginal/degraded lands, wood species required by the industries. It is clear that an agroforestry system would be able to meet the needs of the people and of the industries.

IMPROVE ENVIRONMENT

There is serious concern all over the world due to degradation of environment. Serious adverse ecological manifestation, increase of $CO_2$ in the atmosphere, global warming, serious soil losses, repeated droughts and floods, serious pollution problems, etc., are the results of deforestation. There are concerted efforts to check the process of ecological degradation by increasing the tree cover of the earth.

An agroforestry system helps to increase the tree cover. It also makes available to people the required quantity of timber, firewood, fodder, etc., for which they traditionally depend on forests. Thus, this system helps to reduce the pressure on forests.

Several areas are seriously affected by many kinds of pollutions. Air, water and noise pollutions are common. Trees protect us from different kinds of pollutants. They protect us from dust, dirt and other physical air pollutants. One hectare of a close forest can filter about 50 tonnes of dust and dirt (Durk, 1966). Some of the chemical air pollutants are also absorbed by the trees, protecting human beings from their adverse effects. Noise pollution is also checked considerably by the trees.

Forests and trees increase precipitation and humidity and decrease temperature (Voelcker, 1897; Zon, 1927; Ranganathan, 1949; Nicholson, 1960; Storey, 1966; Dabral *et al.*, 1969; Warren, 1974). Afforestation increases precipitation by about 10 per cent. Air temperature is less in forested areas. Studies indicate that air temperature is 3° to 8°C less in areas covered with trees (Storey, 1966; Dabral *et al.*, 1969; Warren, 1974). In several parts of the country where the air temperature goes very high, reduction in temperature of even 2°–3°C makes life more comfortable and checks various diseases caused by extreme temperature. It also reduces evaporation from the soil which reduces the irrigation requirement. Increase in humidity of 5 to 8 per cent is also reported in forested areas (Warren, 1974). Trees and forests also protect from strong winds and cyclones. A good stand of trees reduces the wind velocity up to 20 to 60 per cent. In frost-affected areas, the trees protect the annual crops from frost damage. Forests and trees help to fight droughts and floods. They are effective in controlling the soil erosion, both by wind and water.

Of the total cultivated area of 164 million ha, more than 100 million ha suffer from various degrees of soil erosion. Agroforestry systems are less prone to soil erosion and hence conserve the environment. Several agro-forestry systems protect and enrich soils. The trees intercept and transpire a large quantity of water and, therefore, a less quantity of water is available for surface runoff. Reduction in erosion and surface runoff helps in reducing flood damage. Subsurface runoff is more which is slow and does not cause damage. The approximate movement of rain-water in an agricultural and wooded ecosystem is indicated in Fig. 2.

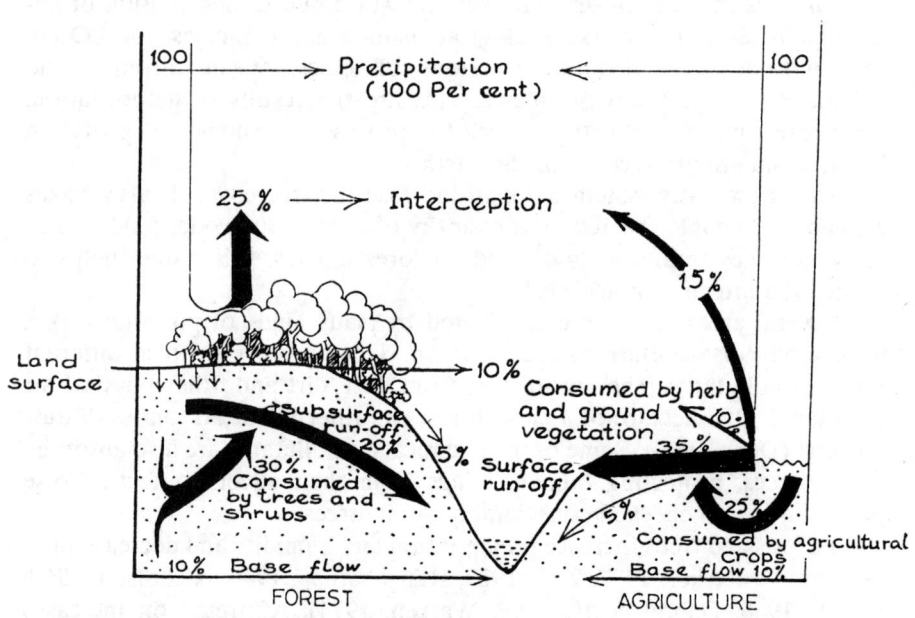

**Fig. 2: Role of forests in hydrology**

Several trees have been found suitable for growing over mined areas and other wastelands. Even in arid areas, some trees can grow and reduce the severity of the desert climate.

Agroforestry systems make available timber, firewood, fodder, and various other products which are usually provided by the forests. Once the needs of the people in respect of forest products is satisfied, the dependence of the people of forests will reduce. This will be of a great help for conservation and development of forests. The forests located in fragile ecological zones will be restored and protected, the degradation will halt and the biological diversity will be preserved. Agroforestry systems increase the farmer's income. Agroforestry, therefore, should be looked upon as a means for improving the economic condition of the rural poor and should be the main plank for integrated rural development programmes. Once the poverty is removed, the ecological restoration and environmental conservation becomes much easier.

# CHAPTER 4
# Classification of Agroforestry Systems

## Introduction

Agroforestry is a new name for old practices. The term now is usually used to include all land-use systems in which woody perennials form a part or are grown deliberately on the same piece of land as agricultural crops and/or animals either in the form of a spatial arrangement or in sequence. The beginning of agriculture and pasture started with the forest being the most dominant component. The present practice of shifting cultivation is indicative where the forestry component dominates over agriculture with regard to time and space. The practice of settled agriculture led the way to the dominance of agriculture. In India, due to the increase in population, food production had to be increased. Hence intensive agricultural practices were adopted. It was, soon realised, however, that such practices lead to increased soil degradation. Also, an acute shortage of wood and other forest products was felt. Therefore, more recently there have been serious efforts to deliberately introduce trees in one form or the other in various agricultural systems. The demand for wood and other products heightened the interest in the concept of intercropping and integrated farming system in the country.

Agroforestry systems refer to such distinct agroforestry practices in which agriculture, forestry and pasture uses of land are combined either temporally or spatially. There may not be distinct pathways of inputs and outputs as required in typical systems. However, the arrangements of different types of crops, e.g., animals, perennials, trees, etc., and the level of interaction between different components are distinct for individual agroforestry systems. Several agroforestry systems are old. Some of them have been considerably modified over the years. However, some of the systems retain their form to a considerable extent. These agroforestry systems vary depending upon the region, terrain, climate, soil, etc.

**Classification of Agroforestry Systems**

Since their inception, a large number of agroforestry systems have evolved in different areas. One system differs from the other in respect of structure, composition, age, intensity, technology, inputs, etc. Several systems could be grouped together as they are similar in one or many of these characteristics. The objectives of classification of agroforestry systems should be: (i) include a logical grouping according to the major factors on which production of the system will depend, (ii) indicate how the system is managed, (iii) offer flexibility in regrouping the information and (iv) be easily understood. It is difficult to work out anyone system of classification of agroforestry systems capable of meeting most of the requirements. The systems can be grouped on the basis of anyone factor or function of the farming system. Nair (1985) classified the agroforestry systems on the basis of structure, function, socio-economic and ecological status. The agroforestry systems can be classified on the basis of the following factors:

  (i) Structure
 (ii) Function
(iii) Physiognomy
 (iv) Floristics
  (v) Ecological considerations
 (vi) Socioeconomic factors
(vii) History

1. STRUCTURAL CLASSIFICATION
    Structure refers to composition, stratification, and dimension of the crop. The classification of agroforestry systems on the basis of the nature of composition is widely recognised and several workers have classified agroforestry systems on the basis of composition into: (i) agrisilviculture, (ii) silvipasture, (iii) agrisilvipasture, and (iv) others (King, 1979; FAO, 1979; FAO, 1981; Nair, 1985; Gholz, 1988). King (1979) considered agroforestry to be a generic term and included: (i) agrisilviculture systems, (ii) silvipastural systems, (iii) agrisilvipastural systems and (iv) multipurpose tree plantation systems as specific components of agroforestry.

(i) *Agrisilviculture*
    Agrisilviculture means use of land for the concurrent production of agricultural crops and forest crops. Agrisilviculture covers all systems in which land is used to produce both forest trees and agricultural crops, either simultaneously or alternately (FAO, 1978). Many forms of agrisilviculture are prevalent. In India, agrisilviculture is used for intercropping of a forest plantation with agricultural crops in the initial years, until the canopy of the forest trees closes. The system has been used to raise forest plantations on

government-owned land using the labour of land-hungry farmers or labourers. Agrisilviculture also includes growing agricultural tree crops with forest trees. Agricultural tree crops include cocoa, coffee, oil palm, coconut, citrus, rubber, papaya, etc. In this form of agrisilvicultural system there is a keen competition between the two types of crops for water, nutrients, light and space.

### (ii) *Silvipasture system*

The silvipasture system means a land-management system in which forests are managed for the production of wood as well as for rearing of domesticated animals (King, 1979). It does not include the destructive over-grazing as practised in Indian forests, grazing lands and pastures. It also does not include the growing of fodder crops that are harvested and fed to stalled animals, which is agrisilviculture. In Indian conditions, silvipasture includes management of fodder grasses in natural forests or in plantations or in grasslands with a view to obtaining the maximum yield of wood, fodder and other products. The trees may include both wood-yielding trees and fodder-yielding trees. The fodder may include natural grasses or improved fodder grasses seeded artificially. In fact there should be no rigidity about grazing by cattle. The system should include even grass-cutting. Under Indian conditions, it is very difficult to manage the area under proper intensity of use with uniform and appropriate livestock use and within carrying capacity of the area.

### (iii) *Agrisilvipasture system*

This system is a combination of the agrisilviculture and silvipasture systems. The land is managed for the concurrent production of agricultural and forest crops and for grazing by domestic animals. If a unit of land is managed under crop rotations or practices which may include production of foodgrains, fodder and wood and has provision for grazing cattle, the system can be called an agrisilvipasture system.

### (iv) *Multipurpose forest tree production system*

In this system, forest tree species are regenerated and managed for their ability to produce not only wood, but leaves and/or fruits that are suitable for food and/or fodder. In this system, the forest is managed to yield multiple products. In addition to wood, the trees may yield fruits, flowers, leaves, barks, roots, gums, honey, medicines, etc., which may be eaten and/or utilised for other purposes. A multiple-product system is practised particularly in areas where tribals live as they derive from forests not only wood, but a large number of other products which they consume themselves or sell in the market. In some parts of the country, villagers cultivate *mahua*, *sal*, jackfruit, etc., not only for wood, but for other products such as fruits, seeds, etc. Mahua flowers are edible. Tribals produce liquor from them. Mahua seeds yield oil which is edible; mahua leaves are utilised for making plates. Similarly, sal seed yields oil; sal gum is used as essence. Jackfruit yields fruits. There are similar other examples.

Combe (1982) suggested 24 classes of agroforestry systems based on mainly the kinds of associated agricultural products, major functions of tree components, spatial arrangements of the trees, rotation of trees and the tree and crop combinations. The International Council for Research in Agroforestry (ICRAF), Nairobi, has tried to enlist all agroforestry systems being followed in different parts of the world. The important agroforestry systems which are recognised and prevalent in different parts of the world are classified and given in Table 35 (Nair, 1985; Gholz, 1988).

(i) *Classification based on dominance of components*: Tejwani (1987) tried to classify these systems on the basis of predominance of the components. Accordingly, if trees are a major component of land use and agricultural crops are integrated with them, the system may be called '*silvoagriculture*', for example, shifting cultivation, taungya cultivation, etc. Here the primary aim of the land use is silviculture. When agriculture is the major component and trees are secondary, the system may be called '*agrosilviculture*'. Several systems, such as multipurpose trees on farm land, hedgerow cropping or alley cropping, intercropping of trees with fruit trees and commercial crops, home gardens, etc., are examples of agrisilviculture. Similarly, in trees with pasture, if trees are the major component and pastures are secondary, the practice may be referred to as the '*silvopasture system*'. Most grazing in forests can be called silvipasture, specifically forest grazing in arid and semi-arid areas and temperate areas. The overgrazing practices beyond the carrying capacity of forests is not a silvipasture system. If pasture is a major component and trees are secondary, the system may be called *pastural silviculture*. This system will include those areas both under government and private control which are primarily used as grazing lands. The combination of crops, trees and pastures are classified as an '*agrosilvopasture*' system. '*Silvoagropasture systems*' are those where silviculture is dominant. '*Agrisilvopastural systems*' include combinations of agriculture, tree crops and pastures, e.g., home gardens, etc. This is one of the most complex and multilayer agroforestry systems. In this system, the agricultural component takes precedence over the other components. '*Silvoagropasture*' could also be identified as a system in which trees are the dominant component among other components.

Nair (1985) tries to differentiate between the words 'agrisilviculture' and 'agrosilviculture'. According to him, 'agrisilviculture' is used to denote those combinations involving only agricultural crops and trees, while 'agrosilviculture' can encompass all forms of agriculture (including animal husbandry) with trees and thus is another word for agroforestry (Nair, 1985). It should be pointed out, however, that such a distinction might create

**Table 35: Classification of agroforestry systems**

| Major systems | Subsystems/practices | Indian examples |
|---|---|---|
| 1 | 2 | 3 |
| 1. Agrisilvi-cultural systems | 1. Improved fallow (in shifting cultivation areas) | Improvements in shifting cultivation practices in several areas of north-east India. |
| | 2. The taungya system | Forest plantations are established and agricultural crops are allowed to be grown until canopy closes. The practice has been popular in several states, e.g. Uttar Pradesh, West Bengal, etc. |
| | 3. Tree gardens | Usually associated with fruit trees mixed with trees yielding timber, firewood, fodder, etc. |
| | 4. Multipurpose trees and shrubs on farm lands | Several forms of plantations, e.g. single row or multiple rows on fields on one side or all sides, scattered trees, etc., e.g., eucalypts, sissoo, babul, poplars, etc., on field bunds in Haryana, Punjab and Uttar Pradesh, and *Casuarina* in South India. |
| | 5. Hedgerow cropping or alley cropping | Intercropping *Leucaena leucocephala, Sesbania* spp., etc., with jowar, maize, and other crops in Maharashtra and western Madhya Pradesh. |
| | 6. Tree crop with plantation crops | Plantation of several multipurpose trees such as *Albizia lebbeck, Terminalia myriocarpa* and *Gliricidia* spp. in coffee and tea plantations in South India and Assam as a combined production system. |
| | 7. Homestead plantations | Homestead agroforestry in which trees of different species, e.g., *Acacia nilotica, Azadirachta indica, Dalbergia sissoo Moringa oleifera,* bamboos, etc., are grown around houses, wells, tanks, paths, etc., to supply small timber, firewood, fruits and fodder. |
| | 8. Shelterbelts and windbreaks | Shelterbelts and windbreaks of *Prosopis* spp., *Eucalyptus, Acacia* spp., *Dalbergia* spp., *Albizia* spp., etc., in arid areas and *Casuarina equisetifolia* in coastal areas are popular in India. |
| 2. Silvipastural system | 9. Cut and carry fodder production | In several areas, particularly in social forestry plantations where grazing is prohibited but grass cutting is allowed. In plantations, fuel and fodder trees mostly predominate. |

| | | | |
|---|---|---|---|
| | 10. | Living fence of fodder trees | Several multipurpose trees and shrubs, e.g., *Sesbania* spp., *Leucaena leucocephala, Acacia nilotica, Dalbergia sissoo, Syzygium cumini, Grewia optiva* (in northern Indian hills) are planted on the boundary of the field. |
| | 11. | Trees and shrubs in pasture | Several tree species, e.g., *Prosopis cineraria* and *Acacia* spp., in arid areas, *Acacia nilotica, Ailanthus excelsa, Dalbergia sissoo, Pongamia pinnata, Aegle marmelos*, etc., in plains and oaks, *Celtis*, etc., in the hills are widely used. |
| 3. Agrisilvipastural systems | 12. | Woody hedges for mulch, green manure, soil conservation | *Sesbania* spp., *Leucaena* and some other species are used in several forms. |
| | 13. | Home gardens | The system is common in almost all areas. |
| 4. Other systems | 14. | Multipurpose woodlots | Practised in areas where acute shortage of fuelwood and small timber is felt. |
| | 15. | Agrisilvifishery | It is being tried in Nagaland where fish ponds are made in paddy fields. |
| | 16. | Shifting cultivation | Common in north-eastern states. |
| | 17. | Agriculture with sericulture, apiculture, etc. | Practised in some parts of the country, particularly in Madhya Pradesh, Bihar, Orissa, etc., in tribal areas. |

confusion as it is improper to equate the terms 'silviculture' and 'forestry'. In this book, therefore, the term agrisilviculture in its true meaning has been used.

(ii) *Classification based on arrangement of components*: Nair (1985) classified agroforestry systems also on the basis of arrangement of components in respect of space and time. In respect of space, arrangement of components may be: (i) *mixed dense* as in home gardens, (ii) *mixed sparse* as in pastures with scattered trees, scattered trees on agricultural lands and (iii) *strip or boundary plantations* as trees on the edges of a field or intercropping of trees and crops.

On the basis of the arrangement of components in respect of time, the agroforestry systems are classified into: (i) *Coincident*: when different crops occupy the land together as coffee under shade trees, and pasture under trees. (ii) *Concomitant*: when components stay together for some part of life as in taungya plantation. In taungya plantation, agricultural crops are allowed to be grown only for a few years. (iii) *Intermittent* (space dominated): when annual crops are grown with perennial crops, e.g., paddy with coconut. (iv) *Interpolated* (space and time dominated): when different components occupy space during different times as in home gardens. (v) *Separate*: when components occupy space during separate times as improved fallow in shifting cultivation.

The different components involved in agroforestry systems are: (i) agricultural crops, (ii) forestry crops which consist mainly of woody perennials, e.g., trees, shrubs, etc., and (iii) pastures or animals. Under Indian conditions, however, only pastures may be considered as practices consisting of regulated grazing is rarely available. Some other components are also included in agroforestry systems. Sericulture, apiculture, pisciculture and lac-culture are usually associated with agriculture and forestry. *Sericulture* is cultivation of silkworms for the production of silk which feed on the leaves of *Morus alba, Terminalia arjuna,* etc. Apiculture is the cultivation of honey bee (*Apis* spp.) which depends for honey on flowers of different species of plants. Some species of trees are known to produce abundant flowers and are considered important for the production of honey. *Pisciculture* is the cultivation of fish. *Lac-culture* is the cultivation of lac insect (*Laccifer lacca*) for the production of lac. The insect feeds on the leaves of palas (*Butea monosperma*), kusum (*Schleichera oleosa*), etc. It can be seen that sericulture, pisciculture, apiculture and lac-cultivation, etc., are also associated with growing of trees and land use. Agroforestry systems should therefore combine production of agriculture, forestry, pasture or animals, sericulture, apiculture, pisciculture, lac-culture, etc. Therefore, possible agroforestry practices may be many depending upon the number of components. For example, agriculture + silviculture = agrisilviculture. Agriculture + silviculture + pasture = Agrisilvipasture. Agrisilvipasture can be combined with sericulture, lac-culture and apiculture also. Paddy cultivation and pisciculture are being tried in some areas of Nagaland. Tree growing and fish culture is also practised in some areas. Sericulture and lac cultivation is already associated with agricultural practices, particularly in the tribal areas of Bihar, Orissa and Madhya Pradesh. Some of the agroforestry systems which are more common in India are shown in Fig. 3.

(iii) *Classification based on stratification of components*: On the basis of vertical stratification, agroforestry systems may be classified into: (i) single stratum or single layered, (ii) double stratum or double layered and (iii) multistratum or multilayered. Some of the examples of these under Indian conditions are as under:

(i) *Single layered*: The crops are usually in one layer or storey, as in hedgerow intercropping, multipurpose tree cropping, tree gardens, etc.

(ii) *Double layered*: The crops are usually in two layers as in agrisilviculture—wheat cultivation under poplar (common in Haryana, Uttar Pradesh), plantation crops under shade trees, silvipasture, etc.

(iii) *Multiple layered*: The crops form several layers as in some systems in agrisilviculture, silvipasture and agrisilvipasture. For example, *Coffee arabica, Piper nigrum* under *Terminalia myriocarpa* or *Albizia* spp. Home gardens where large trees, small trees, shrubs, fruits, vegetables, etc., are grown together.

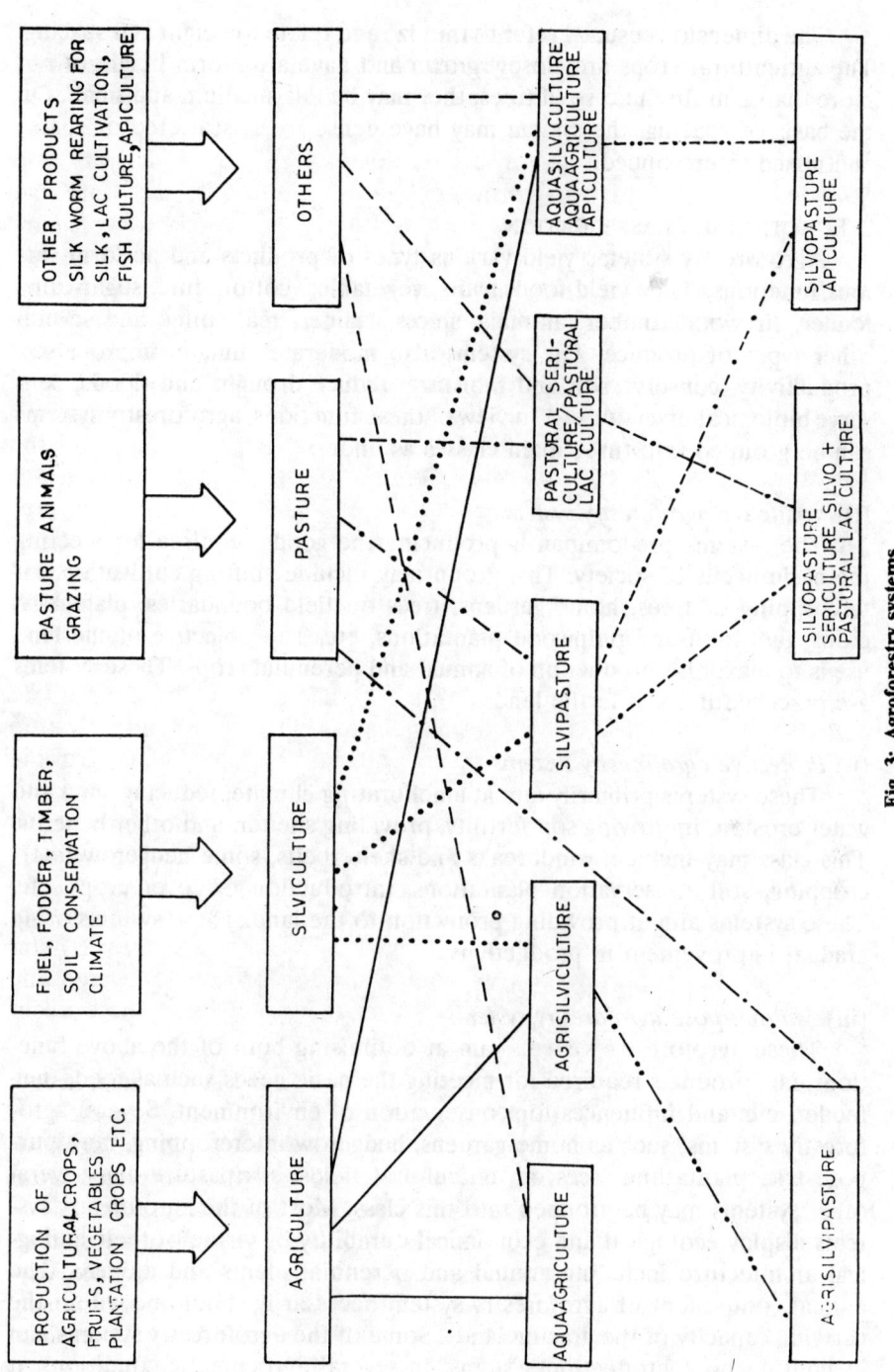

**Fig. 3: Agroforestry systems**

The dimensions usually refer to the size and relate to height and spacing. The agricultural crops are closely grown and have a uniform height of not more than 2 m. In the case of trees, they may be tall, medium and small. On the basis of spacing, the system may have dense trees, scattered or sparse and mixed intercropped.

## 2. FUNCTIONAL CLASSIFICATION

Agroforestry systems yield various types of products and perform various functions. They yield food, fruits, vegetables, cotton, jute, sugar-cane, fodder, firewood, timber, animals, spices, rubber, tea, coffee and several other types of produce. The systems also moderate climate, improve soil productivity, conserve soil and moisture, reduce drought and floods, conserve biological diversity, etc. In view of these functions, agroforestry systems can be grouped into three main classes as under:

### (i) *Productive agroforestry systems*

The systems predominantly producing the goods required for meeting the basic needs of society. This group may include shifting cultivation, intercropping of trees, home gardens, trees on field boundaries, plantation crops, fuelwood and pulpwood plantations, etc. The objective of the land use is to maximise production of annual and perennial crops. These systems are practised in most fertile lands.

### (ii) *Protective agroforestry systems*

These systems primarily aim at ameliorating climate, reducing wind and water erosion, improving soil fertility, providing shelter, and other benefits. This class may include, windbreaks and shelterbelts, some hedgerow intercropping, soil conservation plantations, introduction of cover crops, etc. These systems aim at providing protection to the land. These systems bring gradual improvement in productivity.

### (iii) *Multipurpose agroforestry systems*

These agroforestry systems aim at optimising both of the above functions, i.e., products required for meeting the basic needs such as food, fuel, fodder, etc. and influences for conservation of environment. Several agroforestry systems, such as home gardens, hedgerow intercropping, multipurpose tree plantation, trees on agricultural fields, silvipasture and several other systems may be grouped into this class. Most of the agroforestry systems display ecological and economical durability by virtue of their biological architecture including annual and perennial plants and animals. The animal component of agroforestry system necessarily should be within the carrying capacity of the grazing lands. Some of the agroforestry systems aim at multipurpose production systems. In several agroclimatic conditions, it has been demonstrated that some of these systems bring real increase in the quality and variety of foodstuffs in addition to their role in soil rehabilitation and environmental improvements. The multipurpose agroforestry systems have such productivity which is sustainable.

## 3. PHYSIOGNOMIC CLASSIFICATION

Physiognomy refers to characters of vegetation, such as xeromorphic, mesomorphic or hydromorphic, etc. Paddy cultivation and fish culture and aquasilviculture may be *hydromorphic systems*. The systems are prevalent in some of the north-eastern states. Dry farming in arid and semi-arid areas may be a *xeromorphic system*. The system is characterised by such species of plants, both annuals and perennials, which have low moisture requirements. The system where water is available in sufficient quantity for crop growth may be called a *mesophytic system*. Mesophytic conditions are good for plant growth. These systems are characterised by a large variety of useful plants which are available for cultivation.

## 4. FLORISTIC CLASSIFICATION

Floristic characters can also be used for classification of agroforestry systems. For defining agroforestry systems on the species composition, it is necessary to have an inventory of all existing systems along with their component plant species. Some of the known combinations which are widely adopted in different parts of the country are given in Table 36.

## 5. ECOLOGICAL CLASSIFICATION

Agroforestry practices relate to cultivation and growth of different species of plants. The cultivation and growth of different species of plants depend mainly upon climate, soil, physiography, natural vegetation, etc., of the area. On the basis of these parameters, several agroecological or silviecological regions can be identified and the agricultural and forestry species which are commonly growing or which could be grown can easily be identified. In such agroecological regions, the system may be classified on structural basis as agrisilviculture, silvipasture, agrisilvipasture and others or any other basis.

The agroforestry systems can be classified on the basis of individual ecological parameters also. For example, the agroforestry systems on the basis of climate may be: (i) tropical, (ii) subtropical, (iii) temperate and (iv) subalpine and alpine. On the basis of moisture conditions, each of these groups can be further subdivided into: (i) wet, (ii) moist and (iii) dry. In India with a wide variation in climatic, edaphic and physiographic conditions and with a large biological diversity of flora and fauna, it is difficult to have a perfect ecological classification.

**Table 36: Plant species in agroforestry systems**

| Regions/Area | Agricultural components | Forestry components |
|---|---|---|
| 1 | 2 | 3 |
| 1. Alluvial region | Rice, wheat, sugar-cane, pulses, oilseeds | *Eucalyptus* hybrid, *E. tereticornis, Populus deltoides, Dalbergia sissoo, Morus alba, Azadirachta indica, Acacia nilotica, Bombax ceiba,* etc. |
| 2. Arid and semi-arid areas | Maize, jowar, bajra, small millets, wheat, pulses, etc. | *Acacia nilotica, A. senegal, A. tortilis, Prosopis cineraria, P. chilensis, Eucalyptus camaldulensis, Azadirachta indica, Ziziphus* spp. |
| 3. Northern hill areas | Maize, paddy, wheat, fruits, vegetables, etc. | *Grewia optiva, Morus serrata, Celti. australis, Quercus* spp., *Albizia chinensis, Prunus* spp., *Populus ciliata,* etc. |
| 4. Central region | Wheat, rice, maize, jowar, bajra, pulses, oilseeds, etc. | Bamboos, *Mangifera indica, Dalbergia sissoo, Moringa oleifera, Acacia nilotica, Azadirachta indica, Terminalia arjuna, Albizia* spp., *Tectona grandis, Eucalyptus* hybrid, etc. |
| 5. Southern region | Rice, tobacco, chillis, sugar-cane | *Casuarina equisetifolia, Eucalyptus* hybrid, *E. tereticornis, Acacia* spp., *Albizia* spp., *Dendrocalamus hamiltonii, Tamarindus indica, Santalum album, Anacardium occidentale.* |
| 6. Coastal areas | Rice | *Casuarina equisetifolia, Cocos nucifera, Areca catechu.* |
| 7. Plantation crop areas of south and eastern states | Tea, coffee, cocoa, etc., banana, black pepper, pineapple | *Albizia odoratissima, A. chinensis, Erythrina* spp., *Gliricidia* spp. |
| 8. North-eastern areas | Paddy | *Dendrocalamus hamiltonii, Cocos nucifera, Areca catechu, Artocarpus* spp., *Terminalia myriocarpa, Dipterocarpus macrocarpus, Anthocephalus chinensis.* |

## 6. SOCIOECONOMIC CLASSIFICATION

Agroforestry systems can be classified on the basis of socio-economic considerations. Socioeconomic considerations may include level of production, management systems, technology, etc. Economic criteria may be more relevant in classifying agroforestry systems. On an economic basis, agroforestry systems may be classified into: (i) subsistence systems, (ii) commercial systems and (iii) intermediate systems.

(i) *Subsistence agroforestry systems* are those which aim at meeting the basic needs of the family. The holdings are usually small. These systems are usually managed by the family with a low investment. Usually, no outside labour is employed. There may be some marginal surplus production for sale (but only marginally). Marketing may not be very well developed. The subsistence type of agroforestry systems may include: shifting cultivation, scattered trees on fields, homestead forestry, homegardens, etc.

(ii) *Commercial agroforestry systems* are those where production is on a large scale on commercial lines. Sale of the produce is the main consideration. These systems are worked on the lines of business and require comparatively sizable investment. The systems are managed by individuals but usually by companies, corporate bodies or government. There is a regular requirement for a large labour force. Plantations of tea, coffee, cocoa, etc., under shade trees, plantations of oil palms, coconut with underplanting of food crops, *Eucalyptus* plantations for pulp and paper on agricultural land in combination with agricultural crops, etc., are some examples of commercial agroforestry systems.

(iii) *Intermediate agroforestry systems* are those which are intermediate between commercial and subsistence scales. These are practised on small to medium-sized farms. The systems aim at production of sufficient food, wood, fodder and other products which are not only enough to meet the needs of the family but yield a substantial quantity as surplus which is available for sale in the market. Some examples of this system may include cultivation of fruit trees with agricultural crops, fruit trees with multipurpose trees, *Eucalyptus* plantation for paper and pulp, *Casuarina* plantations with agricultural crops for firewood and poles, coconut with agriculture crops, etc.

On the basis of management, the systems could be grouped into: (i) extensively managed and (ii) intensively managed. Shifting cultivation, silvipasture, or pastural silviculture may be grouped into extensively managed systems. Intensively managed systems include home gardens, trees with agricultural crops, plantation of tea, coffee, cocoa, etc., agricultural crops with coconut, etc. The technology used may also form the basis of classification. However, there may be various aspects of technology, e.g., mechanisation, tillage, water management, post-harvest technology, etc., and all the factors singly or jointly may be used to classify agroforestry systems. On the basis of technology the agroforestry systems are classified into: (i) low technology, (ii) medium technology or intermediate technology and (iii) high technology agroforestry systems. *Low technology agroforestry systems* are those which depend on primitive technology, e.g., shifting cultivation. *High technology agroforestry systems* are those which depend upon modern technology of plant propagation, use of improved seeds, fertilisers, insecticides, weedicides, use of tissue culture, etc. These technologies require much greater investment per unit area. Tea, coffee, rubber plantations, cultivation of trees in several crops, etc., are examples. *Intermediate technology* agroforestry systems are those in between the two systems discussed above. Most of the systems followed in India are of this category.

## 7. HISTORICAL CLASSIFICATION

The various agroforestry systems practised these days have developed through the ages. It is widely accepted that cultivation of agricultural crops started about ten thousand years ago, in the form of *shifting cultivation*. Over the years, due to the spread of knowledge, settlements, development of irrigation and drainage infrastructure, *sedentary farming* systems came into existence. Some of the most intensive agroforestry systems with a combination of agricultural crops and trees are perhaps the most recent systems. It may, however, be quite difficult to classify various systems on the basis of historical background because several systems developed simultaneously depending upon socioeconomic conditions of the community and edapho-climatic conditions of the area.

Of all the classifications of agroforestry systems, the structural basis of classification is more important and commonly adopted. The classification of agroforestry into agrisilviculture, silvipasture, agrisilvipasture and others including the multipurpose forest trees production system is more widely used (King, 1979; FAO, 1981; Nair, 1985; Gholz, 1988). The practices may differ from region to region under the above subsystems. Several systems, such as shifting cultivation, taungya cultivation and other forms of agrisilviculture, silvipasture, social forestry, village woodlots, and some other combinations are important in our country.

# CHAPTER 5
# Shifting Cultivation

Shifting cultivation is also known as 'slash and burn' or 'Swidden' cultivation. It was a remarkable innovation during primitive cultures and a transition between *food gathering* and *food production*. Even in the modern age, the system is widely adopted in different parts of the world. In India, shifting cultivation is known as 'jhuming' in the north-east and 'khallu' or 'kurwa' in Bihar and 'dahiya' or 'podo' in Orissa, Madhya Pradesh and Andhra Pradesh. Early man was basically a *food gatherer*. He gathered food by hunting animals. He collected fruits, roots, rhizomes, seeds and leaves of selected plants and trees. Perhaps shifting cultivation was a first step in the direction of food production during early civilization.

Shifting cultivation is the oldest system of cultivation of crops. About ten thousand years ago man discovered that seeds of some plants and grasses could be eaten. He tried to grow them by sowing seeds after clearing forests. The crops grew very well for a few years but later productivity declined. People used to abandon the old areas and shift to new areas. Thus, they moved from place to place and did not come to the first area until it had regained its productivity. This form of agriculture was prevalent and still is, in many parts of Africa, Latin America, South-East Asia and the Indian subcontinent. The system detail varies from region to region and country to country.

This system has continued for thousands of years and stood the test of time. Therefore, many people conclude that the system has an in-built mechanism of sustenance and conservation. However, due to increase in population, the land available for shifting cultivation does not get enough time for restoration of productivity and ecological system. Therefore, now shifting cultivation has become a source of ecological degradation, soil erosion and converting good forest areas into wastelands. Shifting cultivation is a primitive cultivation technique and is a transition between *food gathering* and *food production*.

### Area under Shifting Cultivation

The total area under shifting cultivation is not precisely available. NCA (1976) estimated that the total area under shifting cultivation in the north-eastern states is about 2.7 million and the practice in Madhya Pradesh and Bihar is being gradually replaced by settled cultivation. Recent information on area, locality and people practising shifting cultivation as available from some reports is given in Table 37. About 2.65 million tribal families depend upon this type of farming system. The approximate location of the area is given in Fig. 4.

Unfortunately most of the area subject to shifting cultivation is not surveyed. The basic information in regard to the practice, such as actual location, the exact extent, population dependent on it, etc., is not available. Various estimates have been made at various times. The Task Force on Shifting Cultivation in its report in 1983 estimated that shifting cultivation was practised in 13 states of India, extensively in Arunachal Pradesh, Assam, Manipur, Meghalaya, Mizoram, Nagaland, Orissa and Tripura, and rather restrictedly in Andhra Pradesh, Bihar, Kerala, Karnataka, M.P., Maharashtra, Sikkim and West Bengal. In other words, the practice is largely confined to the north-eastern hill States and Orissa. The Task Force estimated the forest area affected by shifting cultivation at 4.35 million ha and the number of families practising it at 6.22 lakhs. A report prepared by the North-Eastern Council Secretariat, Shillong, in 1975, indicated that the total area affected by shifting cultivation in the seven states of the north-eastern region was 2.70 million ha. FAO estimated the forest area affected by shifting cultivation in the north-eastern region and Orissa by interpretation of Landsat imagery at 9.00 million ha. Details of these estimates are given in Table 38.

It can be seen that the Task Force, the North-Eastern Council Secretariat, and FAO have given separate estimates of the forest area affected by the shifting cultivation in various states. The FAO figures are obviously overestimates. For example, the figure of 1.66 million hectares, stated by FAO to be affected by shifting cultivation in Orissa, is more than even the recorded forest cover in the state in those districts (Anon., 1988b). Of the three sets of figures quoted, the estimates made by the Task Force appear to be closest to the truth (Anon., 1988b).

The area under shifting cultivation has been on the decrease mainly because of the developmental programmes launched by state governments. The practice of shifting cultivation in Bihar, Madhya Pradesh and Orissa is merging with settled cultivation gradually. More and more cultivators are turning towards settled cultivation sometimes because of stricter land laws. Even in the north-eastern states, the area under shifting cultivation has recently been found decreasing as reported by the Forest Survey of

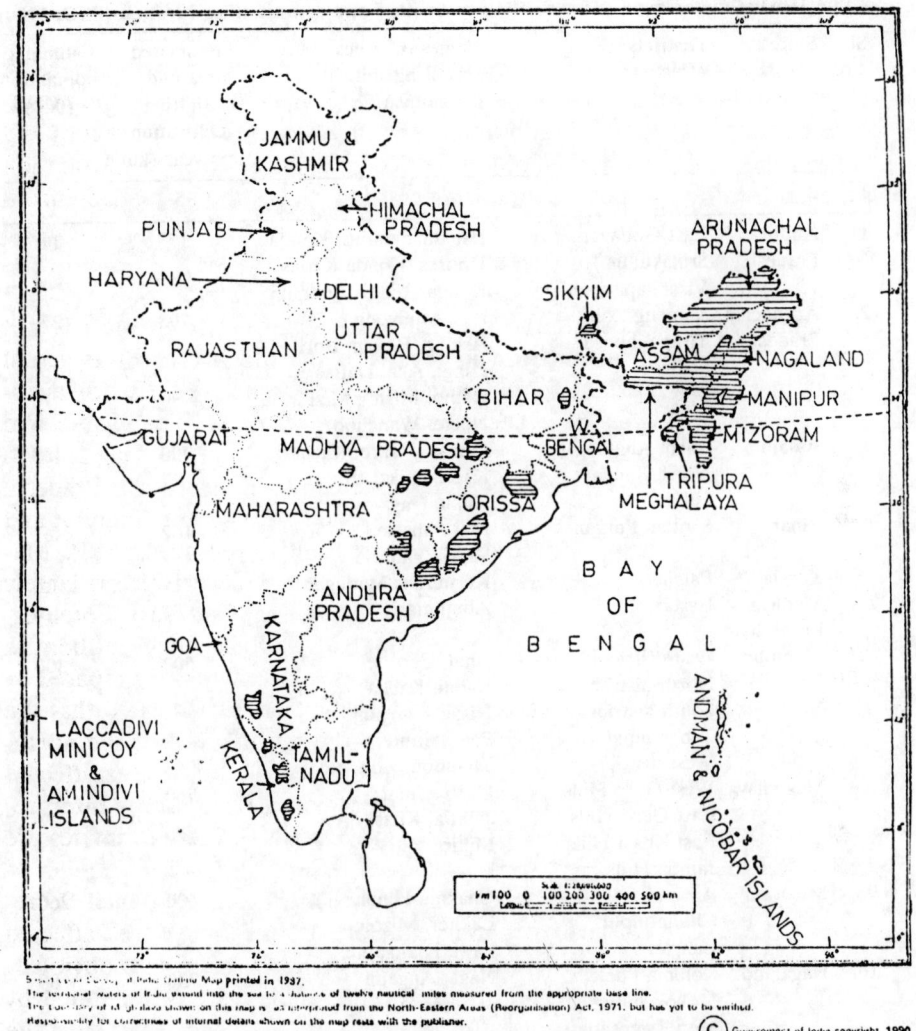

**Fig. 4: Area under shifting cultivation**

India (FSI, 1989). By the visual interpretation of Landsat imagery pertaining to the years 1975 and 1984, the Forest Survey of India has come to the conclusion that during the period of 9 years, the area affected by shifting cultivation in the north-eastern hill states has reduced from 73,410 km$^2$ to 62,854 km$^2$. It is indeed a happy situation that the practice of shifting cultivation is on the decrease. The results of the study are summarised in Table 39 (FSI, 1989).

**Table 37: Estimated area under shifting cultivation and communities involved (FAO, 1980)**

| Sl. No. | State | Districts affected | Names of tribes practising shifting cultivation | Estimated area under shifting cultivation in a year (km²) | Estimated population (000) |
|---|---|---|---|---|---|
| 1 | 2 | 3 | 4 | 5 | 6 |
| 1. | Andhra Pradesh | East Godavari Srikakulum Visakhapatnam | Bagata, Jatapus, Konda Dhores, Konda Kapus, Konda Reddis, Valmiki | 173 | 116 |
| 2. | Arunachal Pradesh | Kameng Lohit Tirap Khonsa Siang | Adi, Miniyong, Ashing, Bokan, Bori, Pangi, Aka, Dafla, Ramo, Mismi, Mizi, Nocte, Wanchoo | 703 | 270 |
| 3. | Assam | Karbi Anglong North Kachar Hills | Dimasa, Garo, Kachari, Karbi, Mikzr, Khasi, Kuku, Naga | 700 | 403 |
| 4. | Bihar | Santhal Pargana | Mal Paharia, Sauria Paharia | 162 | 61 |
| 5. | Kerala | Palghat | Kurumbas Madugar | 19 | 16 |
| 6. | Madhya Pradesh | Bastar | Abujhmaria | 81 | 14 |
| 7. | Manipur | East district North district South district Tengnoupal West district | Anal, Chothe, Kabui, Kacha Naga, Lamganj, Paite, Simte, Thaudou, Zou, etc. | 500 | 300 |
| 8. | Meghalaya | West Garo Hills East Garo Hills East Khasi Hills Jaintia Hills | Garo, Hmar, Jaintia, Khasi, Mikir | 760 | 350 |
| 9. | Mizoram | Aizawal Chhimtuipui Lunglei | Chakma Hmar, Lakher, Mizo, Pawi, Riang, etc. | 600 | 260 |
| 10. | Nagaland | Kohima Phek Mokokchung Mon Twensang Wokha Zunbrebote | Naga, Angami | 730 | 80 |
| 11. | Orissa | Dhekanal, Ganjam Kalahandi Keonjhar Koraput Baudh- Khondmals Sambalpur Sundargarh | Bhuia, Bondo Poraja, Didagi, Gadaba, Juang, Khond, Saora, etc. | 5298 | 706 |
| 12. | Tripura | North district South district West district | Chakma, Halam, Jaintia, Lushai, Mag, Naotia, Riang, Tripuri, etc. | 170 | 100 |
| | | | | 9896 | 2676 |

**Table 38: Estimates of area under shifting cultivation (km²)**

| State/U.T. | Estimate by Task Force, 1983 | Estimate by North Hill Council 1975 | Estimate by FAO |
|---|---|---|---|
| 1. Andhra Pradesh | 1,500 | — | — |
| 2. Arunachal Pradesh | 2,100 | 2,500 | 7,900 |
| 3. Assam | 1,392 | 5,000 | 4,200 |
| 4. Bihar | 810 | — | — |
| 5. Madhya Pradesh | 1,250 | — | — |
| 6. Manipur | 3,600 | 1,000 | 17,800 |
| 7. Meghalaya | 2,650 | 4,200 | 10,200 |
| 8. Mizoram | 1,890 | 6,000 | 16,100 |
| 9. Nagaland | 768 | 6,100 | 11,600 |
| 10. Orissa | 26,490 | — | 16,600 |
| 11. Tripura | 1,115 | 2,200 | 6,200 |
| Total | 43,565 | 27,000 | 90,000 |

**Table 39: Area of shifting cultivation in the north-east during 1975 and 1984**

| State | Geographic area (km²) | Extent of forest affected by shifting cultivation (km²) 1975 | 1984 | %change in 1984 compared to 1975 |
|---|---|---|---|---|
| Arunachal Pradesh | 83,740 | 7,940 | 8,521 | (+) 7.3 |
| Assam | 78,520 | 4,160 | 7,276 | (+) 74.9 |
| Manipur | 22,360 | 17,770 | 13,846 | (−) 22.1 |
| Meghalaya | 22,480 | 10,240 | 6,812 | (−) 33.5 |
| Mizoram | 21,090 | 16,110 | 12,442 | (−) 22.8 |
| Nagaland | 16,520 | 10,970 | 10,641 | (−) 3.0 |
| Tripura | 10,490 | 6,220 | 3,315 | (−) 46.7 |
| Total | 255,050 | 73,410 | 62,854 | (−) 14.4 |
| % | 100.0 | 28.8 | 24.6 | (−) |

## Shifting Cultivation in States

### Andhra Pradesh

Shifting cultivation in Andhra Pradesh is restricted to about 10 blocks of three coastal districts, namely Visakhapatnam, Srikakulam and East Godavari covering an area of about 8392 km² (FAO, 1980). The landscape is rugged and the altitude varies between 300 m to 1700 m. The area is densely forested but the degradation process has already set in. Several tribal communities, e.g., Konda Reddis, Samantha, Konda Kapus, Konda Dhores, Savaras, Bagata, Valmiki, Jatapus, etc., practise shifting cultivation. These communities primarily depend upon shifting cultivation; however, they also practise settled cultivation in valley floors. The soils in the area are mainly réd and mixed red. The rainfall varies from 1100 mm to 1600 mm. Though paddy is a staple crop, it is not cultivated in the shifting cultivation area. People cultivate the land for about 3 years and vacate it for a short period of 5–6 years only. Important crops sown in shifting cultivation area include: millets, pulses, oilseeds, turmeric, etc. The areas are not properly surveyed and land records are not properly maintained. Individual ownership is recorded in all areas under permanent cultivation. But the shifting cultivation area is either owned by the village community or by the state forest department. With increase in population and other developmental programmes, there is continuous increase in permanent cultivation.

### Arunachal Pradesh

Arunachal Pradesh is located in the extreme north-east of the country and covers an area of 83,740 km². The area under forest and permanent agriculture is 51,540 km² and 1366 km² respectively. Shifting cultivation is prevalent over a larger part of the state and covers an area of 7812 km² with an annual cultivated area of 703 km². About 0.27 million population depends on it.

Arunachal Pradesh has a hilly terrain. Altitude increases from South to North where it borders China. Due to montane terrain, the distribution of rainfall is not uniform and varies from 1500 mm to 5750 mm. The soil in forested areas is rich in humus and nitrogen content. Soils in foothills are loams and sandy loams mixed with *Kankar* nodules. Shifting cultivation is practised more in Siang, Lohit, Tirap, Subansiri, Khonsa and Kameng districts. Several tribes, e.g., Adi, Miniyong, Gallong, Pasi, Ashing, Bangni, Bogum, Bokan, Bori, Pangi, Ramo, Aka, Dafla, Khowa, Mismi, Mizi, Nocte, Tangsa, Wanchoo, etc., practise shifting cultivation. The shifting cultivation area is traditionally owned by the village community and is allotted to individual families for cultivation. When the land is not under cultivation, it is possessed by the community. The forest land is cleared and cultivated for 2 years and is vacated for 10 to 12 years. The important crops cultivated are paddy, maize, millets, yam, chilli, tobacco, etc. Sugar-cane is also cultivated in Subansiri district. Cotton is also cultivated in some areas. More and more people are taking to settled cultivation due to the extension efforts by various agencies.

### Assam

In Assam, shifting cultivation is confined to Mikir and North Kachar hills and covers an area of about 4900 km² (FAO, 1980). The terrain is hilly and the altitude varies from 500 m to 1500 m. Rainfall in these areas varies from 1200 mm in the North to 3500 mm in the South. The main communities who practise shifting cultivation include Dimasa, Garo, Kachari, Karbi, Mikir, Khasi, Naga, etc. Paddy is the main crop cultivated in these areas.

Other crops include maize, vegetables, potato, cotton, ginger, chillis, yams, etc. In Mikir hills, people cultivate the area for one year and vacate it for regeneration of forests for 10 years. But in the southern side, people hold the land for 2 to 3 years and vacate it for about 8 years. The area is managed by the District Council. At the village level the village councils distribute the area for cultivation.

### Bihar

In Bihar, the practice of shifting cultivation is restricted to about 9 blocks of Santal Pargana district and covers an annual area of about 162 km². The area is located in Rajmahal hills, dominated by flat-topped hills and plateaus. The average annual rainfall of the area is 1430 mm. The soils are not rich. The undulating topography supports a thin layer of soil which is not fit for sedentary cultivation. The main community practising shifting cultivation is Paharia. Normally the land is cultivated for 2 to 3 years and vacated for 4 to 5 years for rest and regeneration of the forest. After clearing the forest, the area is not usually burnt. In the second and third year, people cultivate millets, pulses, and maize. Paddy is not cultivated in these areas.

### Kerala

In Kerala, the practice of shifting cultivation is restricted to Attapaddy block in Palghat district. The area is hilly and borders the Nilgiri hills. About 83 per cent of the area of the block is covered by forests. Though shifting cultivation is discouraged by the government, the Forest Department has allotted land to some settlements for shifting cultivation. The area is under the ownership of the Forest Department. The headman of each settlement distributes the land to families for cultivation. The area under cultivation is 19 km² per year. The crops sown in these areas include millets, maize, gram, etc.

### Madhya Pradesh

In Madhya Pradesh, shifting cultivation is restricted in Abujhmarh block of Bastar district. However, marginally shifting cultivation is also reported in some areas of Balaghat, Bilaspur, Chindwara, Raigarh and Surguja districts (FAO, 1980). These are fast changing into sedentary cultivation. The area in Abujhmarh is mostly forested. The area is unsurveyed and there is no proper land record. The area is covered with deciduous trees with different species of palms, e.g., sulphi (*Caryota urens*), chind (*Phoenix sylvestris*) etc. The estimated area under annual shifting cultivation is 81 km². Abujhmarias are the main habitants of the area. The soils are red, mixed red and lateritic in nature. Average rainfall may be about 1400 mm. Important crops grown in the shifting cultivation area include millets, such as *Panicum milliaccum*,

*P. trumentaceum*, beans, pulses, sweet potato, potato, chillis, cucumbers, oilseeds, etc.

### Manipur

The central portion of Manipur is agriculturally well developed. However, the hilly areas of surrounding districts, e.g., East, North, West, South and Tengnoupal have areas where shifting cultivation is practised. The hilly areas bear good soils developed from soft rocks. The annual rainfall averages about 2500 mm in the hills. In these areas, more than 75 per cent of the population belong to Naga and Kukichin major groups of the tribes including, Aimol, Anal, Kabui, Kacha, Naga, Paite, Angami, Kom, Zou, etc. For shifting cultivation, people occupy the land for 1–2 years and vacate it for regeneration of forest for 6 to 10 years. Paddy is the major staple crop cultivated followed by maize, potato, yams, chillis, millets, vegetables, etc. Cash crops, e.g., mustard, til, sugar-cane, ginger, turmeric, cotton, etc., are also grown. The total area affected by shifting cultivation is 13,846 km$^2$ with an annual area of 1832 km$^2$ (FSI, 1989).

### Meghalaya

Meghalaya has a geographical area of 22,480 km$^2$ with three distinct parts: the western part or Garo hills is inhabitated by the Garo tribal community; the central part or Khasi hills is dominated by the Khasi tribal community and the eastern part or Jaintia hills is dominated by the Jaintia or Pnar tribal community. Shifting cultivation is the main cultural practice and covers a total area of 6812 km$^2$. These hills receive a very heavy rainfall ranging from 3000 mm to 11000 mm. The soils are red loam and lateritic in the hills and old alluvium in the foothills. The cropping pattern depends upon the altitude. Above 1200 m elevation, potato, paddy, maize, millets, yam, etc., are cultivated. In lower levels, paddy is the dominant crop supported by sweet potato, tapioca, vegetables, oilseeds, millets, cotton, etc. In most of the area, the land is owned by the community chief who distributes for cultivation.

### Mizoram

Mizoram borders Burma and Bangladesh. The whole state is covered with longitudinal parallel ranges running from North to South. The average height of hills in Mizoram is about 1000 m but ranges from 600 m to 2300 m. Average annual rainfall is more than 2500 mm. The soils are medium textured with a good amount of nitrogen and humus. Most of the area is capable of supporting a luxuriant growth of tall and evergreen forests. It is rich in bamboo forests also.

In Mizoram, most of the people depend upon shifting cultivation. Of the three districts, namely Aizawal, Lunglei and Chhimtuipui, the practice of shifting cultivation is more common in Lunglei and Chhimtuipui districts. The land is under the ownership of the village council. Each year the shifting cultivation area is divided into plots and allotted to individual families usually by lottery. The forest growth in these areas which usually consists of various kinds of bamboo and some trees are cut and burnt before April. Sowing is done in the beginning of the monsoon. The main crop is paddy. Besides paddy, people also cultivate maize, millets, sugar-cane, beans, chillis, ginger, turmeric, yam, tobacco, etc.

## Nagaland

This state is situated in the north-east corner of India bordering Burma. The entire area of the state, 16527 km², is hilly and has three distinct regions, i.e. (i) western region forming foothills up to 600 m elevation, (ii) central region consisting of hills of 600 m to 1200 m elevation and (iii) eastern region consisting of high hill ranges of 1200 m or more. The rainfall increases from West to East and averages about 1500 mm in the western, 1500 mm to 2500 mm in the central and 2500 mm in the eastern region. The whole state is dominated by Naga tribes. The people depend primarily upon shifting cultivation. Settled cultivation has started making inroads particularly in the western side.

The total area affected by shifting cultivation is about 6754 km² with an annual cultivated area of 730 km² (FAO, 1980). Land ownership is either with the individual as in the Angamis or with the village community clan as in several cases. The vegetation is cut and burnt and cultivated for 2 years and left for about 8 years. With demand of bamboo by paper mills, the bamboo obtained is now not burnt but sold to paper mills. Paddy is the dominant crop. Other crops cultivated include millets, potato, yam, chillis, maize, etc.

## Orissa

Shifting cultivation in Orissa is practised in a large area extending over eight districts, namely Koraput, Kalahandi, Ganjam, Keonjhar, Dhenkanal, Sambalpur, Sundergarh and Baudh-Khondmals. The areas are generally hilly and forested. The important tribe communities practising shifting cultivation include, Bhuia, Bondo, Poraja, Didagi, Gadaba, Juang, Khond, Koya, Saora, etc. The soils in these areas vary considerably but red and lateritic soils predominate. Average annual rainfall varies from 1000 mm to 2750 mm.

The total area affected by shifting cultivation is 37084 km² with annual cultivated area of about 5298 km² (FAO, 1980). The districts of Keonjhar, Dhenkanal, Sundergarh and Sambalpur have recently been opened up and the Juang and Bhuiya tribes who dominate are fast adopting settled agriculture. The shifting cultivation area of Ganjam, Kalahandi and Baudh-

Khondmals, is located in higher hills and still to be opened up. The people, which mainly belong to the Khond community, largely depend upon shifting cultivation. They cultivate millets, pulses and vegetables. They also produce turmeric, banana, jackfruit, pineapple, etc. At lower altitudes paddy is also cultivated. The district of Koraput, which is dominated by Parojas, has also taken up settled agriculture.

## Tripura

Of the total area of 10477 km² of Tripura state, about 70 per cent is covered by hillocks. The highest altitude is about 900 m and the general altitude is 200 m. The climate is warm and humid. Average annual rainfall is 2100 mm. The soils are good and fertile. Red soils predominate on hills and alluvial soils in valleys. Shifting cultivation is confined in the hilly region and concentrated in nine blocks of West, South and North districts (FAO, 1980). The tribes which practise shifting cultivation include Riang, Chakma, Mag, Jaintia, Lushai, Halam, Tripuri, etc. The total area affected by shifting cultivation is 1360 km² with an annual cultivated area of 170 km² (FAO, 1980). Government of Tripura is trying to convert shifting cultivators into settled cultivators. The important crop is paddy. Other crops include beans, pulses, cotton, jute, mesta and fruits such as banana, pineapple, citrus, jackfruits, etc.

### FEATURES OF SHIFTING CULTIVATION

Shifting cultivation is the most primitive form of cultivation of agricultural crops. The system has been practised over thousands of years. When forest areas were large and the population was small, the practice of shifting cultivation created no adverse effects on forests and the environment and worked in perfect harmony with nature. But increase in population during the last few centuries and reduced forest area shortened the cultivation cycle, which has adversely affected forests and the environment. Some of the important features of the present form of shifting cultivation are as under:

(1) Shifting cultivation is practised in remote forested areas where means of communication, markets, etc. are not developed. The most common system of sale is the *barter system*. The population is undeveloped. Developmental efforts have not been able to make a significant impact because of lack of communication and low level of education among the people. The system has its roots due to comparatively low population density and still sufficient forest area being available.

(2) In some areas, shifting cultivation is due to the land ownership system. In most areas of the north-east, the land is owned by the village council or head of the clan. The council allots the forest land for cultivation. The individual has no ownership over the land. This works as a disincentive

for settled cultivation and making any investment for land development activities.

(3) Shifting cultivation is a sequential agrisilviculture system wherein the trees are usually not utilised for wood. There is emphasis on agriculture. However, there can be shift in the utilisation of wood with improvement of accessibility and creation of a market. In Nagaland and adjoining states, the usual practice was to cut and burn hardwoods and bamboo resources but with the establishment of a paper mill, bamboo clumps are not destroyed but sold as a raw material to the paper mill.

(4) Shifting cultivation generates local subsistence and low economy with little or no surplus. Markets are least developed. The main interest of the cultivator is to produce almost everything that he needs. Therefore, he grows not only a variety of cereals but also vegetables, oilseeds, fruits, spices, cotton, etc. The requirements of timber and fuel are easily met and, therefore, no importance is usually attached to trees. Trees and other vegetation are considered necessary for providing fertiliser in the form of ash, preparing a clean seedbed by burning the cut trees, shrubs, etc., and restoring the productivity of the soil during the fallow period.

(5) Shifting cultivation occurs generally in areas with extensive hilly habitat with steep slopes and no flat areas or valleys, plenty of forest around in the hills, low population density, humid climate and with little development of communication facilities.

(6) The cultivation cycle in shifting cultivation is very important and varies considerably. The longer the cycle, the better it is. The present shifting cultivation cycle varies from 6 to 12 years. These cycles are too inadequate to provide recuperation to the site and repair the ecological damage. Perhaps, a cycle of 50 years or more may be adequate for the purpose. Shifting cultivation on short cycles decreases soil productivity due to excessive loss of soil and nutrients. The loss of nutrients is usually in the form of: (i) cutting of forest growth, either by removal or burning, (ii) accelerated erosion and run-off caused by exposure of surface soil to the erosive forces of rain and wind and (iii) removal of nutrients by annual crops. Longer cultivation cycles gradually build the nutrient store through secondary succession (Fig. 5).

(7) The practice of shifting cultivation has socioeconomic relevance for production of agricultural crops in north-eastern India and some areas of central India. There is no other alternative means of livelihood. The initiative and financial inputs by the individual are completely lacking and are largely coming from the governments. The socioeconomic compulsions of poor tribals generally lead to practising shifting cultivation. They are largely interested in better methods of cultivation, e.g., terracing, ploughing, etc. but these require a large investment and immediate returns do not compensate such investments.

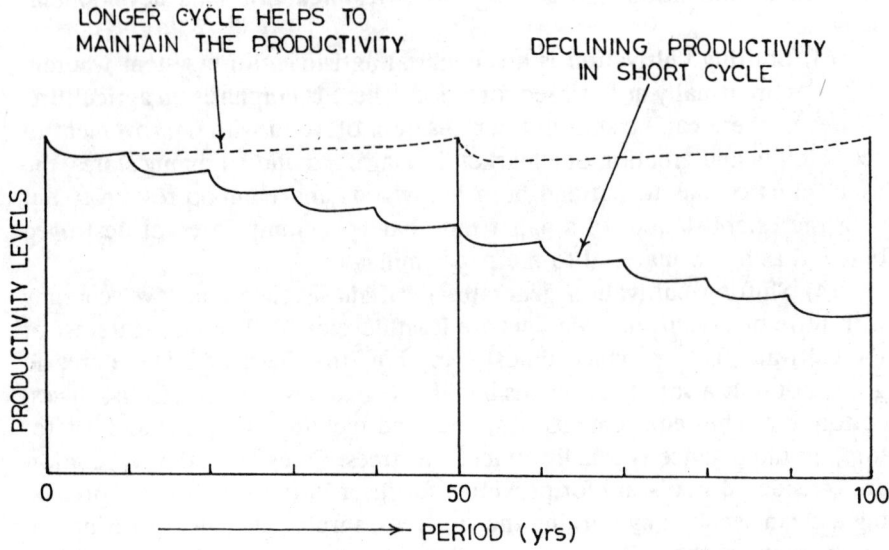

**Fig. 5: Declining productivity in shifting cultivation due to short cycles**

TECHNIQUE OF SHIFTING CULTIVATION

Shifting cultivation areas are mostly hills covered with moist deciduous, evergreen or bamboo forests in their secondary succession stages. There is no permanent plot or field. The land is allotted to the individual cultivators by the village council or head of the village community or clan. For this, the land for cultivation is selected on the basis of soil organic matter, standing forest crop and fertility of the site. The selected area is clear cut during November–December and burnt and the area is distributed among individuals according to the prevailing practice of the community. In Lushai hills of Mizoram, the plots are distributed by the village council by lottery. In Garo hills of Meghalaya, the village chief or *Nokhma* claims rights over certain plots and the remaining plots are distributed to others. In Abhujmarh of Madhya Pradesh, the shifting cultivation area is selected by the village head or *patel* and the villagers are free to cultivate as much land as they want. In Juang hills of Keonjhar district of Orissa, the plots are allotted by the village head but disputes over the land are common due to shortage of land.

If required, second burning is also given before the onset of the monsoon. In some areas of Bihar, the area is not burnt. The areas are not ploughed; the surface soil is not disturbed. No animal or implement is used for preparation of the land. The only implements used in farming are the chopping knife, dibbling stick and a small hoe for weeding (Borthakur *et al.*, 1979).

The most important crop cultivated is paddy, particularly in the north-east. But all other crops, such as millets, pulses, oilseeds, cotton, sugarcane, vegetables, fruits, spices, etc., are also cultivated to make the family self-sufficient in almost all basic needs.

The land is sometimes cultivated in the second year but rarely during the third year.

EFFECTS OF SHIFTING CULTIVATION

(i) Shifting cultivation causes large-scale damage to the forests and has resulted in deforestation and denudation of hill slopes. In the north-eastern region, however, secondary succession is quick to take place but in most cases the area is occupied by reeds, useless shrubs and different species of bamboos. The tree cover, essential for hills due to ecological reasons, is almost negligible. In several areas of Arunachal Pradesh, Manipur and Tripura, the shifting cultivation areas are occupied by tall grasses, reeds, etc. Almost all the shifting cultivation areas in Mizoram are occupied by several species of bamboo, such as *Melocana bamboosoides, Dendrocalamus hamiltonii Bambusa balcoa*, etc. After the land is abandoned by the shifting cultivators, it is rarely occupied by the original vegetation. More commonly, due to xerophytic conditions, evergreen trees and shrubs are replaced by the more hardy reeds, bamboos and coarse grasses. Pioneer species such as *Macaranga* spp., and *Trema* spp., often occupy the abandoned areas in the north-east. In Meghalaya, some of these areas are occupied by useless weeds such as *Eupatorium* spp., *Imperata cylindrica, Andropogon* spp., etc.

Areas in the central peninsula are also devoid of trees and are usually occupied by grasses and shrubs. The destruction of forest takes place over a large area while the effective cultivated area is usually much less. Shifting cultivation reduces the species diversity. Several species of trees have already disappeared due to continuous cutting and burning.

(ii) Shifting cultivation in hill slopes causes soil erosion which leads to: (a) soil and nutrient losses, (b) silting of reservoirs and streams, (c) reduction in water-yield and (d) landslips and landslides. Some of the areas are ecologically very fragile and the soils are loose; shifting cultivation in such areas causes serious soil degradation problems.

(iii) Shifting cultivation leads to considerable loss of soil nutrients through run-off and leaching. Burning and high temperature reduces the organic matter and leads to rapid mineralisation and volatilisation of inorganic nutrients. Studies indicate that shifting cultivation leads to: (a) lowering of organic matter content; (b) decreasing the availability of phosphorus, potassium, calcium and magnesium; (c) lowering the total quantity of sesquioxides, iron, aluminium, calcium, potassium, phosphorus, etc.; (d) affecting adversely the cation exchange capacity and physical properties, e.g., water-holding capacity and field capacity and (e) increasing the pH and reducing microbial activity (FAO, 1981). The soil fertility decreases rapidly and is reflected in crop yield which is good in the first year, reasonable in the second year and poor in the third year. Cultivation during the third year and beyond is usually uneconomical.

(iv) The production of agricultural crops under shifting cultivation area is usually low. The yield of paddy under shifting cultivation varies from 305 kg/ha to 1120 kg/ha against an average production of 3500 kg/ha under settled agriculture with proper terracing (Borthakur *et al.*, 1979; Mathur, 1979; FAO, 1981). Similarly the yields of other crops, e.g., millets, cotton, oilseeds, pulses, vegetables, spices, etc. are also low. Therefore, families practising shifting cultivation live near scarcity or famine conditions.

(v) These days, wood is a very important commodity. The country is importing a large quantity of wood to meet its domestic need. A large number of important tree species such as *Shorea* spp., *Terminalia* spp., *Toona ciliata*, *Schima wallichii*, *Lagerstroemea parviflora*, *Gmelina arborea*, *Duabanga* spp., *Carea arborea*, *Cassia fistula*, etc. occur in the north-east which are cut and burnt. Cutting and burning of such valuable timber species is a national waste. Already trees of larger size are almost absent. Valuable trees are gradually replaced by useless shrubs and more resistant tree species.

(vi) The system of shifting cultivation provides no opportunity for infrastructural development which could be helpful in increasing agricultural production and providing basic amenities, such as public health, roads, education, communication, etc. since there is no permanency of settlements and agriculture.

(vii) Shifting cultivation areas are usually infested by weeds. One of the reasons for poor crop yields during the second and the third years is increased weed infestation. In the first year, due to intensive burning, the weed growth is less but during the second and subsequent years weed infestation increases.

(viii) Shifting cultivation is practised by tribals in remotely located hills and it has become their way of life. The calendar of operations of shifting cultivation controls their social life, festivities and religious observances. Almost all operations are connected with socioreligious functions. There are many dances, songs, folk tales associated with this type of cultivation in tribal areas.

CONTROL MEASURES FOR SHIFTING CULTIVATION

Shifting cultivation is a backward, wasteful and ecologically undesirable farming system. But the system has been going on for several thousand years because the system itself is indicative of a backward culture. The system aims at meeting all family needs of cereals, pulses, vegetables, fibres, spices, oilseeds, narcotics such as tobacco, dyeing material, etc. usually raised as mixed crops. Such mixed cultivation provides some insurance against crop failures. Any effort to improve or replace this system should take into account the sociocultural life of the people, infrastructural development, acceptability by the cultivators, etc. It is difficult to change a traditional shifting farmer who has been born and brought up with it to a modern farmer, even if we provide him all the modern farm inputs. Firstly, he shall have to be motivated for adopting settled agriculture. Already a large area has been affected and degraded due to cutting of trees/shrubs. Rampant soil erosion due to loss of forest cover and cultivation has caused serious soil degradation problems. It is therefore necessary to identify suitable areas for settling the shifting cultivators. In doing so, proper soil conservation measures have to be adopted. For north-eastern areas, control of shifting cultivation programmes includes terracing of the land, afforestation of the upper hill areas, undertaking plantation of cash crop, providing communication facilities, providing economic assistance for erecting houses and for agricultural operations.

In the hills, for undertaking proper soil and water conservation measures preparation of terraces is almost essential. The farming system should include forestry, silvipasture/horticulture and agriculture. It is suggested that the upper one-third of the hill should be developed under forestry the middle one-third portion of the hill should be allocated for horticulture crops or silvipasture, leguminous fodder crops, etc. for providing fruits, fodder and effective cover against soil erosion and the lower one-third portion should be properly terraced for growing agricultural crops such as paddy, pulses, etc. (Borthakur *et al.*, 1979). This system has been found quite effective for soil and water conservation and increasing agricultural yields (Fig. 6).

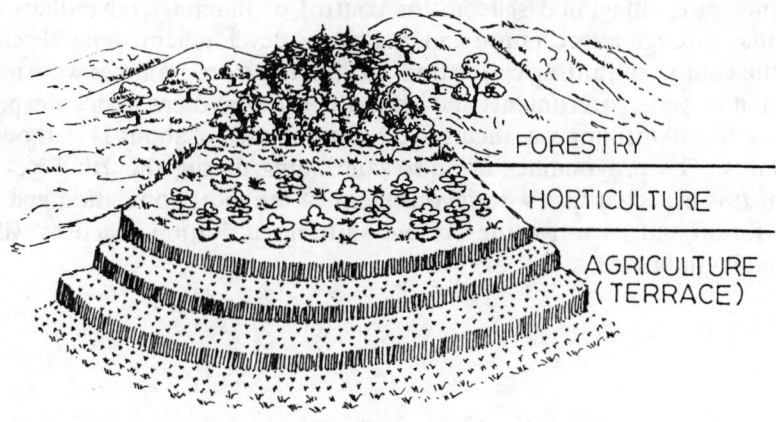

FORESTRY

HORTICULTURE

AGRICULTURE
(TERRACE)

**Fig. 6: Model land use as alternative to shifting cultivation**

While preparing terraces and selecting crops, it is necessary to keep in mind the requirements of the farmers. It must be demonstrated that all crops required by the farmers can be grown in terraced fields. The terrace should be protected against erosion. It has been found that several fodder grasses and legumes such *Stylosanthas hamata*, *S. gnyanasis*, *Serato* spp., *Clitoria*, spp., etc. are quite effective.

The unterraced portion of the land should be allocated for growing trees, shrubs and grasses. There are several trees which yield food, fodder and other valuable products preferred by the shifting cultivators. Only such species should be planted.

Ingty and Goswami (1979) suggest that abrupt change from shifting cultivation to settled cultivation may not succeed due to various problems; therefore, the shifting cultivators may first be involved in taungya cultivation and gradually efforts may be made for settled agriculture. The system may be an alternative to shifting cultivation and work as a useful tool to reclothe the hills.

Shifting cultivation control schemes have not been quite successful because of the following reasons:

 i) The settlements disturb their sociocultural life abruptly.
 ii) They are not used to cultivation on terraces using bullocks and implements.
 iii) They find that production in terraces is low during the first year due to removal of top soil for making the terrace.
 iv) Sometimes there is more weed infestation and damage by insect pests and diseases than in shifting cultivation.

The Government of India and state governments have launched several schemes for the control of shifting cultivation. Some of the important schemes are: integrated scheme for control of jhuming, rehabilitation of jhumias through afforestation and cash crops development, general scheme for the control of shifting cultivation, etc. All these programmes were implemented by state governments and up to the 6th Five-Year Plan an expenditure of Rs. 500 million was incurred covering an area of about 1175 thousand hectares. The programmes are also continuing during the 7th Five-Year Plan. It is expected that with involvement of the local population and with motivation and incentive the present shifting cultivation practices will be gradually improved.

CHAPTER 6

# Silviagriculture Systems

Silviagriculture systems include those systems whereby the land is primarily managed for production of trees and cultivation of agricultural crops is of secondary importance. The management of land may be such that the agricultural component may cover the whole or a part of the rotation period of forest crops. Management practices are adopted so as to cause the least damage to trees. Most of these systems are practised in forest land. These systems may include the following practices.

1. Taungya system
2. Cultivation in established plantations
3. Cultivation of commercial crops in forests
4. Sericulture, lac-culture, apiculture and pisciculture
5. Wood lots

## Taungya System

The taungya system is defined as a method of establishing a forest crop in temporary association with agricultural crops. It is a method of raising forest plantations in which cultivators are allowed to raise agricultural crops for initial periods of a few years and in return they are made to raise forest plantations. Agricultural cropping is usually confined to the period by which the canopy of the forest crop starts closing. In fact, the taungya system is an extension of shifting cultivation. In shifting cultivation, a cultivator clearfells a patch of forest, burns it when dried and broadcasts or dibbles the seed of crops with a minimum of soil working. He takes one or two crops and moves to another patch of the forest. When he leaves the area it is covered by weeds and useless shrubs. In the taungya system, in given forest area, the cultivator is to plant or sow a new forest crop along with his food crops, so that when he moves out, the area is occupied by useful trees and not by inferior shrubs, etc.

The term 'taungya cultivation' means hill cultivation and the system was started for the first time in Burma in 1856. In India, the first taungya plantation was raised in North Bengal during 1863 followed by plantations in Chittagong and Sylhat (now Bangladesh) in 1870 and Coorg in Karnataka in 1890. However, regular taungya cultivation as a system of raising forest plantations started in North Bengal in 1911 when it was used for raising sal (*Shorea robusta*) plantations. It was extended for raising teak (*Tectona grandis*) plantations soon after, in 1912. Gradually, the system was extended to other states. In Uttar Pradesh, it has been adopted for raising sal plantations since 1923. The system spread to other states, e.g., Maharashtra, Kerala, Madhya Pradesh, Andhra Pradesh, etc. It was also tried in the Western Himalayas for raising deodar plantations, but without success.

The taungya system even now is a standard practice for raising forest plantations and for regenerating forests in some states, e.g., West Bengal, Uttar Pradesh and Kerala. It is also practised to a lesser extent in Maharashtra, Andhra Pradesh, Orissa, Karnataka and Tamil Nadu.

PROCEDURE OF TAUNGYA SYSTEM

The taungya system begins with harvesting of the forest. The harvested area is clear cut and burnt. Some trees are often reserved for shade or for fruits, etc. Usually, for the taungya system such areas are selected which are flat and the soil is deep and fertile. Generally, one to two hectares of land are allotted to one family. In some states, taungya cultivators are also allotted additional land other than plantation area. In West Bengal and Assam, taungya cultivators are given additional land for paddy cultivation.

Usually, cultivation of an agricultural crop is permitted for one or two years before forest plantation. This helps the farmers to take one or two good harvests in the highest fertility condition of forest land. Allowing one or two crops before tree plantation is an incentive to the farmer for clearing the land for forest plantation. When tree crops are planted with agricultural crops in the second or third year, they receive the benefits of the intensive cultural operation which the farmer adopted for his agricultural crops. The main objective in this system is to raise forest plantation by utilising the labour of taungya cultivators. When forest plantations get established, the farmers have to leave the area. Sometimes, however, a compromise between the forestry objectives and agricultural objectives is allowed in order to permit the cultivation of the land as long as possible, as in case of poplars in Uttar Pradesh. Sometimes the spacing of trees is increased in order to accommodate cultivation of crops for longer periods.

PROGRESS OF TAUNGYA PLANTATIONS IN STATES

*Andhra Pradesh*

In Andhra Pradesh this system of plantation is called *kumri* and was practised mainly in Vishakhapatnam, Guntur and East Godavari districts. In Vishakhapatnam district, cashew plantations were raised during 1957 to

1964. The area was leased out to landless poor for 3 years. The cultivators uprooted the stumps, burnt the slash and cultivated paddy, groundnut, sweet potato, etc. In Guntur district, the cultivators had to pay lease rent up to Rs. 1500 per ha. Besides cashew, other tree species, e.g., *Tectona grandis*, *Bombax ceiba*, bamboo, *Eucalyptus* spp., have also been raised in these districts. During 1978–79, taungya cultivation was resorted to over 10 km² area by allotting land to poor and landless people. The practice has been given up recently.

*Assam*

In Assam, taungya cultivation has been practised since 1934. *Shorea assamica* has been successfully raised in the foothills in Nowgong and Goalpara districts. This practice has considerably reduced the area of shifting cultivation. Paddy is the main agricultural crop taken by the cultivators.

*Karnataka*

Taungya system was first practised in 1890. It was practised for plantation of teak and sandal. However, the system has never been adopted on a large scale. The system has been given up now.

*Kerala*

In Kerala, plantations of *Tectona grandis* and *Eucalyptus grandis* have been raised by taungya system. In Kerala, the strips between plant lines, in plantations of *E. grandis* are leased for 2 to 5 years. In these plantations, tapioca is the common agricultural crop raised. Tapioca requires intensive soil working and, therefore, soil conservation safeguards are adopted. Common practices include: (i) contour bunding and cross bunding the area and (ii) discouraging tapioca cultivation after 2 years.

*Maharashtra*

Taungya plantations started in Berar area for raising plantations of *Acacia nilotica* and *Tectona grandis*. However, the system has been given up for raising plantations. Cultivation of agricultural crops, such as sunhemp, jute, mesta and oilseeds, etc. in-between the tree lines of plantations has shown success. The material obtained is used as a raw material for paper and pulp.

*Tamil Nadu*

The taungya system was adopted for raising plantations of teak, bamboo, eucalypts, sandal, tamarind, babul, wattles, silk-cotton, cashew and rubber. Food crops grown by cultivators included millets, pulses, tapioca, groundnuts, cotton, medicinal plants, e.g., *Catharanthus roseus*, etc. In the Nilgiri hills, some *Eucalyptus globulus* plantations have been raised through taungya.

## Uttar Pradesh

In Uttar Pradesh, the system was first tried in 1923 and since then has become almost an important system for raising plantations of several species. Taungya plantations have been raised in several areas, particularly in Gorakhpur, Saharanpur, Dehra Dun, Haldwani, Gonda, Bahraich and Pilibhit areas. Several tree species, e.g., *Shorea robusta*, *Acacia catechu*, *Dalbergia sissoo*, *Bombax ceiba*, *Eucalyptus hybrid*, *Populus deltoides*, etc. have been raised successfully. So far, about 44000 ha of plantations have been established by the taungya system. The important species being raised are poplar in the Terai, semal, khair, sissoo, teak and other miscellaneous species in the bhabar area and eucalypts in the intermediate zone. Many of these plantations have shown a very good growth. In Uttar Pradesh, cultivation is allowed for 4 to 5 years. Intensive taungya cultivation is prevalent in the Terai area for poplar plantation. Poplar is usually planted at 5 m × 4 m or 5 m × 5 m. Soil working by tractor, irrigation by tube-well and use of fertilisers are common. The average lease rent for a few years for the Terai and Bhabar area of Haldwani Division is given in Table 40 (Rawat, 1990).

**Table 40: Average lease rent received for terai and bhabar area (Rs./ha)**

| Year | Terai | Bhabar |
|------|-------|--------|
| 1985–86 | 3000 | 1500 |
| 1986–87 | 7000 | 5000 |
| 1987–88 | 8000 | 6000 |

## West Bengal

West Bengal is perhaps one of the pioneers for adopting the taungya system. Both sal and teak have been raised through taungya since 1912. The system is even now practised for raising sal and teak plantations in moist and humid areas where natural regeneration does not occur. Besides sal and teak, several other species, such as *Michelia champaca*, *Terminalia myriocarpa*, *Chukrasia velutina*, *Schima wallichii*, *Toona ciliata*, *Gmelina arborea*, *Bombax ceiba*, *Dalbergia sissoo*, etc., have also been raised. The system has also been tried in hills where tree species, e.g., *Symingtonia populanea*, *Betula alnoides*, *Alnus nepalensis*, *Cryptomeria japonica*, etc., have been raised.

Taungya plantation is one of the best systems for raising forest plantations as it is done without cost and at the same time provides employment and sustenance to poor people. However, the system has been given up gradually due to the serious difficulties that forest departments have encountered from the cultivators.

### CULTIVATION OF CROPS AND PLANTATIONS

Usually the cultivator sows agricultural crops soon after the onset of monsoon and does planting of seedlings during the monsoon. In some areas, e.g., Gorakhpur (U.P.) cultivation for one year is permitted without forest plantation. The period for which the cultivation is allowed varies with area. In Garo hills, it is usually 2–3 years. In Gorakhpur (U.P.), it continued for 4 years and in Saharanpur (U.P.), it goes up to 7–8 years. It has been observed that if crops are allowed for more than 2–3 years, the adverse effect of agricultural crops on trees becomes visible.

Agricultural crops are selected depending on the area. In the north-eastern hills, it is commonly a mixture of hill rice, maize, cotton, vegetables and some tobacco. A wide variety of crops, such as cereals, millets, pulses, oilseeds, root crops, cotton, sugar-cane, medicinal crops, vegetables, fodder grasses, banana, etc. are grown in taungya plantations. The taungya cultivator has to sign an agreement which details what crops he can raise and for how many years. The choice of agricultural crops to be raised in taungya plantations are left to the discretion of the cultivator, although some restrictions on raising certain crops, such as climbing pulses, sugar-cane, paddy, etc. are sometimes imposed on the grounds that these crops damage the forest plantation. Important tree crops and agricultural crops grown in the taungya system in India are summarised in Table 41. (Champion and Seth, 1968; FAO, 1981; Tejwani, 1987).

Taungya cultivation as a practice of regenerating forests or establishing plantations is now being practised mainly in Uttar Pradesh and West Bengal and to a lesser extent in Tripura, Maharashtra and Karnataka. The total area of plantation raised up to 1978 with annual targets for taungya plantation in some of the states is given in Table 42 (FAO, 1981).

More recently, there have been problems with cultivators in raising plantations and leaving the area and the tendency is to drop the traditional taungya methods.

The compatibility of crops and forest trees to be grown together has not been properly studied. In the first year almost any crop can be grown but during subsequent years when interaction between forest plantations and crops grows more intense, selection of suitable crops becomes important. Saukat Hussain (1925) reported that for Gorakhpur area, wheat, barley and arhar were not good companions for sal except where there is danger of frost. Jowar and bajra are also not considered good because these crops cast shade on forest plants. Stewart (1933) reported that cotton with sal and khair killed weeds resulting in better growth of tree species. According to Howard (1939) arhar crops reduced the frost damage in sal seedlings in Dehra Dun area. In Garo hills, millets, maize, sweet potato, cotton, vegetables, hill paddy, etc., are cultivated in sal plantations (De, 1932). Lahiri (1972) reported that intercropping of turmeric with two-year-old sal and teak plantations in north Bengal gave additional net profits by more than Rs. 375 per ha and increased the height growth of teak. Experiments conducted in Palghat Forest Division (Kerala) indicated that many agricultural

Table 41: **Important tree and agricultural crops grown in taungya**

| State/Union Territory | Tree crop | Agricultural crops |
|---|---|---|
| 1 | 2 | 3 |
| Andhra Pradesh | *Anacardium occidentale, Tectona grandis, Bombax ceiba,* bamboo, *Eucalyptus* spp. etc. | Hill paddy, groundnut, sweet potato. |
| Assam | *Shorea robusta, S. assamica* | Paddy |
| Karnataka | *Tectona grandis, Santalum album, Cassia siamea* | Paddy, tapioca, etc. |
| Kerala | *Eucalyptus* spp., *Tectona grandis, Bombax ceiba.* | Paddy, tapioca, ginger, turmeric, etc. |
| Maharashtra | *Tectona grandis, Acacia nilotica* | Sunhemp, jute, mesta, sunflower, castor, etc. |
| Tamil Nadu | *Acacia mearnsii, A. nilotica, Tectona grandis,* bamboo, *Santalum album,* cashew, rubber, etc. | Millets, pulses, ground-nut, cotton, tapioca, potato, etc. |
| Tripura | *Shorea* spp., *Schima* spp., *Michelia* spp. | Paddy, maize, etc. |
| Uttar Pradesh | *Shorea robusta, Tectona grandis, Acacia catechu, Dalbergia sissoo, Eucalyptus* spp., *Populus* spp. | Paddy, maize, sorghum, pigeon pea, soyabean, wheat, barley, rape-seed, etc. |
| West Bengal | *Tectona grandis, Shorea robusta, Schima wallichii, Cryptomeria japonica, Michelia* spp. | Paddy, maize, millet, turmeric, ginger, pine-apple, hemp, etc. |
| Andaman & Nicobar Isls. | *Pterocarpus dalbergioides.* | Maize, sugar-cane, etc. |

crops, e.g., hill rice, chillis, cotton, millets, tapioca, gram, ginger, etc., can be grown with teak with no detrimental effect on teak growth (Anon., 1947, 1949). In Nilgiri, plantations of black wattle (*Acacia mollissima*) and blue gum (*Eucalyptus globulus*) have been successfully raised with potato as the taungya crop. Experiments conducted in Tamil Nadu indicated that raising of tree species in association with field crops was not only cheaper but also beneficial to the tree crops (Anon., 1955). According to Chaturvedi (1981) agricultural crops, such as soyabean, maize, urad, moong, mustard, wheat, potato, peas, gram, millets, etc. are grown successfully in block plantations of poplars raised by the forest department of Uttar Pradesh. The yield of crops decreases in second and subsequent years as tree shade increases, but a wheat crop can be grown successfully up to the 5th year of plantation. These observations and studies are made keeping in view the goal of af-forestation or reforestation of the area. Here, the primary aim is to afforest the areas as early as possible as these lands are forest lands.

Table 42: Total area of taungya plantation up to 1978 and annual target for some states

| Sl. No. | State | Total area (ha) | Annual target (ha) |
|---------|-------|-----------------|--------------------|
| 1. | West Bengal | 127,400 | 4000 |
| 2. | Uttar Pradesh | 43,800 | 2900 |
| 3. | Tripura | 15,400 | 480 |
| 4. | Maharashtra | 7,900 | 210 |
| 5. | Karnataka | 810 | 100 |

TAUNGYA AGREEMENT

Usually, in every state an agreement is made with the taungya culti-vators. The terms of the agreement include the extent of land, possible inclusion of additional dry or wet area, period of cultivation, the crops to be grown, the nature of work required to be done on the forest crop, ques-tion of payments, etc. if any. In West Bengal some amount was paid to the cultivator at the end of a specified period if the area was adequately stocked. For poor work, however, there was a provision of penalty. But if there is acute land hunger as in Uttar Pradesh, the cultivators are willing to pay the rent for raising agricultural crops. For raising a plantation, staking of the area is done by the forest department with or without the help of the cultivator. Nursery-raised seedlings are supplied to the cultivator and plan-tation is done under the supervision of the staff of the forest department. The cultivator is helped to raise a small nursery for replacement of casualty. If a plantation is to be raised by seed, the collected seed may be supplied to the cultivator or the cultivator himself collects seed of selected species and sows them at recommended spacings. In order to ensure success, planning of various operations is essential.

CROP YIELDS

The emphasis on this system is on the regeneration of the forest. The growth and yield of agricultural crops is secondary for the forest department. However, for the farmer, yield of agricultural crops is important as it is going to decide the wages for him for raising plantations. The yield of agricultural crops depends upon area. However, in the first year if good burn is given, the yields are good. In the second year, yield declines and in the third and subsequent years, the yields are poor without fertilisers. The farmers are attracted only for better sites. For poorer areas usually no cultivator comes forward. In some of the areas of Uttar Pradesh where subsoil water is cheaply available, progressive taungya cultivators practise intensive cultivation with irrigation, etc. This substantially increases the yield of agricultural crops. However, sometimes serious damages are caused to planted trees due to tractor ploughing and harrowing.

MERITS AND DEMERITS OF TAUNGYA SYSTEM

The taungya system provides an opportunity for proper use of land and maximises production. Due to prevailing land shortages, the system enables the poor people to cultivate the land temporarily and obtain employment and food. The forest department, which incurs a huge sum for raising plantations, can save this cost and divert the funds for other needy sectors. In this system plantation, protection, etc., is guaranteed. The intensive soil working, irrigation, fertiliser application done in agricultural crops boosts the growth and yield of trees.

In practice, however, several problems have been encountered. The cultivators' main interest is cultivation of agricultural crops. They do not take an interest in plantation and care of forest trees. Sometimes a serious problem was faced by the forest department as cultivators did not leave the area and the area had to be allotted for agriculture. Due to these reasons the taungya system has been given up in several states.

## Cultivation in Established Plantations

In Kurseong division of West Bengal, in the foothills where annual rainfall is 3000–4000 mm, intercropping of plantation of *Tectona grandis, Chukrasia, velutina, Shorea robusta*, etc., has been tried mostly with turmeric (*Curcuma longa*), ginger, black pepper, etc., with a good degree of success (FAO, 1981; Lahiri, 1972; 1983).

Turmeric, ginger, etc., are shade-bearing crops and require a warm humid climate; therefore these species were selected for growing as intercrop. Soil between the lines is worked manually. Two ridges are made between the tree rows three metres apart. Ridges 30 cm in height and 30 cm in width are made 100 cm apart leaving 70 cm from either side of the forest crop line. In areas where a plantation is spaced at 4 m, three ridges are made between the tree lines. Planting is done in the sides of these ridges. The yield obtained from intercrop in established plantations is given in Table 43 (Lahiri, 1983).

**Table 43: Yield of intercrop in older plantations**

| Sl. No. | Locality | Age of plantation | Tree species | Intercrop | Yield Qtl/ha |
|---------|----------|-------------------|--------------|-----------|--------------|
| 1. | Badamtam | 2 | *Leucaena leucocephala* | Ginger | 25 |
| 2. | Champta | 2 | *Shorea robusta* | Turmeric | 12 |

**Cultivation of Commercial Crops in Forests**

Several commercial crops and spices are grown in forest plantations and natural forests. Some of these are small cardamom (*Elettaria cardamomum*), large cardamom (*Ammomum aromaticum*), black pepper (*Piper nigrum*), etc.

Cardamom has a good aromatic odour and is largely used for flavouring and medicinal purposes. In India, cardamom grows wild in many parts of the Western Ghats. It is also cultivated in this region at an altitude of 750 m to 1500 m. The plant likes a warm and humid climate and thrives best in the shade provided by lofty forest trees. It is cultivated mostly in moist evergreen forests of Kerala, Karnataka and to a small extent in Tamil Nadu and Maharashtra.

For cultivation of cardamom a good piece of virgin forest or some old plantation of the evergreen forest type is selected and cleared of all undergrowth, leaving only tall trees to provide shade. Small pits are dug in cleared areas usually 3 m × 3 m before the onset of the monsoon. Usually, 2-year seedlings or rhizomes from the old cardamom crop with their aerial shoots are planted at the rate of two rhizomes per pit. During August–September one or two weedings are given. The crop starts yielding fruits from the third year onwards. The Malabar variety of cardamom is widely cultivated. In the initial stages, cardamom plants require light shade for proper growth. If trees are not enough, suitable quick growing trees are planted. Several species, e.g., *Toona ciliata, Artocarpus heterophyllus, Mesopsis emini*, etc., are usually planted. Observations however, indicate that tall trees with uniform and light crown are the best. The basic objective of managing the area is the development of forest. The trees are owned by the state governments, although growers can collect dead and fallen branches and trees for fuelwood. The normal yield of cardamom is about 100 kg/ha.

Large cardamom is cultivated in the north-eastern Himalayas mostly in Sikkim, Darjeeling district of West Bengal and Arunachal Pradesh. The large cardamom is also grown under the shade of natural forests as well as in flat or sloping land by planting temporary and permanent shade trees. Large cardamom takes 4 to 5 years for production and is reported to yield 200–300 kg cardamom per ha per year (Gupta, 1982).

The pepper (*Piper nigrum*) is another important plant which largely grows in warm humid climate. It is cultivated in forests in Kerala, Karnataka and Tamil Nadu. The pepper plant is a woody perennial climber which is grown with supports of trees or other supports. It is introduced in existing plantations or natural forests where trees provide natural support for climbing and growth. Sometimes *Erythrina indica* is planted to give necessary support to the pepper vines. The vines grow rapidly on these support trees and cover them with dark green foliage. The vines flower during July and the fruit, i.e., berries ripen during February/March.

In these systems, the species preference for shade, the quantity and quality of shade requirement under different microsites, effects of inputs on the trees, effect of continued growing of these species on the bio-diversity and species composition of forests, etc. are some of the common questions on which sufficient information is not available. The forests are basically managed keeping in view the general objectives of forest management, i.e., for the production of goods and services. The evergreen forests are ecologically very sensitive and therefore there is need to develop this combination as a management system on sound ecological principles.

## Sericulture, Lac-culture, Apiculture and Pisciculture in Forests

SERICULTURE

Sericulture today is a well-recognised practice combining silviculture, agriculture and cottage industry. The main trees raised are mulberry (*Morus alba*) and arjun (*Terminalia arjuna*). The silkworm (*Bombyx mori*) is reared on the leaves of these trees. The total area under mulberry is 240 thousand ha in the country and it plays an important role in socioeconomic development of rural poor in certain areas.

A national sericulture project has been started with the view to increasing the productivity of mulberry silk in the country from the present level of 9000 tonnes to 15,000 tonnes annually. Traditional silk-producing states in the country are Karnataka, Andhra Pradesh, Tamil Nadu, West Bengal and Jammu and Kashmir. But the programme has been taken up in 12 other. states, viz., Kerala, Maharashtra, Gujarat, Rajasthan, Uttar Pradesh, Orissa, Madhya Pradesh, Bihar, Assam, Himachal Pradesh, Punjab and Haryana. The project consists of establishment of a basic seed farm, plantation of mulberry, cocoon testing and grading, establishment of technical service centres, development of marketing facilities, etc.

There are several varieties of *Morus alba*. The important varieties are L 146, RSS 175, RSS 135, S 54, TR 10, K 2, etc. Six- to nine-month-old seedlings prepared from cuttings are planted. The tree grows at a fast rate. About 2000 seedlings are planted per ha. Sometimes, these trees are also planted on bunds and other places in a scattered manner. After about three years, a tree starts producing about 10–15 kg of leaves. Agricultural crops can easily be taken in the interspaces of tree lines by adjusting the spacing of trees. The caterpillar of the silkworm moth feeds on the leaves of the mulberry and about 50 kg cocoon are obtained in about 100 trees, which gives an income of about Rs. 3000 from these trees. The silk insect is reared in insect chambers. The leaves are collected and fed to the insects. Some persons having mulberry trees sell the leaves to others who are rearing the silkworm. The leaves are usually sold at Rs. 0.50 per kg. The leaves are good fodder for cattle. Branches are used for making baskets. The timber is in great demand in sport goods industry.

In several states, e.g., Madhya Pradesh, Bihar, Orissa, West Bengal, etc. forests are used to provide host plants for tasar silk insects. The tasar silk has uneven tan filaments which are coarser, shorter and stronger than the normal cultivated silk. The wild tasar silk is obtained from the silkworms of the genus *Antheraea* which comprises 36 species secreting silk, which feed on *Terminalia alata, Terminalia arjuna, Shorea robusta, Syzygium cumini,* etc. *A. mylitta* is the only species exploited commercially at present. *Terminalia alata* is usually preferred. Forest areas where density of *Terminalia alata* is good, are selected for tasar cultivation. The tasar worm is reared on these trees to form cocoons which are collected by the villagers. The temperate tasar insect, *A. proylei,* which feeds on *Quercus serrata, Q. leucotrichophora, Q. himalayana,* etc. produces the finest tasar silk.

LAC-CULTURE

Lac is the resinous substance secreted by an insect called the lac insect (*Laccifer lacca*) which generally feeds on *Schleichera oleosa, Butea monosperma, Ziziphus mauritiana, Shorea talura, Samania saman, Acacia catechu, Ficus religiosa,* etc. There are two strains of the lac insect: (i) kusumi and (ii) rangini; the former is considered superior. Lac insects secrete a resinous material which forms a hard continuous crustation over the twigs. The resinous crustation called stick-lac, is scraped from the twigs and branches of the host trees. The removal of insect bodies, twigs, lac dye, and dirt from the stick-lac gives commercial seed-lac. 'Seedlac', which is used in plastics, electrical adhesives, leather and wood finishing, printing ink, sealing wax, etc. is obtained from seed-lac. At present, lac culture is done by about 3 million tribal cultivators mainly in Bihar, Madhya Pradesh, West Bengal and Orissa.

The selection of host trees is important for production of lac. The main hosts are *Butea monosperma* and *Ziziphus* spp. for rangini and *Schleichera oleosa* for the kusumi strain. Other factors include: proper management of host plants, use of pestfree brood, time and manner of inoculating host plants and control of pests and predators which control lac production. Important areas where sufficient work is still lacking include: pruning intensity of host plants, time of collection of brood lac, selection and use of brood lac, artificial and natural infestation methods, precautions against predators and parasites, systematic cultivation, preservation of brood lac, etc. The lac production could be increased substantially if forest areas rich in the host plants are allowed to be used for the purpose. The cultivators may be supplied with brood lac and proper technology for pruning, inoculation, harvesting, etc.

APICULTURE

Two important non-fibre products in our forests are honey and beeswax. Several species of bees are found in our country, e.g., *Apis indica, Apis dorsata*, etc. Some forest areas can be used for beekeeping. A knowledge of the nectar and pollen source plants in the area is necessary before initiating a beekeeping programme. Several tree species, e.g., *Acacia* spp. *Tilia* spp., *Nephelium litchi, Prosopis chilensis, Eucalyptus* spp., etc. are supposed to support beekeeping. However, knowledge in the field is limited. Forest areas could be developed as good apiary sites provided: (i) species and provenances known to produce plentiful, high-quality nectar are selected and enriched, (ii) species which bloom at different periods are planted adjacently which may extend the honey production period and (iii) some forest plantations or natural forest rich in suitable species are developed into apiary sites for beneficial effects of bees in pollination.

SILVIPISCICULTURE

Silvipisciculture technique has been tried in the Sundarban deltaic area of West Bengal. The region is criss-crossed by creeks and channels of varying width and depth. The settled area consists of more than 50 islands which are protected by earthen embankments. People in this area largely depend on cultivation and fishing. The mangrove areas are rich in marine life. Lahiri (1987) reports that about 42 per cent area has been retained as a national park and the remaining area is under systematic forest management where selective fellings and enrichment planting is usually carried out. Several non-mangrove tree species, e.g., *Acacia nilotica, Prosopis chilensis, Leucaena leucocephala, Casuarina equisetifolia* and mangrove species, e.g., *Avicinia* spp., *Bruguiora gymnorhiza, Rhizophora mucsonate*, etc. were planted and have given good success. For cultivation of fish, the area is enclosed with earthen embankments. Inside the embankment, a system of ridges and canals has been created. Several varieties of fish were cultivated in these areas. It was observed in the initial trials that the fish yield suffered due to the presence of a number of predatory fishes which preyed on shrimps and prawns but subsequently mixed cultures of prawns and mullets were established with the fish yield varying from 150 to 200 kg/ha (Lahiri, 1987).

The land receives intensive treatment. The rain-water is collected by making bunds which helps in developing non-mangrove species. In the lower part of the bund, the mangrove species are grown. Trenches are provided for pisciculture. This land use creates more job opportunities and helps the local population to improve their earnings.

Another example of silvipisciculture is seen in central and southern India where tank-shore afforestation has been taken up on a large scale under a social forestry programme. The tank fills up during the monsoon but the water level recedes during the summer. *Acacia nilotica* is widely planted in these tanks, utilising about 1/3 to 1/2 of the submerged area. Some of these tanks have been taken up recently for pisciculture also. The tanks are therefore providing area for water storage, forestry and fishery. The trees help to meet the timber and firewood needs and fishery improves the nutritive value of the diet.

## Wood Lots

In many parts of the country farmers grow trees in separate fields as wood lots. The field is managed basically for the production of wood until the tree crop is harvested. The system is spreading rapidly due to the acute shortage of fuelwood and timber in rural areas. India is already importing a large quantity of industrial and constructional timber but import of firewood is not feasible.

Most of the firewood collected from the forests by the neighbouring rural population goes unrecorded. However, the recorded production of firewood is only about 18.5 million $m^3$. According to one report the maximum permissible removal of firewood from forests might only be about 40 million $m^3$ against the estimated consumption of about 235 million $m^3$. The gap between the demand and production is met through pilferage and illicit removals from the forests. Pilferage from forests constitutes a removal in excess of the silviculturally permissible limit. This leads to continuous depletion of our forests. Firewood is no longer available within the vicinity of village. People have to walk long distances to collect firewood from the forests. There is an acute shortage of timber for various household works.

In order to surmount this problem, people have started cultivation of trees in small fields for production of firewood and timber. Such tree plantations are usually referred to as wood lots. Several fast-growing species have become popular for growing in such plantations.

*Casuarina equisetifolia* is extensively grown in Andhra Pradesh, Tamil Nadu and Karnataka, on such lands which are usually poor and not suitable for profitable agriculture. The trees are planted very close, usually 1 m × 1 m and harvested in 7 to 10 years. The clearfelling of such plantations yields about 120 tonnes of firewood (Reddy, 1981; Ram Parkash and Hocking, 1986). The produce consists of firewood and small timber which are in demand in the local markets. After 7 to 10 years, the crop is clearfelled and stumps, which also fetch a good price, are uprooted. Often the land is subsequently brought under agriculture for a few years and then is converted into a *Casuarina* wood lot. Other fast-growing species are *Eucalyptus tereticornis* and *Eucalyptus hybrid*, which have been grown extensively in wood lots in several parts of the country, particularly in Punjab, Uttar Pradesh, Haryana, Gujarat, Karnataka, Madhya Pradesh, etc. The species is usually grown at spacing of 2 m × 2 m but narrower spacings of even 1 m × 1 m, 1 m × 1.5 m, 1.5 m × 1.5 m have also been tried. Several wood lots are irrigated which show much faster rate of growth than unirrigated ones. In narrower spacings most trees can be used only as firewood. Only a small portion can be utilised as poles and small timber. Apart from *Eucalyptus* spp. other species which are planted include *Dalbergia sissoo, Acacia nilotica,* etc. Under irrigated conditions the growth of *Eucalyptus tereticornis* in wood lots of Punjab and Haryana is given in Table 44 (FAO, 1981; Rawat, 1988; Dwivedi and Sharma, 1990).

**Table 44: Growth data of *Eucalyptus* under irrigated conditions**

| Age | Average height (m) | Average diameter (cm) |
|---|---|---|
| 4 | 11.1 | 10.8 |
| 5 | 12.0 | 12.1 |
| 6 | 13.5 | 14.2 |
| 7 | 14.0 | 17.5 |
| 8 | 14.8 | 19.0 |
| 9 | 15.0 | 24.0 |
| 10 | 16.8 | 26.0 |

Subabul (*Leucaena leucocephala*) is another important multipurpose tree yielding fodder and fuelwood and has been grown in Maharashtra, Madhya Pradesh and Andhra Pradesh in certain areas. Plantation of subabul at the spacing of 1 m × 1 m at the age of 5 years gives about 100 $m^3$ to 200 $m^3$ of wood/ha (Vaishnav, 1981; Anon., 1984 b; Hegde, 1987).

Under unirrigated conditions, at Gandhi Nagar, Gujarat experiments for five years on the biomass production under high-density plantation and intensive culture of different species yielding firewood, indicated that the mean annual increment of firewood biomass is maximised at three years of age and the same is summarised in Table 45 (Gurumurti and Bhandari, 1987).

Farmers in several states have started diversifying their farming practices due to market pressures and economic returns. Due to acute shortage of timber and firewood in the country, plantations of *Eucalyptus tereticornis* or *Eucalyptus* hybrid could be commercially more viable than agriculture. Farmers in Assam grow wood lots of bamboo along with paddy fields. Bamboos are in demand in the local market and by paper-mills.

In Madhya Pradesh, wood lots of *Bambusa arundinacea* and *Dendrocalamus strictus* are being grown owing to their multipurpose use and greater financial returns. In Andhra Pradesh, plantations of *Bambusa arundinacea* have been raised in depressed and waterlogged areas. In Cuddapah district, farmers grow *Pterocarpus santalinus* (Tejwani, 1987). In semi-arid areas of

**Table 45: Mean annual increment of firewood biomass (stem and branch wood) at 3 years of age**

| Species | Trees per ha | MIA of firewood T/ha (oven dry) |
|---|---|---|
| *Eucalyptus* hybrid | 5587 | 14.5 |
| *Dalbergia sissoo* | 5848 | 28.0 |
| *Acacia nilotica* | 5264 | 22.8 |
| *Acacia tortilis* | 4695 | 24.0 |
| *Cassia siamea* | 2611 | 21.2 |
| *Prosopis chilensis* | 5917 | 29.6 |

Rajasthan, Andhra Pradesh and Uttar Pradesh, farmers have taken to planting ber (*Zizyphus* spp.) for fodder, fruit and firewood.

In Bhavnagar district of Gujarat, it is reported that *Eucalyptus tereticornis* and *Leucaena leucocephala* are planted at a close spacing of 45 cm × 45 cm and 60 cm × 60 cm respectively. NPK fertilisers and insecticides are added to ensure a good growth and higher survival per cent. At the fourth year about one-third of the best stems are harvested. In the 5th year, one-third of the second best stems and in the 6th year the remaining one-third of the stems are harvested. In the seventh year after planting, trees cut in the fourth year become ready for harvesting again. This system provides a continuous production of 90 tonnes of marketable wood and 22.5 tonnes of firewood per ha per year (FAO, 1989). The system has become popular and is being adopted by a large number of farmers in Gujarat.

# CHAPTER 7

# Agrisilviculture Systems

Agrisilviculture systems include those combinations of agriculture and forest crop where agricultural components predominate over the forestry components. These systems are primarily used for managing lands classified as agricultural lands. Important agrisilviculture systems may include the following:

1. Growing of multipurpose trees on farm lands
2. Homestead plantations and home gardens
3. Trees with plantation crops
4. Multipurpose trees with horticultural crops
5. Alley cropping or hedgerow cropping

The agrisilviculture systems aim at production of enough foodgrains, timber, fodder, firewood and other products. The agricultural systems require heavy inputs and the production cannot be increased beyond certain limits. The trees provide further opportunity to increase production. The agrisilvicultural practices are closer to agricultural practices in comparison to silviculture. The trees are grown on agricultural land for security against adverse climatic factors, supplementing agricultural income and increasing the total productivity of the agricultural lands. The agrisilviculture systems are more productive and sustainable than agriculture. Agronomic practices under agroforestry systems should be worked out taking into consideration the needs of the perennial crops. Though trees find place in agricultural cropland with annual crops traditionally, no work has been done so far to harmonise the crop mix under various agroecological conditions with a view to maximising production and increasing farm income. Some of the forms of agrisilviculture are discussed.

## Multipurpose Trees on Farm Lands

Trees have always been grown on farm lands in every region of our country. Some trees were grown to meet the family needs in respect of firewood and timber, some were grown for fruits and seeds, others were grown for fodder, fibre, leaves, etc., and still others were grown for shade, shape, flowers, etc.

More recently, trees were grown on farm lands for money as they fetched a handsome income and increased the total income of the farm. In some areas, trees were grown on farm lands to protect the site, for conservation of soil and moisture and to avoid damage by wind, frost, etc. People also grew trees of some species, e.g., *Morus* spp., *Terminalia* spp., etc., so that they could practise *sericulture*. Some trees were grown because they were important for meeting the raw material needs of industries. Some trees, such as bamboos, babul, etc., provided raw material to rural artisans. In other words, people have always found some use for the trees planted on their farm lands. Also, Indian farmers planted trees on their farm lands because they liked them and worshipped them. In fact, growing of trees is linked traditionally and religiously with Indian farmers.

The trees are planted on farm lands in several types of plantation geometry. Some of the more common geometrical types include:

  (i)  Scattered trees
 (ii)  Trees on field bunds or borders
     (a) One side
     (b) Two sides
     (c) Three sides
     (d) All four sides
(iii)  Row plantation or strip plantation

The trees may be planted anywhere on the farm land. They may occur in the border, middle or anywhere in the field. In such cases, the farmers usually do not plant the trees but nurse them at those places where they grow naturally. In most regions, farmers plant trees on farm bunds or borders. When fields are small, planting of trees on all four sides may have serious competition with agricultural crops, adversely affecting them. In such cases, usually planting of trees is resorted to only on one or two sides. Generally in fields larger than 0.5 ha, planting is done on all four sides (Fig. 7a). Even in larger fields, if the field shape is rectangular with lesser width, planting is avoided on both of the long sides. However, these practices depend on the region, water availability, characteristics of species to be planted, etc. Row or strip plantation of some tree species along with agricultural crops is also common in several areas (Fig 7c). Planting is also done in the form of blocks (Fig. 7d). Planting geometry and species of trees usually depend on several factors, such as the region and its edapho-climatic conditions, availability of irrigation, physiography, etc. Details of plantation of multipurpose tree species on farm lands in some regions of the country are discussed below.

## 1. ARID AREAS

The arid areas cover about 32 million ha, spread in Rajasthan and part of Gujarat and Haryana. The rainfall is usually less than 500 mm. In some areas of Rajasthan, the average annual rainfall is below 250 mm. The soils

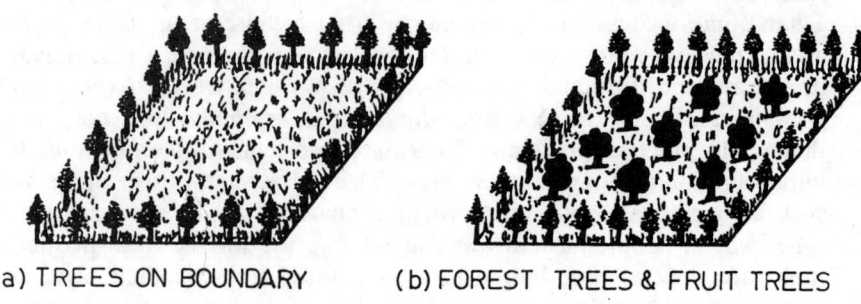

(a) TREES ON BOUNDARY        (b) FOREST TREES & FRUIT TREES

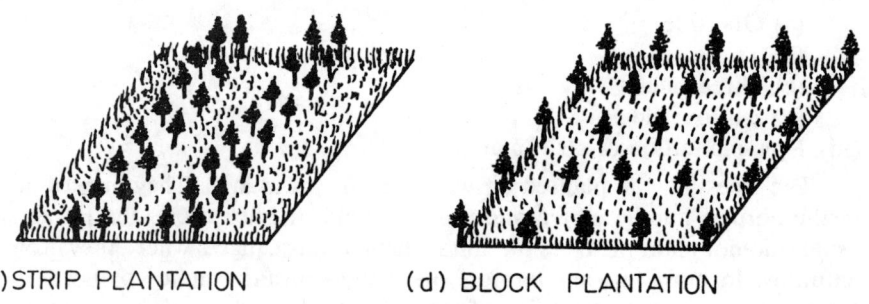

(c) STRIP PLANTATION        (d) BLOCK PLANTATION

**Fig. 7: Trees on farm lands**

are mostly sandy with an alkaline nature. These are low in organic matter and have a low water-holding capacity. In the arid zone, about 45 per cent of the land is cultivated but only about 2 per cent is double cropped. Several studies have revealed that crop production in arid areas is unstable and risky, leading to uneconomic yields (Mann *et al.,* 1977; Gupta and Mohan, 1982). The density of human population is 50/km$^2$, which is one of the highest in the world in such areas.

Arid areas consist of four distinct soil types: (i) sandy plains where soil depth varies from 70 cm to 150 cm with kankar pan beneath, (ii) sandstone rocky sites usually devoid of soils, (iii) saline areas and (iv) shifting sand dunes with deep sandy soils. Agriculture is most common in sandy plains and less common in other types. Important agricultural crops grown in Rajasthan include: wheat, jowar (*Sorghum* spp.), maize (*Zea mays*), bajra (*Pennisetum typhoides*), other small millets, oilseeds, pulses, etc.

Livestock plays an important role in the economy of the region. Cattle, sheep, goat and camel are important livestock. Production of enough fodder is one of the main objectives of land use.

The practice of raising trees with agricultural crops has been prevalent in desert areas for centuries as trees provided insurance against crop failures during droughts. The most important tree species for desert areas are *Prosopis cineraria* and *Acacia nilotica*. Other important species which are widely grown are *Acacia senegal, A. tortilis, Tecomella undulata, Tamarix articulata, Prosopis chilensis, Capparis decidua, Ziziphus nummularia*, etc. In comparatively higher rainfall areas, *Albizia lebbeck, Eucalyptus camaldulensis, Azadirachta indica, Hardwickia binata, Cassia siamea, Ailanthus excelsa, Holoptelea integrifolia*, etc. are also cultivated. *Prosopis cineraria* is the most important and widely planted species. It is a medium-sized tree with tiny dark leaflets. It is evergreen with thorny branches. It is drought hardy and frost resistant but relatively slow growing. It is a very useful tree and almost every part of it finds some use (Muthana, 1980). The leaves produce a good fodder and are extensively lopped. They are rich in protein (15–20 per cent) and minerals (Ca = 2.5 per cent; P = 0.25 per cent). A tree of 25 cm diameter, having a height of 7 to 8 m produces about 10 kg of dry leaves (Paroda and Muthana, 1979). Pods of this tree produce vegetable beans for human consumption. The wood has a high calorific value and is excellent fuel. About 50–70 trees/ha are maintained in agricultural fields.

*Acacia nilotica, A. leucophloea, A. senegal, Ziziphus nummularia* and *Salvadora persica*, etc. are also commonly grown. In Barmer, Bikaner and Jodhpur belt, the *P. cineraria* and *Z. nummularia* are grown with crops. In Jaisalmer region, *Ziziphus nummularia* predominates. In Pali area, *Acacia nilotica* is more common and in Jalore area, *Salvadora* predominates (Shankarnarayan *et al.*, 1987). These species are planted on the bunds and also inside the field. In slightly better areas, several other species such as *Albizia lebbeck, Azadirachta indica, Acacia nilotica, A. jacquimontii, Tecomella undulata, Dalbergia sissoo, Eucalyptus camaldulensis, Hardwickia binata, Parkinsonia* spp., *Dichrostachys nutans, Colophospermum mopane*, etc. are widely grown (Paroda and Muthana, 1979; Shankarnarayan *et al.*, 1987).

Planting of trees in form of wood lots is also becoming popular in several areas of the desert region. *Ziziphus nummularia* is becoming very popular particularly its improved varieties. The improved cultivars, namely 'gola', 'seb', and 'mundia' are popular in low rainfall areas and these can be budded on the rootstocks of any species of *Ziziphus*, e.g., *Z. nummularia, Z. rotundifolia, Z. jujuba* and *Z. spinachristic* (Pareek and Vashishtha, 1978).

## 2. ALLUVIAL PLAINS

Alluvial plains comprise the Indo-Gangetic plains and plains of Assam. The Indo-Gangetic plains comprise plains below the foothills of Himalayas running parallel and covering most of Punjab, Haryana, Delhi, Uttar Pradesh, Bihar and West Bengal. These plains consist of five sections, namely (i) Punjab plain, (ii) Upper Gangetic plain, (iii) Middle Gangetic plain, (iv) Lower Gangetic plain and (v) Assam plain. These areas consist of deep alluvial soils with elevation ranging from 150 m to 600 m from mean sea level. The total area under alluvial plains in India is 70 million ha. The characteristics of the area have important bearing on the agrisilviculture systems being followed in the area.

### (i) Punjab plain

It covers parts of Punjab, Haryana and Delhi, comprising an area of about 10 million ha with about 76 per cent area under agriculture and only about 3.5 per cent area under forests. The remaining area is under uncultivable wastelands, viz. saline-alkaline soils, waterlogged areas and arid areas.

About 50 per cent of the agricultural area is irrigated. Important agricultural crops include: rice, wheat, maize, pulses, oilseeds, sugar-cane, cotton, etc. *Agrisilviculture* has been widely adopted as a form of land use in this region due to assured price of wood products. During the last 10–15 years *Eucalyptus* hybrid or *E. tereticornis* has been widely adopted due to its fast rate of growth. Other species which have been planted include *Dalbergia sissoo, Acacia nilotica, Populus deltoides, Leucaena leucocephala, Sesbania* spp., *Morus alba, Ziziphus* spp., etc. The most common system of agrisilviculture practices followed in Punjab and Haryana are: (i) boundary planting and (ii) block planting (Dhanda, 1983; Shrivastava, 1983). The planting of *Eucalyptus* spp. on the field boundaries started during the sixties and picked up subsequently. The trees are usually planted at the spacing of 1 m to 2 m. Though adverse effects of trees on the annual crops have been recorded near the trees, still farmers have planted trees on the field boundaries. On average, about 75 to 100 trees per ha are supported due to boundary plantations.

Some farmers, especially absentee land-owners have taken up block planting of *Eucalyptus* spp. and *Populus deltoides* to avoid difficulty in share-cropping and tenancy disputes (Dhanda, 1983). *Eucalyptus* spp. are usually planted at a spacement of 1 to 2 m × 1 to 2 m and *Populus* spp. at 4 to 5 m × 4 to 5 m. When agricultural crops are also taken up in block planting, the spacing is wider usually, 5 m × 5 m, so that tractor ploughing can be done.

Most of the *Eucalyptus* wood is consumed by the local paper-mills. The wood of *Dalbergia sissoo* is utilised by the sport goods and furniture industries. Since *Dalbergia sissoo, Acacia nilotica* and several other species take usually 40 to 50 years for harvesting, these species are not preferred by the farmers. *Eucalyptus* spp. and *Populus deltoides* which are harvested at about 10 years, are the most popular species for *agrisilviculture*. A large production of *Eucalyptus* wood has saturated the market, resulting in a slump in prices recently. This has discouraged eucalypts planting.

More recently, several industries have started supporting tree planting in farm lands. WIMCO is supporting plantation of *Populus deltoides* under a buy-back guarantee scheme. One or two paper-mills have also come forward to support eucalypts plantation under a similar scheme.

### (ii) Upper Gangetic plain

It covers about half of Uttar Pradesh, comprising an area of about 15 million ha. More than 65 per cent of the area is under agriculture, 8.5 per cent under forest and the remaining area under wastelands and other uses. About 60 per cent of the cropped area is under cereals, 20 per cent under pulses and about 10 per cent under sugar-cane. About 35 per cent of the cropped area is irrigated. Wheat and rice are important cereal crops.

Agrisilviculture is a popular practice similar to that in the Punjab plains. The main systems of tree planting include scattered trees, boundary plantation and block plantation. The system of planting scattered trees on the farm lands is most common in unirrigated areas. The species common in unirrigated areas are: *Acacia nilotica, Dalbergia sissoo, Ziziphus* spp., *Acacia auriculiformis, Tamarindus indica, Syzygium cumini, Ailanthus excelsa, Azadirachta indica, Moringa oleifera, Artocarpus heterophyllus, Albizia lebbeck, Leucaena leucocephala, Mangifera indica, Emblica officinalis, Bauhinia variegata*, etc. These trees are planted on the boundaries and elsewhere in the field.

In irrigated areas, the more common species are: *Eucalyptus* hybrid, *Dalbergia sissoo, Acacia nilotica, Populus deltoides, Artocarpus heterophyllus, Syzygium cumini, Albizia lebbeck*, etc. Several other species, e.g. *Bombax ceiba, Tectona grandis, Morus alba, Grevillea robusta, Melia azedarach*, etc. are of recent introduction. In this area, the farmers have taken to tree planting for supplementing their income. The produce is usually sold in the market or supplied to meet the demand of industries. Of all species *Eucalyptus* hybrid is perhaps the most widely planted tree species in irrigated area. Planting of trees along field boundaries is the most common system. *Eucalyptus* hybrid is planted at a spacing of 1 m to 1.5 m. *Populus deltoides* is the most common tree in the terai area. It is planted in rows as well as in blocks.

Other species such as *Dalbergia sissoo, Acacia nilotica, Albizia lebbeck, Syzygium cumini*, etc., are planted at a wider spacing of 3 m to 4 m apart. Block plantation of *Eucalyptus* hybrid or *Eucalyptus tereticornis* and $G_3$ and $G_{48}$ clones of *Populus deltoides* is also common. In block plantation, *Eucalyptus* is planted at spacings of 1 m × 1 m to 2 m × 2 m. *Populus deltoides* is usually planted at 4 m × 5 m or 5 m × 5 m. Intercultivation of agricultural crops is done with poplars, however, it is not common with *Eucalyptus* spp. at advanced stages.

Several industries have come forward to support agroforestry planta- tions. This has helped farmers to plant more trees.

In this part of Uttar Pradesh, fruit orchards are very common where forestry trees are extensively grown. The most common practice is to grow a dense one or two row of trees on the four boundaries of the orchard. These trees work as a windbreak and also supplement wood production. They also help in better bearing of fruits. The most common fruits which are being cultivated are mango, guava, litchi, etc. The most commonly planted tree species in the boundaries of fruit orchards are *Eucalyptus* hybrid, *Dalbergia sissoo*, *Syzygium cumini*, *Acacia nilotica*, etc.

*(iii) Middle Gangetic plain*

It comprises approximately an area of 14.4 million ha covering east- ern Uttar Pradesh and Bihar. About 85 to 90 per cent land area is sown. Cultivable wastelands constitute about 5 per cent. About 35 per cent area is irrigated. The holdings are small. Rice is the main cereal crop. Other important crops include wheat, pulses, oilseeds, sugar-cane, etc.

Growing of trees in agricultural land is common but in comparison to the Punjab plains and the upper Gangetic plains, the proportion is less. This is mainly because of small holdings and the general poverty situation prevailing in the area. The common agrisilvicultural practices include grow- ing of scattered trees, field boundary plantation and to a lesser extent block plantation. The species most commonly planted are *Acacia nilotica*, *Eucalyp- tus* hybrid, *Leucaena leucocephala*, *Dendrocalamus strictus*, *Tectona grandis*, *Bambusa vulgaris*, *Bambusa tulda*, *Dalbergia sissoo*, *Mangifera indica*, *Aegle marmelos*, *Grevillea robusta*, *Artocarpus heterophyllus*, *Albizia lebbeck*, *Syzy- gium cumini*, *Moringa oleifera*, etc. Tripathi (1983) observes that in Bihar, trees are being raised on field bunds, fruit orchards and tree orchards. The species common in field bunds are *Dalbergia sissoo* and *Wendlandia exserta*. The mango orchards have peripheral planting of litchi, jackfruit, jamun, guava, anola, bel, etc.

*(iv) Lower Gangetic plain*

It comprises the eastern part of the Gangetic plains, covering almost the whole of West Bengal and encompassing a total area of about 8 million ha. About 70 per cent of the area is under agriculture. Irrigation facilities are inadequate although average rainfall is comparatively high. The main crops in this region are rice and jute. A large variety of useful species are being planted in agricultural croplands. Some of them are *Eucalyptus* spp., *Tectona grandis*, *Michelia champaca*, *Lagerstroemia speciosa*, *Dalbergia sissoo*, *Gmelina arborea*, *Bombax ceiba*, *Terminalia myriocarpa*, *Leucaena leucocephala, Sesbania* spp., *Bambusa tulda, Dendrocalamus hamiltonii, Artocarpus heterophyllus*, etc.

Evaluation of the social forestry project of West Bengal has indicated that the most widely planted species are *Eucalyptus* spp. Other species planted by farmers include: *Dalbergia sissoo, Michelia champaca, Tectona grandis, Gmelina arborea, Acacia auriculiformis*, etc. (Anon., 1983). It has also been indicated that small and marginal farmers have gone for agrisilviculture systems.

### (v) Assam plains

The Assam plains consist of the Brahmaputra valley which is not more than 80 km wide and surrounded by high mountains on all sides except the west. The valley consists of plains and numerous isolated hillocks; the land is fertile and paddy is the main crop. The trees which are planted include *Artocarpus heterophyllus, Cocus nucifera, Areca catechu*, bamboos, canes, etc.

### 3. MADHYA PRADESH

The total geographic area of Madhya Pradesh is 44.2 million ha. Of this, about 22.2 million ha is agricultural cropland, 15.0 million ha is forest and the remaining area is under other uses and wastelands. There are five distinct agricultural regions depending on soil and climatic conditions. The *Rice zone* consists of the eastern part of the State, mainly occupied by Chhatisgarh plateau. Rainfall is comparatively higher. Rice is the main crop. Other crops are negligible. Since forests are plentiful in this region, cultivation of trees on farm land is not very common. Trees on farm land include isolated trees and row/bund planting. *Acacia nilotica* is the predominant tree species. Other species which are common include *Terminalia arjuna, Madhuca indica, Bambusa vulgaris, Eucalyptus* hybrid, *Moringa oleifera, Albizia lebbeck, A. procera*, etc.

The *rice-wheat* zone is in the northern part of the State. The other agricultural crops include pulses, oilseeds, jowar, bajra and smaller millets. The tree species which are widely grown are *Mangifera indica, Madhuca indica, Dendrocalamus strictus, Acacia nilotica, Bambusa arundinacea, Azadirachta indica, Ziziphus* spp., *Eucalyptus* spp., *Albizia lebbeck, Dalbergia sissoo*, etc.

The *wheat zone* is predominantly an area of black cotton soils. The main crop is wheat. Other crops include jowar, pulses, oilseeds, etc. The trees which occur on agricultural cropland in various forms include *Acacia nilotica, Azadirachta indica, Tectona grandis, Butea monosperma, Eucalyptus* hybrid, *Albizia lebbeck*, etc. The cotton-wheat zone is the area adjoining Maharashtra and western Madhya Pradesh. Important trees which are traditionally grown include: *Acacia nilotica, A. leucophloea, Dendrocalamus strictus, Madhuca latifolia, Tectona grandis, Azadirachta indica*, etc. Recently, fast-growing species such as *Eucalyptus* spp., *Leucaena leucocephala*, etc., have become more popular.

The *jowar-wheat* zone is in the north-west part adjoining Rajasthan and Uttar Pradesh. Important agricultural crops are jowar, wheat, pulses, oilseeds, millets, etc. Paddy is also cultivated in pockets. The important trees on farm land include: *Dalbergia sissoo, Albizia lebbeck, Dendrocalamus strictus, Artocarpus heterophyllus, Mangifera indica, Acacia nilotica, Prosopis* spp., *Hardwickia binata*, etc.

## 4. GUJARAT

The total area of Gujarat state is 19.6 million ha. Of this, agricultural cropland area is 10.18 million ha and irrigated area is only 2.3 million ha. Forests cover only 1.9 million ha and the remaining area is covered by different categories of wastelands and other lands. Agrisilviculture has long been practised in Gujarat but real progress made only in recent years. The practice of agroforestry consists of scattered trees on farm lands, boundary plantation and wood lot establishment in the form of block plantation. The farm forestry programme, which is really agrisilviculture, began in a systematic way during 1972 with the distribution of seedlings to the farmers free of cost. During 1972, only 0.65 million seedlings were distributed while during 1984–85, the number of seedlings distributed rose to 194.67 million (Anon., 1990). Most of the farmers prefer planting trees on field bunds (Table 46).

Table 46: **Pattern of agroforestry plantation in Gujarat (Anon., 1990)**

| Type of system | Percentage of farmers by holdings | | | |
|---|---|---|---|---|
| | Marginal | Small | Big | Total |
| 1. Row or bund planting/ water channels homesteads | 89 | 76 | 72 | 82 |
| 2. Block planting | 11 | 24 | 28 | 18 |

However, the seedlings planted in block plantation are about 59 per cent in comparison to only 41 per cent seedlings planted in row/bund planting (Anon., 1990).

The most dominant species is *Eucalyptus* hybrid. In Gujarat, the percentage of species planted during 1980 to 1984 were: *Eucalyptus* spp. = 84 per cent, *Leucaena leucocephala* = 6 per cent, *Tectona grandis* = 5 per cent, *Acacia nilotica* = 1 per cent, *Acacia auriculiformis* = 2 per cent, Bamboo = 1 per cent and others = 1 per cent (Anon., 1990). Most farmers plant trees for production of small timber, firewood and economic benefits.

## 5. MAHARASHTRA

Maharashtra is the third biggest state in India with an area of 30.77 million ha. Of this, about 20 million ha are agricultural cropland, 6.5 million ha forest land and the remaining area is wasteland and other category lands. The state can be divided into three distinct physiographic regions: (i) *Tapi trough*: It lies between Satpura in the North and Balaghat range in the south. It is a transition zone between central India and the Deccan. (ii) *Deccan plateau*: The greater part of Maharashtra between Sahyadri and Satpura. This region is intercepted by rivers forming the valleys of Godavari, Bhima and Krishna. (iii) *Konkan*: A narrow strip of land about 500 km in length between the Arabian Sea and Sahayadri. The Konkan region receives heavy rainfall of over 2000 mm. The western Maharashtra lies in the rain-shadow area with an annual rainfall of about 600 mm. About 62 per cent land area is under agriculture. Important crops in Vidarbha region are cotton, groundnut, bajra, pulses, jowar, oilseeds, wheat, etc. In Marathwada area, particularly, where irrigation facilities are available, cash crops such as sugar-cane, vegetables, etc., are also grown. Cultivation of oranges in Vidarbha is common. Several agrisilviculture systems such as scattered trees, row/bund planting and block plantations are common. The most important tree species are *Acacia nilotica*, *A. leucophloea*, *Eucalyptus* hybrid, *Tectona grandis*, *Azadirachta indica*, *Leucaena leucocephala*, *Hardwickia binata*, *Mangifera indica*, *Moringa oleifera*, etc. In Konkan area, species such as *Mangifera indica*, *Artocarpus heterophyllus*, *Terminalia arjuna*, *Bambusa arundinacea*, etc., are more common.

## 6. ORISSA

The total geographic area of Orissa is 15.54 million ha. Of this, agricultural cropland is 8.4 million ha, forest land is 6.6 million ha and the remaining area is under other uses. The important agricultural crop in Orissa is rice. Other crops include oilseeds, pulses, etc. Trees are widely grown on farm lands in various forms. In some areas of the state, shifting cultivation is practised. State Government has made attempts to rationalise *podu* cultivation by helping farmers to plant trees of economic importance along with agricultural crops. Several species, such as, *Tectona grandis*, *Gmelina arborea*, *Bombax ceiba*, *Terminalia alata*, *Terminalia arjuna*, etc., are widely planted (Das and Mahapatra, 1983). In other areas, on field bunds, the species commonly planted are *Acacia nilotica*, *A. auriculiformis*, *A. leucophloea*, *Eucalyptus* spp., *Leucaena leucocephala*, *Casuarina equisetifolia*, etc., *Agave sislana* and *A. americana* are planted on field bunds in the western part of the state.

## 7. SOUTHERN PLATEAU REGION

The southern plateau region comprises plateau areas of the southern states, e.g., Andhra Pradesh, Karnataka, Tamil Nadu and Kerala. Andhra Pradesh has an area of 27.67 million ha, of which 14.38 million ha are agricultural land, 6.17 million ha forest land and the remaining area is under other uses. The total number of holdings are reported to be 6.2 million indicating average area per holding to be 2.3 ha. The state has semi-arid areas where farmers have taken to tree planting in the form of tank foreshore plantation, village wood lots, shelterbelt plantation, row/bund plantation, etc. The important species which are planted by the farmers include: *Eucalyptus* spp., *Dendrocalamus strictus*, *Bambusa arundinacea*, *Anacardium occidentale*, *Leucaena leucocephala*, *Tamarindus indica*, *Albizia lebbeck*, *Acacia nilotica*, etc. In dry areas, some farmers have taken up tree farming where they are growing mainly *Eucalyptus* hybrid, bamboos and *Casuarina equisetifolia*. Such plantations are more common in the vicinity of paper and pulp-mills which require a large quantity of raw material (Rao, 1983). The farmers raise agricultural crops during initial years until the tree canopy closes.

In semi-arid areas of the state, *Acacia nilotica*, *Azadirachta indica*, *Holoptelea integrifolia*, *Mangifera indica*, *Madhuca indica*, *Borassus flabellifer*, *Phoenix sylvestris*, etc. are grown on field bunds (Rao, 1983).

Tamil Nadu with a total geographic area of 13.0 million ha has an agricultural cropland of 6.03 million ha and forest land 2.03 million ha. The remaining area is under other uses. Important agricultural crops are rice, groundnut, sorghum, pulses, cotton, millets, sugar-cane, spices, etc. Various forms of agrisiviculture are practised in this state. In the southern part of the state, where the north-east monsoon is about 600 mm, two species, e.g., *Acacia planifrons* and *Prosopis chilensis* constitute the tree component in the croplands (Shanmugasundram, 1983). These species regenerate themselves profusely. *Prosopis chilensis* comes up like a weed and sometimes becomes a menace in wetland agriculture. These species are harvested on a coppice rotation of 5 years.

In the districts of Coimbatore, Tirchy and Madurai *Acacia leucophloea* is usually interplanted with jowar, gram and other crops at an interval of 5 to 12 m. *Acacia leucophloea* provides good fodder and yields poles within 10–15 years. The yield of jowar is also increased due to protection afforded by *Acacia leucophloea* (Shanmugasundaram, 1983). Some trees, such as *Borassus flabellifer*, *Tamarindus indica*, *Cieba pentandra*, *Delonix alata* and *Prosopis chilensis* are common in dry unirrigated areas (Jambulingam and Fernandes, 1986).

Several other tree species such as *Albizia lebbeck*, *Albizia falcataria*, *Ailanthus excelsa*, *Artocarpus* spp., *Casuarina equisetifolia*, *Bambusa* spp., *Leucaena leucocephala*, *Annona squamosa*, *Ailanthus excelsa*, *Calophyllum innophylum*, *Ficus* spp., *Dalbergia sissoo*, *Pongamia pinnata*, *Pterocarpus marsupium*, *Azadirachta indica*, etc. are also planted because farmers are supplied up to 500 seedlings free of cost. Fruit trees are also widely planted. Since the State is deficient in wood production, there is considerable demand for firewood and small timber produced by farmers. In the hills of Nilgiri, *Eucalyptus globulus* and wattles are usually planted.

Karnataka has a total land area of 19.10 million ha. Agricultural cropland is 11.15 million ha and forest land is 3.83 million ha. The remaining area is under other uses. The main agricultural crops include rice, jowar, bajra, ragi, smaller millets, pulses, oilseeds and to a smaller extent, wheat, sugar-cane, etc. Several tree species are traditionally planted in farm land. These species include: *Tamarindus indica, Anacardium occidentale, Santalum album, Tectona grandis, Eucalyptus* hybrid, bamboos, *Leucaena leucocephala, Phoenix sylvestris, Casuarina equisetifolia, Cocos nucifera*, etc. Most recently, block plantations of *Eucalyptus* hybrid, *Leucaena leucocephala* and bamboos have been taken up by farmers, particularly in those areas where there is acute demand.

Kerala covers an area of 3.88 million ha. Approximately 2.9 million ha are agricultural cropland, 1.08 million ha forest land and the remaining area is under other uses. The main agricultural crops in Kerala are rice, spices, etc. The main tree species found in the farmland include: *Cocos nucifera, Areca catechu, Anacardium occidentale*, bamboos, *Tectona grandis, Gmelina arborea, Eucalyptus* spp., *Casuarina equisetifolia*, etc.

## 8. COASTAL AREAS

India is surrounded by seas on three sides: the Indian Ocean in the South, the Arabian Sea in the West and the Bay of Bengal in the East. The total length of the coastline is approximately 5600 km. There are about 247 islands which also form a part of the country. The coastal area starts in Gujarat at Kachchha near Kathiawad and ends at the delta of the Ganga-Brahmaputra in West Bengal. The climatic conditions are quite divergent from place to place. In coastal areas, there are several problems. Some of these problems are: (i) soil erosion by water, (ii) soil erosion by wind, (iii) salinity, (iv) periodic cyclones, (v) waterlogging etc. In coastal areas, the climate is generally hot and humid. Agricultural crops which are able to tolerate some salinity are widely grown; however, paddy is the most dominant.

Several species of trees are grown with agricultural crops either in mixture or as wood lots. Some of the important trees for coastal areas are: *Casuarina equisetifolia, Acacia auriculiformis, Anacardium occidentale, Borassus flabellifer, Cocos nucifera, Areca catechu*, etc.

In dry coastal areas, where serious aridity and salinity problems exist, species such as *Prosopis chilensis, Salvadora persica* and *Prosopis cineraria*, are usually grown.

## 9. HIMALAYAN REGION

The Himalayan region can be distinctly divided into two, namely, western Himalayas and eastern Himalayas. The western Himalayas comprise the Himalayan ranges distributed in Uttar Pradesh, Himachal Pradesh and Jammu & Kashmir. Annual precipitation varies between 1000 mm to 2500 mm. There is considerable variation in climate, physiography, soil and vegetation between the outer and inner Himalayas. Vegetation is largely controlled by altitude. The western Himalayas are classified into: (i) the lesser Himalayas and (ii) the great Himalayas. The lesser Himalayas lie in the North of the Siwalik hills. The hill ranges of the lesser Himalayas are usually 50 to 100 km wide and 1000 m to 5000 m high. The terrain in the entire area is rugged and montane. Dhauladhar range in Himachal Pradesh, Pir Panjal in Jammu & Kashmir and Mussoorie in Uttar Pradesh are some of the important hill ranges. There are several important valleys where intensive agriculture is practised. Important valleys are Kangra and Kullu in Himachal Pradesh and Kashmir valley in Jammu and Kashmir. In these hills agricultural fields are terraced. Several agricultural crops such as paddy, maize, pulses, wheat, oilseeds, potatoes, vegetables, etc., are grown. Cultivation is common up to 2500 m elevation.

Fruit orchards of several species are found in the hills of Himachal Pradesh, Jammu & Kashmir and Uttar Pradesh. The fruits of plum, peach, apricot, etc. are grown up to 1500 m elevation. Apple orchards are found above 1500 m elevation. The terraced agricultural fields are in the form of narrow strips whose width varies from 5 m to 15 m. Plantation of most trees on agricultural cropland was not common in the past as enough forests were available in the vicinity. However, deforestation in the hills during the recent past has created acute shortage of firewood and fodder. This has compelled the farmers to grow trees on their farmlands.

The plantation of trees in these hills usually depends upon the altitude. Singh (1986), on the basis of a survey in Himachal Pradesh, found the preference for different species as under:

(a) 500 m elevation: *Eucalyptus spp., Acacia catechu, Toona ciliata, Dalbergia sissoo, Leucaena leucocephala, Terminalia billerica, Albizia* spp.

(b) 500–1000 m elevation: *Eucalyptus* spp., *Grewia optiva, Bauhinia* spp., *Dalbergia sissoo, Acacia catechu, Artocarpus lakoocha, Morus alba, Toona ciliata, Terminalia alata, Mangifera indica.*

(c) 1000–1500 m elevation: *Grewia optiva, Robinia pseudacacia, Sapindus mukorossi, Melia azedarach, Eucalyptus* spp., *Bauhinia* spp., *Juglans regia, Toona ciliata,* etc.

(d) 1500-2000 m elevation: *Robinia pseudacacia, Grewia optiva, Juglans regia, Quercus leucotrichophora, Populus* spp., *Prunus* spp., *Cedrus deodara, Melia azedarach, Quercus himalayana, Morus serrata,* etc.

(e) Above 2000 m elevation: *Robinia pseudacacia, Quercus himalayana, Jug-

*lans regia, Salix* spp., *Morus serrata, Populus* spp., *Prunus* spp., *Celtis australis,* etc.

The great Himalayas form the innermost and loftiest continuous Himalayan ranges. In this part of the country, cultivation is rare as climatic conditions are difficult.

The eastern Himalayan hill ranges constitute humid regions from Sikkim eastwards up to Arunachal Pradesh. The eastern hills run through the northern part of West Bengal, Sikkim, and north-eastern states of Arunachal Pradesh, Nagland, Manipur, Tripura, Mizoram, Meghalaya, etc.

These areas have, depending on the altitude, tropical, subtropical and temperate zones. Agriculture is limited to lower elevations and in valleys. In the higher elevations, shifting cultivation is mostly practised. In areas where settled agriculture is done, only useful trees such as bamboos, canes, *Artocarpus heterophyllus, Areca catechu, Cocus nucifera, Terminalia myriocarpa, Dipterocarpus* spp., *Anthocephalus chinensis,* etc. are preferred.

**Homestead Plantations and Home Gardens**

It is almost a common practice to plant trees around habitations, courtyards, wells, threshing yards and in the fields in every region of the country (Fig. 8). The trees are planted for fruits, wood, seeds, flowers, shade, etc. They are planted as they are beautiful to look at. They harbour different kinds of birds and other animals. In India, the commonest homestead plantation system includes: plantation of trees around houses, courtyards, wells, compounds, etc. Several agricultural crops, vegetables, spices and other crops are grown under them.

The species planted depends on climate, soil and preference of the farmer. The species which are widely planted in the alluvial plains of Punjab, Haryana, Uttar Pradesh and Bihar include: *Acacia nilotica, Azadirachta indica, Syzygium cumini, Albizia* spp., *Dalbergia sissoo, Artocarpus heterophyllus, Moringa oleifera, Mangifera indica, Eucalyptus* hybrid, etc. In arid areas of Rajasthan commonly planted species are: *Acacia nilotica, A. leucophloea, Azadirachta indica, Prosopis* spp., *Parkinsonia aculeata, Ailanthus excelsa, Ziziphus* spp., etc.

In central India, commonly planted species in homesteads are: *Mangifera indica, Acacia nilotica, Azadirachta indica, Moringa oleifera, Tamarindus indica, Dendrocalamus strictus, Bombusa aroundinacea, Dalbergia sissoo,* etc. Fruit species such as citrus, guava, mango, bel, etc. are preferred in homestead plantations. In alluvial plains and the central Indian plateau, important species in homestead plantations are as under:

(i) Trees: Mango (*Mangifera indica*), sissoo (*Dalbergia sissoo*), babul (*Acacia nilotica*), jack fruit (*Artocarpus heterophyllus*), neem (*Azadirachta indica*), bamboo (*Bambusa arundinacea*), anola (*Emblica officinalis*), munga

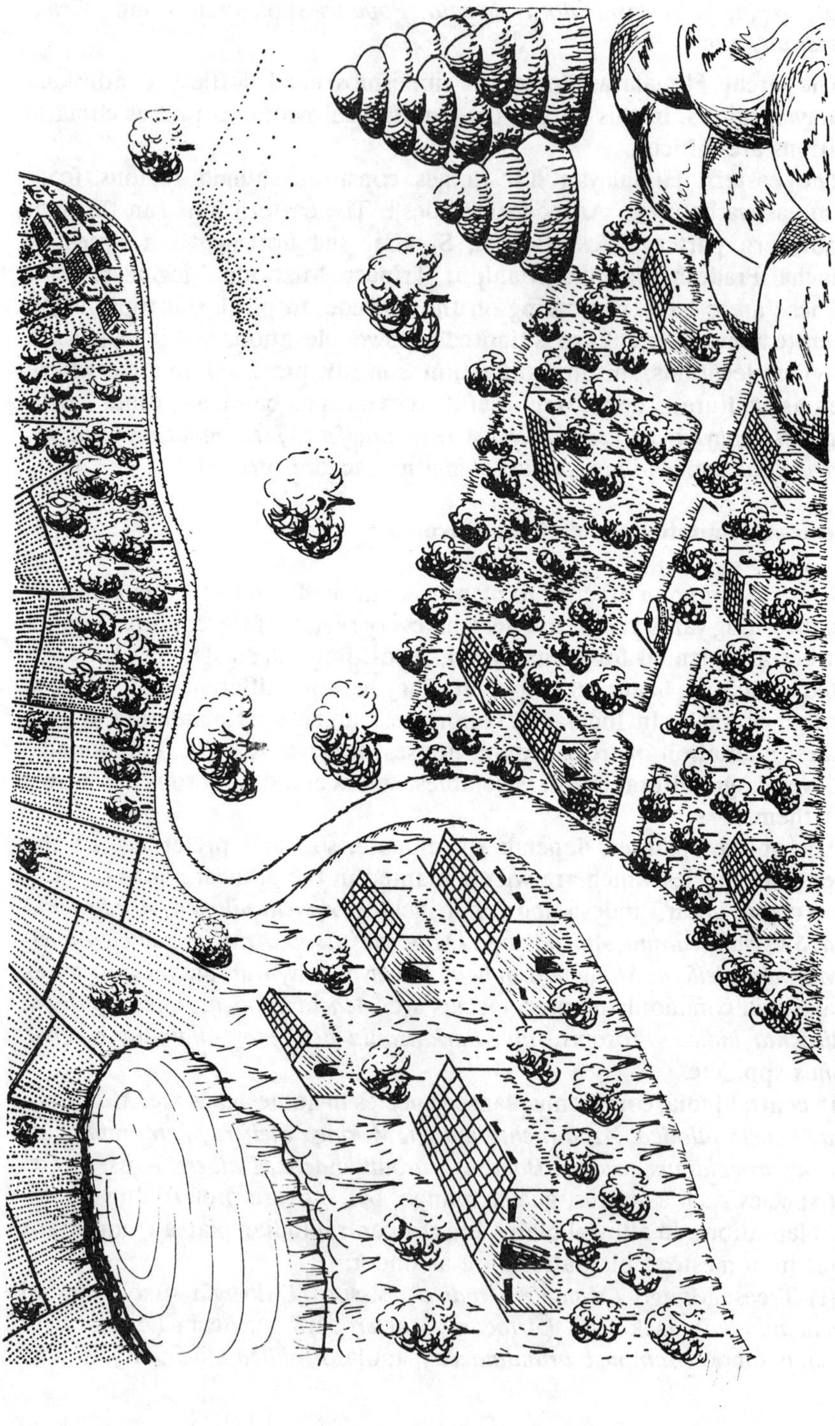

Fig. 8: Home gardens

(*Moringa oleifera*), ber (*Ziziphus* spp.), teak (*Tectona grandis*), etc. (ii) *Small trees*: Citrus (*Citrus* spp.), guava (*Psidium gujava*), papaya etc. (iii) *Food crops*: Rice, wheat, pulses, oilseeds, etc. (iv) *Vegetables*: Tomato, bhindi, brinjal, arbi, haldi, potato, cucumber, beans, etc. (v) *Spices*: Turmeric, ginger, chillis, etc.

In the southern plateau, the common tree species are: *Tamarindus indica, Cocos nucifera, Areca catechu, Gmelina arborea, Azadirachta indica, A. nilotica*, etc. For coastal and valleys of eastern states, *Cocos nucifera, Areca catechu, Artocarpus heterophyllus*, etc. are important species. In some areas of the Western Ghats and Assam, particularly *Dendrocalamus hamiltonii, Bambusa tulda, Bambusa arundinacea*, etc. are also common. In hills, the common species for homestead plantations may include *Grewia optiva, Ficus* spp. *Juglans regia, Punica granatum*, etc.

In rural areas, fruit and commercial trees predominate. However, multipurpose trees such as *Acacia nilotica, Azadirachta indica*, etc. are of common occurrence in most regions of the country. There is a *multilayered cropping* system in homestead plantations. Bigger trees occupy the top layer, small trees and shrubs yielding fruits and other useful products occupy the middle layers and shade tolerating vegetables, and spices, e.g., garlic, ginger, etc., find a place in the ground layer. A large variety of climbing and twining plants yielding vegetables and spices, e.g., beans, cucumbers, are commonly grown in home gardens.

In high rainfall areas, the system of home gardens is more prevalent. In the states of Kerala and Tamil Nadu, the practice is more pronounced where *Cocos nucifera* and *Areca catechu* are the main crops.

An inventory of homestead plantations in Kerala indicated that a large number of cereals, pulses, fruits, oilseeds, spices, vegetables, trees, etc., are grown (Nair and Sreedharan, 1986). Some of the common crops are: (i) *Food crops*: arrowroot (*Maranta arundinacea*), cassava (*Manihot esculenta*), Chinese potato (*Coleus parviflorus*), dioscorea (*Dioscorea* spp.), rice (*Oriza sativa*), sweet potato (*Ipomea batatus*), taro (*Calocasia* spp.), elephant yam (*Amorphophallus companulatus*), etc. (ii) *Pulses*: cow-pea (*Vigna unguiculata*), horse gram (*Dolichos uniflorus*), mung bean (*Vigna radiata*), pigeon-pea (*Cajanus cajan*), etc. (iii) *Fruit*: anona (*Annona* spp.), banana (*Musa* spp.), breadfruit (*Artocarpus utilis*), garcinia (*Garcinia indica*), anola (*Emblica officinalis*), guava (*Psidium gujava*), jackfruit (*Artocarpus heterophyllus*), mango (*Mangifera indica*), papaya (*Carica papaya*), passion-fruit (*Passiflora udilis*), pomegranate (*Punica granatum*), tamarind (*Tamarindus indica*), etc. (iv) *Oils and fats*: coconut (*Cocos nucifera*), groundnut (*Arachis hypogaea*), sesame (*Sesamum indicum*), etc. (v) *Beverages*: Cacao (*Theobroma cacao*), coffee (*Coffea* spp.). (vi) *Spices*: Cardamom (*Eletaria cardamomun*), cinnamon (*Cinnamomum zeylanicum*), clove (*Syzygium aromaticum*), ginger (*Zingiber officinale*), pepper (*Piper*

*nigrum*), turmeric (*Curcuma longa*), etc. (vii) *Vegetables*: Lady's finger
(*Abelmoschus esculentus*), bitter gourd (*Momordica charantia*), brinjal
(*Solanum melongena*), cucumber (*Cucumis sativus*), watermelon (*Citrullus*
spp.), etc. (viii) *Trees*: bamboo (*Bambusa arundinacea*), erythrina (*Erythrina
indica*), gliricidia (*Gliricidia sepium*), subabul (*Leucaena leucocephala*),
mahogany (*Swietenia macrophylla*), moringa (*Morinda tinctoria*), portia tree
(*Thespesia populenea*), teak (*Tectona grandis*), wild-jack (*Artocarpus hirsuta*),
areca-nut (*Areca catechu*), cashew (*Anacardium occidentale*), palmyrah palm
(*Borassus flabellifer*), rubber (*Hevea brazillensis*), etc.

Most home gardens also support a variety of animals. Some trees produce fodder. The waste materials from crops and homes are used as fodder
and feed for the animals and birds. In backwater areas of coasts, mangroves
form an essential part in home gardens. Important species grown in backwater areas include: *Acanthus illicifolius*, *Avicennia officinalis*, *Carbera odollam*,
*Rhizophora conjugata* and *R. maicronata*. Fish and prawns are also grown
in farm ponds, paddy fields and canals and form essential components of
home gardens in Kerala (Nair and Sreedharan, 1986).

Home gardens consist of a number of crops having several functions.
For protection, there are large trees, bamboos, etc. at the periphery of
the home gardens. Under them are various types of fruit trees, e.g., citrus, guava, mango, papaya, etc. Various kinds of cereals and vegetables are
also grown depending on the season, soil and locality, providing different
kinds of goods. The home gardens aim to satisfy the needs of fuel, fodder, small timber, etc. They also help to meet the needs of food, fruits,
vegetables, spices, etc. Some trees yield medicines and raw material for industries.

The homestead plantations and home gardens usually occupy a small
area close to houses, wells, etc. The cropping intensity is considerably high.
The valuable and faster growing trees are preferred to miscellaneous species
having no immediate benefits. The size of home gardens is small and is
usually not more than 0.15 ha. Smaller home gardens are more common.
However, rich farmers can afford to maintain larger homestead plantations.
The home gardens involve a higher degree of labour input compared to
monocropping. Therefore, smaller home gardens are well looked after. Information regarding cropping pattern in different regions, relative yields
of agricultural and tree crops and support to animals, cropping intensity,
suitable management systems, etc. in home gardens has not been docu-
mented. The system is very complex and would need study of several parameters involving several physical and biological parameters including edapho-
climatic conditions, plant to plant interactions, socioeconomic complexities,
etc.

**Trees with Plantation Crops**

Plantation crops have an important place in the country's economy. Although these crops occupy only a small area, their production plays an important role in the country's export. Important plantation and garden crops include, tea, coffee, cocoa, rubber, coconut, areca-nut, etc. The total area under tea, coffee and rubber plantations in India during 1985—86 was 398, 235 and 350 thousand ha respectively (Anon., 1986). Coconut is grown over 1150 thousand ha.

Tea is grown mainly in Assam, West Bengal, Tamil Nadu, Kerala and Karnataka. Tea (*Camellia sinensis*) is a shade-loving plant and thrives well in a humid climate. It does not grow well in poor soils and exposed to the full sunlight. Several tree species are grown in lines to provide necessary shade in a tea plantation. Leguminous trees such as *Albizia chinensis, Acacia* spp., *Dalbergia serica, Derris robusta,* etc., are grown as shade trees in northeast India, while *Grevillea robusta, Acacia* spp., *Erythrina lithosperma,* etc. are used in south India.

The quality and the quantity of shade determines the tea leaf production. It is now concluded that medium to light shade gives a better yield of leaves than no or heavy shade. Therefore, such trees which produce uniform and light shade are preferred. Leguminous trees benefit the tea crops by providing light shade and adding nitrogen. About 90 kg/ha per year of nitrogen is added in the soil with planting of *Albizia chinensis.* Tea is mycorrhizal plant and light intensity also affects the mycorrhizal formation.

The trees planted for shade are usually put in lines of 12 to 15 m apart. For small trees, this distance could be suitably reduced. Bamboos and some of the dense crowned trees are reported to cause damage to the tea crop.

Coffee is cultivated in the hilly tracts of the Western and Eastern Ghats in the states of Karnataka, Tamil Nadu, Kerala and Andhra Pradesh. *Coffea arabica* and *Coffea robusta* are two important species of coffee extensively planted. The coffee plant is delicate and fastidious in its requirements. It thrives well in a warm and humid climate. It cannot withstand frost and extreme heat. In India, coffee is also grown under shade. Shade protects the plants from extremes of temperature and rainfall. Trees also protect the coffee plants from hailstorms, check soil erosion, and add leaf litter in the soil.

Two kinds of shade trees are provided in coffee plantations, viz., permanent and temporary. As permanent shade trees several species, e.g., *Ficus glomerata, F. nervosa, F. tsjekela, Albizia chinensis, A. lebbeck, A. moluccana, A. sumatrana, Artocarpus* spp., *Grevillea robusta,* etc., are preferred. Temporary shade almost always consists of *Erythrina lithosperma* and leguminous shrubs such as *Crotolaria* spp., *Tephrosia* spp., and *Indigofera* spp. These are grown in young plantations to cover the ground between the rows and to keep down the weed growth. They also provide shade and green manure to young coffee plants. Coffee has been successfully grown under the shade in plantations of *Terminalia myriocarpa* and *Dipterocarpus dalergioides* in Arunachal Pradesh. It could also be introduced in natural forests by clearing the ground vegetation and leaving some trees for shade.

Valuable tree species are also planted with rubber, cocoa and other plantation crops.

### Multipurpose Trees with Horticultural Crops

The total area under fruit trees in India is over 2.5 million ha. Mango is the most dominant fruit occupying about 40 per cent of the total area. The other important fruit trees include: apple, citrus, litchi, date palm, almond, apricot, peach, plum, guava, ber, banana, pineapple, etc. Several fruit trees such as banana, pineapple, papaya as well as strawberries are cultivated as field crops and sometimes grown as intercrops among the trees. Several fruit trees such as mango, citrus, apple, plum, litchi, almond, etc. take longer period for fruiting and are usually spaced out to allow intercropping of cereals, pulses, oilseeds, vegetables, etc.

Mango is grown in almost all the regions of the country except the temperate Himalayan region. Mango trees are sufficiently spaced to allow intercropping of cereals, pulses, oilseeds, vegetables, spices, etc. Those varieties are preferred which do not grow tall. In the mango orchards of Uttar Pradesh, the common practice is to cultivate vegetables, pulses, beans, berseem etc. Wheat, mustard, etc. are grown during formative years.

Most varieties of mango are planted at a spacing of 10 to 15 m. During the initial 10 to 15 years there are enough interspaces for cultivation of agricultural crops. The canopy closes after a period of 15 years depending on spacement. Mango trees cause dense and low shade which is heavy and usually nothing grows under them. Sometimes the branches are too low and they touch the ground. However, if trees are properly trained and do not cast a heavy low shade several crops can be grown.

In areas of central India, particularly in Maharashtra and adjoining Madhya Pradesh, cultivation of oranges is common. The trees of oranges are small and their growth is such that it allows intercropping of a variety of crops. However, if spacing is narrow, intercropping of cereals, pulses, vegetables, etc. becomes difficult. In most cases, however, the spacing is so adjusted that it allows intercropping of cereals, vegetables, pulses and other crops in the interspaces available.

In areas where high winds blow during summer, plantation of windbreaks of tall trees around an orchard has been a common practice. The trees which are grown around an orchard are multipurpose trees and include *Dalbergia sissoo*, *Syzygium cumini*, *Acacia* spp., etc. Orchards of citrus, guava, etc. are also intercropped with cereals, pulses, fodder crops, etc. The growing of multipurpose trees either on the boundary or as an intercrop is not very common. However, more recently, a greater number of multipurpose trees have gradually been planted along the boundary of orchards and also as an intercrop. Similarly, other fruit orchards, e.g., apple, plum, peach, almond, litchi, etc. are intercropped with cereals, pulses, vegetables, fodder crops, etc. which do not grow tall, at least not for the first 4–5 years.

With some fruit trees, *filler trees* are also planted. The purpose of filler trees is to grow multipurpose trees and at the same time provide revenue after harvesting. These filler trees are usually removed as the main fruit trees grow up demanding more space. Sometimes fruit trees themselves are used as filler trees. For example, the filler trees of guava, papaya, phalsa, peach, plum, etc. are planted in mango and litchi orchards.

Coconut (*Cocos nucifera*) is an important commercial fruit tree grown in southern India and Assam. Kerala, Karnataka, Tamil Nadu and Andhra Pradesh account for almost 90 per cent of the area and production of coconut. It is a perennial tree. It is tall and has a small crown at the top. Several crops can be grown with coconut. Up to about 8 to 10 years of planting the crown size is small and the area can be intercropped with annuals. It is better if such crops do not compete with developing plum. During the age period of 10 to 25 years the crown size increases and covers almost 80 per cent of the ground. After 25 years, intensive intercropping can be done as the trunk height increases and more light falls on the ground. Also, the root system at this stage is such that feeder roots are below 30 cm (Kushwaha *et al.*, 1973; Nair, 1979).

Coconut is essentially a crop of small growers. Coconut holdings in southern India are small and average only 0.22 ha. A wide variety of crops can be grown with coconut trees. Important crops include rice, pulses, pepper, ginger, turmeric, areca-nut, jackfruit, papaya, etc.

## Alley Cropping or Hedgerow Cultivation

The practice of *alley* cropping in which agricultural crops and trees are cultivated together usually in lines or strips, has not been streamlined in India. However, it is widely practised in Africa and other countries. Several species, such as *Leucaena leucocephala*, *Sesbania grandiflora*, *Gliricidia sepium*, *Cassia siamea*, etc. are used in this system. These species are usually planted at a spacement of 50 cm in rows spaced at 4, 5 and 6 m apart. In-between the rows of trees, agricultural crops such as maize, jowar, pulses, oilseeds, etc. are grown. The trees are often pruned and lopped. The fodder obtained from these trees is either used as green manure in the field or as fodder for animals. *Leucaena leucocephala* is often pruned at the height of 1.0 metre and maintained as a hedge in the field.

Alley cropping and hedgerow cultivation is very useful for sloping land. In the Philippines, sloping agricultural land technology has been developed which is primarily a soil conservation-oriented farming system involving hedgerow cultivation of *Leucaena leucocephala* on contour lines. The system is claimed to be culturally appropriate, ecologically fit, economically sound and a technically astute development tool (Harold and Warlito, 1985). The system consists of growing crops between contoured rows of *Leucaena leucocephala* usually set up 4 to 6 metres apart. The trees of *Leucaena leucocephala* are planted in dense double rows in order to make hedgerows which control erosion. When the trees are 1.5 to 2.0 metres tall, they are cut back to about 40 centimetres and the tops are placed in 3 to 5 metre alleys where crops are growing. The leaves of subabul make very good nitrogen-rich fertiliser and also add organic matter to the soil. In-between the hedgerows, agricultural crops such as soyabean, mung, maize, coffee, banana, citrus, etc. are grown in the Philippines.

Use of *Leucaena* under alley cropping is very prevalent due to its easy management. The leaf is used as green manure and fodder, twigs as firewood and the stem as a pole. It was found that wheat, sorghum and groundnut under subabul gave maximum return (Gill *et al.*, 1986). In alley cropping, the orientation of a hedgerow has important bearing on the yield of fodder and grains. In one experiment it was found that NE-SW orientation of the hedgerow was the best and gave maximum return in comparison to East-West, North-South and NE-SW orientations (Pathak, 1989).

The hedgerow system of cropping has several advantages. It is effective in controlling erosion. Green manuring by the leaves of trees may add organic matter to the extent of 2.0 to 5.0 tonne/ha and 150 kg of nitrogen, 15 kg of phosphorus and 75 kg of potassium per hectare annually (Kang *et al.*, 1981; 1985; ICRAF, 1988). Addition of organic matter in the soil helps in improvement of physical, chemical, and biological properties of the soil. Soil fertility is restored which gives good yield of agricultural crops. The crop yield is reported to be 3 to 4 times more than normal farming system (Harold and Warlito, 1984; ICRAF, 1988). This system is self-sustaining, particularly for sloping lands.

The alley cropping and hedgerow cultivation system is not practised on a large scale. However, more recently, farmers in Maharashtra and Gujarat have taken up hedgerow cultivation. *Leucaena leucocephala* has been popular as a fodder and green manure crop.

# CHAPTER 8

# Pastural Silviculture and Silvipasture Systems

## Livestock and Fodder Needs

India has 416 million livestock population which is about 16 per cent of the total livestock population of the world. The livestock constitutes a very important component in rural economy as agriculture in our country largely depends on livestock for manure and draught power. Though the importance of livestock in the economy of the farmers needs no emphasis, the desirability of having such a large livestock population is questionable, particularly when the quality of the majority of the livestock is poor due to undernourishment and inferior breed.

The availability of fodder is short of the requirement. The availability and requirement of green fodder, crop residue and concentrates is as per figures given in Table 47 (ICAR, 1980).

**Table 47:  Estimated availability and requirement of fodder**

| Type of fodder | Availability (million tonnes) | Requirement (million tonnes) |
|---|---|---|
| Green fodder | 224.08 | 611.99 |
| Crop residue | 231.05 | 869.79 |
| Concentrates | 31.60 | 95.40 |

There are several other estimates of fodder availability and requirement which indicate a huge gap between availability and requirement (Anon., 1976; Anon., 1986a). One of the main concerns of land use is to produce enough fodder.

The present sources of fodder availability are: (i) agricultural crop residue, (ii) agricultural green fodder, (iii) grass and grazing and (iv) tree leaf fodder. The total production of green and dry fodder from various sources has been estimated at 250 million and 441 million tonnes respectively as per details given in Table 9 (Anon., 1986a).

Shah *et al.* (1980) estimated the availability of feeds (concentrates),

**Table 48:  Estimated area under pastural silviculture**

| Category | Area in million ha |
|---|---|
| 1.  Barren and uncultivable land | 20.17 |
| 2.  Other uncultivated land (excluding fallow land) | |
|     (i) Permanent pasture and grazing land | 12.00 |
|     (ii) Land under miscellaneous tree crops and groves | 3.49 |
|     (iii) Cultivable wasteland | 16.73 |
| 3.  Fallow land other than current fallows | 9.82 |
| Total | 62.21 |

green and dry fodder for 1984–85 as 26.69 million, 723.93 million and 334.21 million tonnes respectively. The estimates of fodder availability made by various workers vary considerably. The fodder produced and consumed depends upon the edapho-climatic factors, cropping pattern, areas available for grazing and grass production, type of livestock, etc. Cattle and buffaloes are normally fed on the fodder produced on cultivated areas supplemented by grass and grazing and tree leaf fodder. Sheep and goats are normally maintained on grazing. The grazing and harvested grass constitute the main fodder for donkeys, ponies, mules, etc.

The production of grass and leaf fodder is due to the land use which involves managing the land under grass or tree cover or both. A combination of trees and grasses is referred to as a *silvipasture* system, if grazing land is in the forest. The silvicultural aspect of management takes precedence over pasture management. The majority of the grazings in the forest can be classified as *silvipastural systems*. When a land is managed primarily for the production of grasses or as grazing land and the trees may be found scattered in the pasture/grazing land, the practice is called *pastural silviculture*. The practice of pastural silviculture is followed in government forests and lands classified as barren and uncultivable, permanent pasture and grazing lands, lands under miscellaneous tree crops, cultivable wastelands and fallow land other than current fallows.

**Pastural-silviculture**

AREA

Indian farmers in almost all regions tend to keep a large herd of animals. These animals are integrated with the farming systems. Only in limited areas, the livestock depends on cultivated agricultural fodder. In most parts of the country, livestock depends either partially or completely on grasses produced through pastural silviculture. The area available for pastural silviculture is given in Table 48.

Some of the lands classified as forests are also sometimes managed under pastural silviculture but the extent of such area is not precisely available.

## MAJOR TYPES OF GRASSLANDS

The grass production depends on the type of areas, soil conditions, type of grass, climatic conditions and biotic factors. Depending upon the edapho-climatic conditions, the following eight major types of grasslands are recognised (Table 49).

**Table 49: Major grassland types of India (Anon., 1986a)**

| Grassland type | Environment | Distribution |
|---|---|---|
| *Sehima/ Dichanthium* | Black soils | Western Andhra Pradesh, Maharashtra, Madhya Pradesh, South-Western Uttar Pradesh, Karnataka, Tamil Nadu |
| *Dichanthium/ Cenchrus* | Sandy loams | Punjab, Delhi, Rajasthan, Eastern Uttar Pradesh, Gujarat, West Bengal, Bihar, Orissa, Northern Andhra Pradesh, Eastern Madhya Pradesh, Western Maharashtra, Kerala, Eastern Tamil Nadu |
| *Phragmites/ Saccharum* | Marshy localities | Terai areas of Uttar Pradesh, Bihar, West Bengal and Assam, swamps of Sunderbuns, Kaveri delta |
| *Bothriochloa* | Paddy tracts and high rainfall belt | Lonavala tract of Maharashtra |
| *Cymbopogon* | Low hills | Low hills of Western Ghats, Western Ghats, Vindhya, Satpura, Aravali, Chhota Nagpur Plateau up to about 1000 m elevation |
| *Arundinella* | High mountains | High hills of Western Ghats, Nilgiris, lower Himalayan region in Himachal Pradesh, Uttar Pradesh, Bihar, West Bengal and Assam up to about 2000 m elevation |
| *Deyeuxia/ Arundinella* | Mixed temperate climate | Upper Himalayan region in Himachal Pradesh, Jammu & Kashmir, Uttar Pradesh, West Bengal, Assam |
| *Deschampsia/ Deyeuxia* | Alpine and subalpine climate | Alpine and subalpine regions of the main Himalayas in Jammu & Kashmir, Himachal Pradesh, Uttar Pradesh, West Bengal |

## CONDITIONS OF GRASSLANDS IN INDIA

The grasslands in India suffer from several problems. These problems are: (i) higher incidence of grazing than the carrying capacity, (ii) adverse site factors, (iii) poor management, (iv) degraded conditions, (v) poor productivity, etc. Due to the high number of livestock, grazing pressure is very high on grasslands. Stall-feeding is not prevalent in many areas. Higher incidence of grazing leads to reduced grass cover, increased compaction of soil, reduced porosity and increased soil erosion. In rural areas, people keep a large herd of animals. Due to absence of stall-feeding, most animals are forced to stand in common grazing lands. The grasslands in private ownership are managed, however, and the number of livestock is regulated as a limited number of animals are allowed to graze.

Most of the grazing lands do not have good soils. Almost all good and fertile lands are utilised for agricultural purposes. Only such soils which have serious limitations for agriculture are used for grasslands. The soils for grasslands are generally poor in nutrients, shallow in depth, rocky in nature, sandy in texture, poor in moisture, and are associated with adverse climatic conditions. The soils are usually deficient in nutrients and poor in physical, chemical and biological properties. Continuous management of land under grass production helps in improving the soil conditions provided the grazing is restricted to the optimum number of animals.

The grasslands in India are mostly under three types of ownership, namely, (i) private grasslands, (ii) community grasslands and (iii) government-owned grasslands. Private grasslands are restricted to only those farmers with comparatively larger size holdings. Private grasslands are found in central and southern Indian plateau area and in the Himalayan hills. Such grasslands are restricted to saline-alkaline areas of the alluvial region. These grasslands are comparatively better protected and well managed. Fallow lands are good lands left to regain fertility; they are also used for grass production in some areas. In lands which are of land capability class V and above, pasture use can play a vital role. In class V lands, which are marginal croplands, a phase of 4 to 5 years of pasture can restore the cropping potential to some extent. The community grasslands are those owned by the village community or panchayats. Some grasslands are under the ownership of government. The community and government-owned grasslands offer several problems in management. Most of these lands are encroached upon by individuals. The grasslands are in a severely degraded state. In several areas, grass production is negligible and the areas are used as excercise grounds for animals. There are serious social, human and economic implications in the management of these grasslands. The main problem is to educate the farmers who use these lands for livestock grazing that everything will not be right in the near future, if the present land-use pattern on these lands continues (Albrecht, 1976; 1979). Several problems, such as protection, controlling livestock numbers, lack of investment, ownership, etc., beset community and government-owned grasslands.

Most of these grasslands are in a highly degraded condition. The proportion of palatable and perennial grasses is considerably less. Some of these grasslands have become infested with notorious weeds, e.g., *Lantana, Carissa, Dodonea, Eupatorium, Partheneum,* etc. which threaten the existence of all kinds of grasses as these weeds are gregarious in nature.

Overgrazing has led to deterioration of almost all grasslands in the country. The semi-arid region, which is occupied by the *Dichanthium/Cenchrus* type of grass community, is characterized by perennial grasses: *Dichanthium annulatum, Cenchrus ciliaris, C. setigerus, Bothriochloa pertusa, Heteropogon contortus, Cynodon dactylon,* and annual grasses: *Eragrostis tremula, E. viscosa, E. ciliaris, Cenchrus biflorus, Aristida odscensionis, A. redacta,* etc. When these grasslands are subject to heavy grazing by cattle and sheep, they exhibit all stages of deterioration through a gradual disappearance of the perennial species to sparsely populated annual types and ultimately to almost bare soils (Whyte, 1964). In black cotton soils the grass community of *Sehima/Dichanthium* predominates, which consists of these species: *Sehima sulcatum, S. nervosum, Ischaemum* spp., *Chrysopogon montanus, Dichanthium annulatum* and *Heteropogon contortus.* In overgrazed areas, these grasslands are dominated by *Saccharum spontaneum.*

The Terai area lying in Uttar Pradesh, Bihar and West Bengal is characterised by moist to swampy conditions and is occupied by the *Phragmites/ Saccharum* type of grasses. This community consists of mesophilous and hygrophilous grass species, such as *Phragmites karka, Narenga* spp., *Arundo donax, Saccharum bengalense, S. spontaneum, Thameda arundinacea, Imperata cylinderica* and *Vetiveria zizanioides.* Under the influence of heavy grazing and fire, the conditions become drier which change the grass composition. The first species to disappear under these conditions are *Phragmites karka* and *Arundo donax.* Further biotic pressure leads to the gradual disappearance of *Narenga* spp. and *Saccharum* spp. *Vetiveria zizanioides* becomes the most dominant grass and represents the last stage of degradation. Any further biotic pressure leads to complete loss of soil cover. Sometimes these grasslands are invaded by species, e.g., *Eleusine, Paspalam* and *Cynodon,* which are prostrate growing and capable of sustaining biotic pressures to a large extent.

Similarly, other grassland types deteriorate under biotic pressures. The degree of deterioration depends upon the quantum of biotic pressure. Continuous heavy biotic pressure leads to complete disappearance of the grass cover and a high rate of soil erosion.

The grasslands and grazing lands under the control of government, panchayats and other village institutions are very poor, degraded and overgrazed. The productivity of these grasslands is, therefore, very low. The production of dry grass in such grasslands ranges from 0.5 to 6.0 tonnes/ha /annum depending upon the condition of grasslands (Dabadghao and

Shankarnarayan, 1973). However, in grasslands which are subjected to serious biotic pressure and overgrazing, the annual yield may always be less than 1 tonne/ha. This is one-eighth of a well-managed grassland in the same area.

## IMPROVEMENT OF GRASSLANDS

Of all the classes of lands in India, perhaps grasslands and grazing lands are the most neglected. Most of these lands are ecologically most sensitive and excessive biotic pressure leads to development of wastelands and degradation of these lands. These grasslands need to be improved for optimum production. Their improvement would require a large investment and effective organisation. Grasslands have to be managed so that they have full ground cover and provide a sustained yield of fodder grasses. Important steps which may be essential for improving these lands include: (i) regulating the number of animals and practising controlled grazing, (ii) soil and water conservation, (iii) protection, (iv) introduction of improved species and leguminous fodder grasses, (v) application of manures and fertilisers, (vi) control of obnoxious weeds and (vii) planting of fodder trees.

### (i) *Controlled grazing*

The need for regulating the numbers of livestock on grasslands has been recognised for several decades and has been repeatedly emphasised in several reports (Anon., 1952; Whyte, 1964; NCA, 1976; Albrecht, 1976; 1979). There is an immediate need for controlling the number of livestock grazing on the deteriorated grazing lands. The over-riding principle of good management of grasslands is to determine the average long term and safe carrying capacities of location specific sites. The number of all livestock should be restricted to the *carrying capacity* of the grasslands. The carrying capacity of the grasslands means the number of unit animals which may be allowed to graze with no adverse effect on the grassland.

The grazing systems should be based on the principle of restricting the number of animals and giving rest to some portion of the grazing land. The important systems include: (i) continuous grazing, (ii) seasonal grazing, (iii) deferred grazing and (iv) rotational grazing. Continuous grazing means round-the-year grazing. The practice creates an adverse effect on grasslands unless the number of animals allowed to graze remains much less than the carrying capacity. *Seasonal grazing* has an advantage over continuous grazing as the grassland receives rest for some period. In *deferred grazing* also, the grassland is given rest for some period. In *rotational grazing*, the grazing land is divided into blocks and grazing is allowed on a rotational basis.

Several advantages are afforded by controlled grazing. These advantages include: (i) the vegetation including grasses show better growth, (ii) the proportion of nutritious grasses becomes more, (iii) the grass yield increases leading to increased grazing period, (iv) inferior and course grasses give way to superior types, (v) soil erosion is reduced and (vi) the condition of the livestock improves. These advantages have been confirmed in almost all edapho-climatic conditions (Malik, 1958; Whyte, 1964). Data on the carrying capacity of different types of grasslands is completely lacking. Site-specific research needs to be undertaken in important grasslands.

### (ii) *Soil and water conservation*

Most of the grasslands owned by community and government are in a seriously eroded condition. Large herds of livestock graze and move in these grasslands, reducing the surface cover and exposing the mineral soil. During movement, the animals follow different paths which is evidenced by loss of vegetation. Due to soil compaction, the permeability is reduced considerably, resulting in higher surface run-off. Continuous higher run-off causes different types of erosion in different intensities. Therefore, it is necessary that all grassland improvement programmes should invariably include soil and water conservation activity.

Experience in the drought-prone area programme (DPAP) indicates that soil and water conservation helps in increasing the growth of all vegetation including grasses. Soil and water conservation operations include: contour furrowing and strip ploughing of land, plugging of gullies and nallas, preparation of bunds, etc. Sowing of seeds of *Cenchrus ciliaris*, *Cenchrus setigerus* and *Stylosanthes hamata* in 'U'-shaped furrows and closure for six months have increased the yield of grasses substantially in these areas (Kulkarni, 1979). In Rajasthan, shallow contour trenches (22.5 cm deep and 60.5 cm wide with an interval of 10 m), contour bunds (60 cm high, lower base 187.5 cm, upper base 15.2 cm, height interval, 60.5 cm) and deeper contour trenches (30.5 cm deep, 60.9 cm wide, height interval of 60.9 cm) were made in grasslands and it was found that construction of shallow contour trenches resulted in several-fold increased grass production (Verma, 1975).

### (iii) *Protection*

Grazing in the grassland has to be restricted well within its carrying capacity. Also, the grassland has to be closed for some period. Therefore, it is necessary to protect the grassland from livestock. For protection, grasslands need to be fenced. In Madhya Pradesh, Uttar Pradesh, Maharashtra, etc., *cattle-proof trenches* are usually prepared around the grassland which provide effective protection against cattle. Cattle-proof trenches of different sizes are used but trenches 1 m deep, 1 m wide at the base and 1.5 m wide at the top are more common. In Rajasthan and in areas where stones are easily available, *stone-wall fencing* is common. Smaller areas can be effectively fenced with *brush wood* of thorny species, e.g., *Acacia nilotica*,

*A. catechu, Carissa* spp., etc. *Live hedges* of *Agave sislana, Acacia nilotica, Prosopis chilensis, Euphorbia* spp., etc. can also be planted and maintained as an effective fence. *Wire fencing* is more effective but it is more costly and, therefore, it is not generally recommended. Four-strand barbed wire fencing is effective against most animals. It is much better if fencing can be avoided. A watchman could protect and drive away animals. Also, it is not feasible to provide fencing in all grasslands as most of these are scattered and found in small blocks. *Social fencing* should also be tried in areas where villagers are co-operative and helpful.

(iv) *Introduction of improved fodder grasses and legumes*

Excessive grazing and fire destroy the palatable perennial fodder grasses in the grasslands. Only coarse annual grasses grow in such grasslands. Introduction of high-yielding, palatable and perennial grasses, therefore, becomes an essential component of any grassland development programme. The grasslands are ploughed and the seeds of improved grasses are sown. The seed rates for some of the improved fodder grasses are given in Table 50.

**Table 50: Seed rate of some fodder grasses**

| Species | Seed rate kg/ha |
|---|---|
| *Cenchrus ciliaris* | 8 |
| *C. setigerus* | 12 |
| *Chrysopogon fulvus* | 8 |
| *Sehima nervosum* | 15 |
| *Dichanthium annulatum* | 7 |
| *Heteropogon contortus* | 5 |
| *Panicum antidotale* | 6 |
| *Sateria* spp. | 7 |

Some of the leguminous fodder species yield valuable fodder and their introduction in the grassland helps in improving the grass cover and soil conditions. These fodder species are capable of growing in poor and eroded sites as they fix nitrogen from the atmosphere. The seed rate and method of sowing of some of the important leguminous fodder species are given in Table 51.

The fodder from some leguminous grasses is more nutritive. These species can be grown without irrigation. In arid and semi-arid areas also, some species, such as *Stylosanthes* spp. and *Phaseolus* spp. can be easily grown. Some species, e.g., *Purarea* provide an excellent soil cover and protect the soil from erosion.

The non-availability of improved seeds of fodder grasses and legumes is one of the major constraints in the grassland development programme. In the case of fodder grasses and legumes, the vegetative part is more

Table 51: **Method of sowing and seed rate of important leguminous species**

| Species | Method of sowing | Seed rate kg/ha |
|---------|------------------|-----------------|
| 1. *Microptelium artopurpureum* | Line sowing in 45 cm interval | 10–15 |
| 2. *Stylosanthes hamata* | Line sowing in 45 cm interval | 6–7 |
| 3. *S. gracilis* | -do- | 6–8 |
| 4. *S. scabra* | -do- | 7–10 |
| 5. *S. humilis* | -do- | 4–6 |
| 6. *Atylosia scaraboides* | Line sowing in 30 cm interval | 12–15 |
| 7. *Desmenthis vergatus* | -do- | 6–8 |
| 8. *Glycine javanica* | -do- | 4–7 |
| 9. *Phaseolus* spp. | -do- | 10–12 |
| 10. *Purarea thumbergiana* | Stem planting | — |

important and, therefore, seed production is often neglected. For most grasses, the pre-flowering stage is the most optimum time for harvesting from a fodder nutrition point of view. In several species, seed setting is erratic and poor. In several species, seed viability is also low. Therefore, the seeds of fodder grasses and legumes are often in short supply. There is no definite requirement for seed and, therefore, bulk seed production has not been given sufficient attention. The requirement of seed will depend upon the area to be brought under fodder production. The seed rate of most fodder grasses varies between 5 to 10 kg and, therefore, the seed requirement for improved fodder grasses and legumes is likely to be substantial.

(v) *Use of manures and fertilisers*

Grasslands are usually degraded sites. The introduction of improved varieties of grasses will not show the desired success without the application of manures and fertilisers. Before applying fertilisers, it has to be ensured that only desirable fodder grasses are growing. Most grasslands are deficient in nitrogen and its application alone is reported to increase the grass production by 1.5 to 2 times. The response to fertiliser application depends on the soil, the climate and the grass. Phosphorus and potassium, if added with nitrogen, help to increase the growth of grasses.

(vi) *Weed control*

A large number of undesirable plant species which have no fodder value invade the grasslands if they are not maintained properly. *Lantana camara*, *Partheneum* spp., *Eupatorium* spp., *Dodonea viscosa*, *Cassia tora*, *Argemone mexicana*, *Ageretum* spp., etc., are some examples of weeds which invade the grasslands and suppress the grass production. Also, some grasses are not palatable and need to be eradicated.

In order to have maximum production of desirable grasses, undesirable plant species have to be eradicated. For eradicating them, several methods, e.g., physical, mechanical, chemical or biological and a combination of these methods may be tried. One of the best methods for removing weeds is to uproot them before flowering. Some machines, e.g., shrub cutters, stalk pullers, etc., are also available which can be used for cutting and uprooting of woody weeds. Annual plants, e.g., *Cassia tora, Parthentum* spp., etc. may be removed by tractor ploughing. When ploughing is done to introduce improved grasses and fodder legumes, it must be ascertained that the area is free from unwanted plants. Several *weedicides* are also available which may kill a group of plant species. For example, 2-4 D kills dicotyledonous plants and is used for controlling weeds in wheat, rice, etc. However, the usefulness of applying these weedicides has to be examined taking several factors into consideration. Some biological agents such as insects have also been used for controlling weeds. The lantana lace bug has been found effective in controlling *Lantana camara*.

### (vii) *Trees in grasslands*

A large number of trees yield nutritious fodder for livestock. The chemical composition and nutritive values of some of the tree fodders are indicated in Table 52 (Sen, *et al.*, 1978; Gulati, *et al.*, 1982; Singh, 1982).

Fodder trees can be grown in grasslands to maximise fodder production. The practice of pastural-silviculture ensures a higher yield of better quality

**Table 52: Nutritive value of some fodder trees**

| Tree leaves | Crude Protein % | Digestible Crude Protein % | Total Digestible Nutrients % | Crude Fibre % | Calcium % | Phosphorus % |
|---|---|---|---|---|---|---|
| 1    2 | 3 | 4 | 5 | 6 | 7 | 8 |
| 1. Khair *Acacia catechu* | 12.0–18.7 | 2.9 | 46.3 | 21.9–22.6 | 1.6–2.7 | 0.1–0.2 |
| 2. Babul *Acacia nilotica* | 7.0–15.0 | — | — | 20.1–33.3 | 1.2–2.6 | 0.1–0.2 |
| 3. Israeli babul *Acacia tortilis* | 18.1 | — | — | — | 3.1 | 0.2 |
| 4. Haldu *Adina cordifolia* | 8.7–15.3 | 2.7–2.8 | 50.9 | 12.1–14.0 | 1.7–3.2 | 0.1–0.5 |
| 5. Bel *Aegle marmelos* | 15.1–15.3 | 10.8 | 56.7 | 16.5–18.1 | 4.2–4.8 | 0.1–0.3 |
| 6. Mahrukh *Ailanthus excelsa* | 16.3–19.9 | — | — | 12.8–21.9 | 1.5–2.4 | 0.2–0.3 |
| 7. Ohi *Albizia chinensis* | 15.1 | 4.9 | 40.2 | 31.6 | 1.2 | 0.1 |

| 1 | 2 | 3 | 4 | 5 | 6 | 7 | 8 |
|---|---|---|---|---|---|---|---|
| 8. | Siris<br>*Albizia lebbeck* | 14.9–29.2 | 11.6 | 49.3 | 25.3–37.5 | 1.1–2.7 | 0.1–0.3 |
| 9. | Allmania nodiflora | 10.3 | — | — | 25.8 | 1.0 | 0.1 |
| 10. | Zamin–kand<br>*Amorphophallus<br>campanulatus* | 15.3 | — | — | 14.2 | 1.5 | 0.8 |
| 11. | Bakli<br>*Anogeissus latifolia* | 7.5–11.5 | 0.6 | 47.8 | 16.4–24.2 | 2.7–4.0 | 0.1–0.6 |
| 12. | Kardahai<br>*Anogeissus pendula* | 7.6 | — | — | 19.0 | 3.5–3.7 | 0.1 |
| 13. | Kadam<br>*Anthocephalus chinensis* | 21.3 | — | — | — | 2.7 | 0.3 |
| 14. | Kathal<br>*Artocarpus heterophyllus* | 11.2–14.2 | — | — | 18.7–22.8 | 0.5–2.2 | 0.1–0.3 |
| 15. | Neem<br>*Azadirachta indica* | 12.4–18.3 | 8.4–9.3 | 42.8–53.3 | 11.4–23.1 | 0.9–4.0 | 0.1–0.3 |
| 16. | Bans<br>*Bambusa arundinacea* | 18.6 | 13.5 | 46.5 | 24.1 | 0.6 | 0.2 |
| 17. | Kachnar<br>*Bauhinia variegata* | 10.7–15.9 | 5.0–9.2 | 47.9–55.5 | 20.7–33.0 | 1.4–4.1 | 0.2–0.4 |
| 18. | Bishop wood<br>*Bischofia javanica* | 18.0 | — | — | — | 2.6 | 0.2 |
| 19. | Baddlleia<br>*Buddleia<br>madagascariensis* | 13.0 | — | — | — | 1.1 | 0.2 |
| 20. | Dhak<br>*Butea monosperma* | 17.5 | — | — | — | 3.3 | 0.2 |
| 21. | Kumbhi<br>*Careya arborea* | 10.4 | 0.2 | 43.1 | 25.9 | 1.6 | 0.3 |
| 22. | Amaltas<br>*Cassia fistula* | 15.8 | — | — | — | 1.3 | 0.2 |
| 23. | Kassod<br>*Cassia siamea* | 13.5 | — | — | — | 1.9 | 0.6 |
| 24. | Jangli Saru<br>*Casuarina equisetifolia* | 9.5 | — | — | — | 2.1 | 0.1 |
| 25. | Khark<br>*Celtis australis* | 12.0–15.5 | 5.1–9.7 | 41.2–59.4 | 19.5–22.0 | 2.8–4.9 | 0.1–0.2 |
| 26. | Lasora<br>*Cordia dichotoma* | 12.4–15.1 | 5.1–5.4 | 26.9 | 16.5–26.8 | 2.4–4.3 | 0.2–0.3 |
| 27. | Shisham<br>*Dalbergia sissoo* | 2.7–24.1 | 3.7–9.1 | 20.9–52.2 | 12.5–32.0 | 2.0–2.3 | 0.2 |
| 28. | Bamboo<br>*Dendrocalamus strictus* | 14.2–15.1 | 9.3 | 48.9 | 23.5–15.6 | 1.1–1.6 | 0.2–0.3 |
| 29. | Vayni<br>*Delonix elata* | 25.6–26.5 | — | — | 7.2 | 3.5 | 0.2 |
| 30. | Tendu<br>*Diospyros melanoxylon* | 7.1 | 0.0 | 34.1 | 25.3 | 1.8 | 0.2 |
| 31 | Chamror<br>*Ehretia laevis* | 13.5 | 8.5 | 54.9 | 11.0 | 1.5 | 0.3 |
| 32. | Jamun<br>*Syzygium cumini* | 8.8–10.2 | 0.1 | 43.8 | 19.8 | 1.3 | 0.1–0.2 |

| 1 | 2 | 3 | 4 | 5 | 6 | 7 | 8 |
|---|---|---|---|---|---|---|---|
| 33. | Bargad *Ficus bengalensis* | 7.7–11.5 | 1.9 | 44.5 | 13–30.0 | 1.2–4.1 | 0.1–0.4 |
| 34. | Rubber *Ficus elastica* | 12.0 | — | — | — | 1.3 | 0.2 |
| 35. | Gular *Ficus glomerata* | 11.2–15.2 | 6.6 | 53.8 | 12.3–16.5 | 1.7–3.0 | 0.2–0.5 |
| 36. | Pakar *Ficus lacor* | 7.3–16.0 | 0.4–6.3 | 32.2–62.0 | 12.8–27.7 | 1.1–3.5 | 0.1–0.4 |
| 37. | Pipal *Ficus religiosa* | 9.0–25.6 | 5.5–7.0 | 38.2–39.2 | 14.7–20.0 | 2.3–6.3 | 0.2–0.3 |
| 38. | Timla *Ficus roxburghii* | 12.3–13.4 | — | — | 7.7–17.8 | 1.3–2.2 | 0.2 |
| 39. | Fig *Ficus sycomorus* | 11.4–15.2 | — | — | 13.7–27.4 | 2.6–3.0 | 0.2–0.3 |
| 40. | Bhimal *Grewia optiva* | 15.6–20.6 | 12.4–16.3 | 57.4–61.7 | 18.9–22.1 | 2.7–3.5 | 0.1–0.3 |
| 41. | Dhamni *Grewia tiliifolia* | 13.2 | — | — | — | 1.5 | 0.1 |
| 42. | Anjan *Hardwickia binata* | 9.0 | — | — | 30.4 | 2.3–3.3 | 0.1 |
| 43. | Marorphali *Helicteres integrifolia* | 13.3 | 9.7 | 58.4 | 19.8 | 2.3 | 0.3 |
| 44. | Kanju *Holoptelea integrifolia* | 13.7 | 5.2 | 58.1 | — | 4.3 | 0.2 |
| 45. | Madhavilata *Hiptage benghalensis* | 13.2 | — | — | — | 2.1 | 0.2 |
| 46. | Sausage *Kigelia pinnata* | 12.3 | — | — | — | 2.4 | 0.2 |
| 47. | Pula *Kydia calycina* | 11.1–13.6 | 7.9 | 45.2 | 14.5–23.7 | 3.1–5.2 | 0.4–0.6 |
| 48. | Dhaura *Lagerstroemia parviflora* | 7.8 | 0.9 | 49.5 | 17.3 | 2.6 | 0.3 |
| 49. | Jhingan *Lannea coromandelica* | 11.4–16.7 | 4.9 | 55.2 | 16.1–19.8 | 1.8–2.4 | 0.2–0.3 |
| 50. | Subabul *Leucaena leucocephala* | 15.2–27.6 | 12.6–16.4 | 57.1–70.2 | 10.2–17.2 | 2.7–3.1 | 0.2 |
| 51. | Banda *Loranthus falcatus* | 14.8 | 6.2 | 58.9 | 17.9 | 1.8 | 0.2 |
| 52. | Mahua *Madhuca indica* | 9.4–10.0 | 0.0 | 37.0 | 19.5–20.3 | 1.6 | 0.1–0 2 |
| 53. | Karuala *Mallotus philippensis* | 12.5–23.1 | 7.9 | 42.1–16.6 | 21.6–29.7 | 1.3–1.6 | 0.2 |
| 54. | Am *Mangifera indica* | 9.3 | — | — | 23.7 | 1.9–2.2 | 0.2–0.3 |
| 55. | Cassava *Manihot esculenta* | 23.0 | — | — | 24.4 | 1.6 | 0.2 |
| 56. | Khirni *Manilkara hexandra* | 9.4 | — | — | 23.1 | 1.5 | 0.4 |
| 57. | Bakain *Melia azedarach* | 13.3 | — | — | — | 2.6 | 0.2 |

| 1 | 2 | 3 | 4 | 5 | 6 | 7 | 8 |
|---|---|---|---|---|---|---|---|
| 58. | Gauj<br>*Millettia extensa* | 22.7 | 15.5 | 44.9 | 32.5 | 1.9 | 0.3 |
| 59. | Kaim<br>*Mitragyna parvifolia* | 7.7 | 1.5 | 49.9 | 19.6 | 2.3 | 0.2 |
| 60. | Al<br>*Marinda coreia* | 15.1 | — | — | 22.7 | 1.7 | 0.2 |
| 61. | Sainjana<br>*Moringa oleifera* | 15.3–20.7 | 11.1 | 61.5 | 7.1–17.9 | 1.7–3.8 | 0.1–0.5 |
| 62. | Mulberry<br>*Morus alba* | 15.0–27.6 | 10.7 | 59.6 | 9.1–15.3 | 2.4–4.7 | 0.2–1.0 |
| 63. | Sandan<br>*Ougeinia oojeinensis* | 11.6–15.1 | 3.7 | 45.6 | 21.9–28.5 | 0.8–3.6 | 0.2–0.4 |
| 64. | Karanja<br>*Pongamia pinnata* | 17.6 | — | — | — | 2.2 | 0.2 |
| 65. | Jand/Khejra<br>*Prosopis cineraria* | 13.9–15.3 | — | — | 17.5–22.1 | 1.9–3.6 | 0.2–0.5 |
| 66. | Tilonj<br>*Quercus dilatata* | 9.6 | 4.2 | 43.2 | 29.1 | 1.6 | 0.3 |
| 67. | Phaliant<br>*Quercus glauca* | 9.6 | 4.6 | 39.8 | 29.0 | 1.9 | 0.2 |
| 68. | Banj<br>*Quercus leucotrichophora* | 8.9–15.9 | 4.2–5.8 | 43.2–43.8 | 13.7–42.1 | 1.0–1.7 | 0.1–02 |
| 69. | Robina<br>*Robinia pseudoacacia* | 17.1–25.5 | — | — | 16.6–23.1 | 1.1–3.5 | 0.1–0.3 |
| 70. | Gogina/Ratendu<br>*Saurauia napaulensis* | 12.3 | 1.2 | 34.0 | 18.4 | 2.7 | 0.2 |
| 71. | Kusum<br>*Schleichera oleosa* | 10.4 | 3.4 | 47.5 | 32.3 | 1.7 | 0.3 |
| 72. | Sal<br>*Shorea robusta* | 10.1 | 0.1 | 42.7 | 27.4 | 0.8 | 0.1 |
| 73. | Padri<br>*Stereospermum personatum* | 8.1–11.4 | — | — | 22.0–28.9 | 1.1–2.7 | 0.2–0.6 |
| 74. | Imli<br>*Tamarindus indica* | 11.2–15.4 | — | — | 14.2–22.0 | 1.7–3.2 | 0.1–0.6 |
| 75. | Laurel<br>*Terminalia alata* | 8.9–12.7 | 0.0 | 34.9 | 12.4–21.8 | 2.7–3.2 | 0.2–0.4 |
| 76. | Bahera<br>*Terminalia bellirica* | 7.7–17.2 | 0.9 | 54.5 | 7.4–18.6 | 1.6–3.6 | 0.2–0.3 |
| 77. | Amrita<br>*Tinospora cordifolia* | 11.2 | — | — | 17.5 | 1.1 | 0.6 |
| 78. | Mored<br>*Ulmus wallichiana* | 18.1 | — | — | — | 0.7 | 0.3 |
| 79. | Ber<br>*Ziziphus mauratiana* | 8.6–16.9 | 3.1 | 30.7 | 13.5–30.1 | 1.3–3.6 | 0.2–0.3 |
| 80. | Jharber<br>*Ziziphus nummularia* | 11.5 | 5.5 | 51.1 | 33.8 | 1.9 | 0.3 |

fodder besides maintaining the productivity of the sites. The leaf fodder of some trees is almost as nutritious as that of leguminous fodder crops. The trees are capable of producing almost as much green fodder per unit area as agricultural fodder. The tree fodder production does not suffer during drought when most annual grasses dry up. The leaf fodder production offers another advantage of producing fuelwood as a by-product which helps in meeting the energy requirements of the rural population. The trees and annual fodder grasses together utilise the site better. Leguminous fodder trees enrich the site through fixation of nitrogen. The practice also helps in effective soil and water conservation.

Except for scattered trees naturally occurring in the grasslands, the practice of pastural silviculture has not been adopted in a systematic way in the country. Experiments conducted at different areas indicate the efficacy of grass + fodder trees in maximising fodder production per unit area. Such trees should be given preference which can withstand repeated loppings.

It is necessary to identify suitable grasses and trees which are compatible for a given edapho-climatic condition. Preliminary results of experiments conducted at various research institutes indicate the compatibility of several trees and grasses under different conditions (Table 53).

In almost all forest types, the grass community is an important ground flora. The important forest types are given in Table 54. In tropical wet evergreen and semi-evergreen forests, the understorey vegetation is dense and therefore, grasses are present only in blanks or in pockets where trees are fewer.

IMPORTANT PASTURAL-SILVICULTURAL PRACTICES IN INDIA

The pastural-silviculture system is practised in several areas, particularly in areas with a deficient water supply. The farmers allot some areas of their holding for production of grasses where they graze their animals. The grasslands owned by the village community, organisations or government are by and large not well managed. Some of the specific systems are discussed below:

(i) *Grassland and tree management in arid areas*

The grasslands in arid areas of Rajasthan, Gujarat and Haryana consist of several types of grasses, namely, *Urochloa* spp., *Eragrostis* spp., *Dichanthium annulatum*, *Lasiurus sindicus*, *Panicum antidotale*, *Cenchrus setigerus*, *C. ciliaris*, *Aristida* spp., *Chloris* spp., *Heteropogon contortus*, *Sehima nervosum*, *Tetrapogon* spp., *Eremopogon* spp., etc. The important fodder tree species which occur scattered naturally in these grasslands are *Prosopis cineraria* and *Ziziphus nummularia*. These two species play a vital role in the pastural-silviculture system followed in these areas. These village grazing lands in Rajasthan are called *oran* or *bir*. In individual *birs* and also

**Table 53: Suitable mixture of grass and tree species for different conditions**

| Sl. No. | Conditions | Tree species | Grass species |
|---|---|---|---|
| 1. | Arid climate, sandy soil | *Prosopis cineraria, Acacia tortilis, A.nilotica, Albizia lebbeck, Ziziphus* spp. | *Cenchrus ciliaris, Sehima nervosum, Microptelium artopurpurium, Atylosia scarboides* |
| 2. | Semi-arid climate, sandy soil | *Acacia nilotica, Leucaena leucocephala, Anogeissus pendula, Azadirachta indica, Dalbergia sissoo* | *Cenchrus ciliaris, C. setigerus, Dolichos lablab, Microptelium artopurpurium, Stylosanthes hamata, Phaseolus lathyrus* |
| 3. | Semi-arid, red soil | *Albizia lebbeck, A. procera, Leucaena leucocephala, Acacia auriculiformis* | *Cenchrus* spp., *Sehima nervosum, Chrysopogon fulvus, Microptelium* spp., *Stylosanthes hamata, S. humilis.* |
| 4. | Semi-arid, black cotton soil | *Acacia nilotica, A. lebbeck, A. procera, Hardwickia binata, Sesbania* spp., *Azadirachta indica* | *Dichanthium annulatum, Panicum antidotale, Pennisetum pedicellatum, Sehima nervosum, Chloris* spp. |
| 5. | Medium climate, black soil | *Moringa oleifera, Leucaena leucocephala, Sesbania grandiflora, Dendrocalamus strictus, Albizia lebbeck,* etc. | *Cenchrus* spp., *Sehima nervosum, Dichanthium annulatum, Stylosanthes* spp., *Cymbopogon* spp., *Chloris* spp. |
| 6. | Temperate climate | *Acacia dialbata, Alnus* spp., *Fraxinus excelsa, Grewia, optiva, Celtis australis, Morus serrata, Quercus leucotrichophora,* etc. | *Phalaris tuberosa, Agropyron* spp., *Bromus inermis, Poa pratensis, Dactylis glomerata, Trifolium* spp. |
| 7. | Saline-alkaline soils | *Acacia nilotica, Dalbergia sissoo, Pongamia pinnata, Albizia procera* | *Cenchrus ciliaris, Ischaemum muticum, Chloris* spp., *Lopholepis* spp., *Sporobolus, marginatus, Dicanthium annulatum, Cynodon dactylon Eriochlora procera,* etc. |
| 8. | Ravines | *Acacia nilotica, Prosopis cineraria, Albizia lebbeck, Dalbergia sissoo, Leucaena leucocephala,* etc. | *Cenchrus ciliaris, Dicanthium annulatum, Bothriochloa glabra, Stylosanthes hamata, Eulaliopsis binata,* etc. |

in community *birs*, these two species are encouraged and lopped for fodder. *Prosopis cineraria* is a very good fodder tree for arid areas. The tree can be lopped and the firewood and timber can also be used for household purposes. If it is cut at the ground level, it coppices profusely, giving rise to about 6 to 12 coppice shoots. If grazing or browsing is allowed, the tree forms a bushy structure. Under severe grazing intensities, the branches spread horizontally, which facilitates grazing for sheep. The shrubs of *Ziziphus nummularia* grow in grasslands in arid areas. It regenerates usually through the root suckers. It has been found that 14 per cent density of *Z. nummularia* bush is optimum for obtaining the highest yield of leaf fodder and grass (Kaul and Ganguli, 1963).

**Table 54: Area and occurrence of different forest types (Anon., 1989)**

| Forest types | Area (million ha) | Occurrence |
|---|---|---|
| 1. Tropical wet-evergreen and semi-evergreen forests | 6.4 | Assam, West Bengal, Karnataka, Kerala, Orissa, Andaman & Nicobar Islands |
| 2. Tropical moist deciduous forests | 22.6 | Karnataka, Orissa, Tamil Nadu, Uttar Pradesh, Bihar, Madhya Pradesh, Maharashtra, Kerala, Andaman & Nicobar Islands |
| 3. Tropical dry deciduous forests | 29.2 | Almost all over India |
| 4. Littoral and swamp forests | 0.7 | Deltas of big rivers |
| 5. Tropical thorn forests | 5.2 | Punjab, Haryana, Rajasthan, Upper Gangetic plains, plateau of peninsular India |
| 6. Subtropical forests | 4.2 | Slopes of Himalayas up to 1500 m and Nilgiris and Malabar |
| 7. Temperate forests | 4.5 | Himalayan hill ranges between 1500 m to 3000 m |
| 8. Alpine and subalpine forests | 1.8 | Himalayan hill ranges above 3000 m elevations |

## (ii) *Grassland and tree management in semi-arid zones*

The system is practised in the plateau area of Andhra Pradesh, Karnataka, Tamil Nadu, Maharashtra, Madhya Pradesh, Uttar Pradesh, and Bihar. The average annual rainfall varies from 750 mm to 1250 mm with 50—70 rainy days by the South-west monsoon. The system is known as *Kanchas* or *ghas birs* locally. These *kanchas* or *ghas birs* are owned by individuals although some are with the government forest department and have been taken up for afforestation under the social forestry programme. The system consists of protecting the grasses and trees against grazing, fire, etc. The grasses are cut or grazed, the number of animals remaining controlled. In the *kanchas* under the control of the forest department, free grass cutting is allowed to local villages. Earlier these were auctioned to individuals or panchayats for cutting of grasses or for grazing.

The grasses in these grasslands fall mainly in the category of *Sehima-Dicanthium* type. The important grasses occurring in these areas are *Sehima nervosum, Dichanthium annulatum, Pennisetum pedicelatum, Chloris gayana, Cenchrus ciliaris, C.setigerus, Heteropogon contortus, Chrysopogon fulvus, Iseilema laxum, Cymbopogon* spp., *Eremopogon* spp., *Saccharum* spp., *Bothriochloa* spp., etc. The tree species found scattered in these areas include: *Tectona grandis, Acacia nilotica, Acacia catechu, Diospyros melanoylon, Terminalia alata, Madhuca latifolia, Lagerstroemia parviflora, Bauhinia variegata, Anogeissus latifolia, Adina cordifolia, Hardwickia binata, Emblica officinalis*, and several others. Palms such as *Borassus flabellifer* and *Phoenix sylvestris* are common. Several fodder and fuelwood species have been planted in these areas under the social forestry programme. The commonly planted species include: *Leucaena leucocephala*, *Bauhinia variegata*, *Acacia nilotica*, *Hardwickia binata*, *Dendrocalamus strictus*, *Azadirachta indica*, *Ougeinia oojeinensis*, *Casuarina equisetifolia*, *Mangifera indica*, *Madhuca indica*, etc. Introduction of improved varieties of grasses and legumes is also done. Management consists of closing the area for grazing and introduction of improved fodder grasses, legumes and fodder trees. Usually a cattle proof trench is provided in government kanchas. In private kanchas, social fencing is adopted. The yield of grass depends upon the stage of succession and management. In a well-protected kancha on red soils, 3 tonnes of air-dried grass per ha per annum may be obtained; in gravelly soils, the yield is 0.5 to 1.0 tonne air-dried grass per ha per annum; a well-protected kancha on black soils can yield 10 tonnes/ha/annum (Tejwani, 1987).

(iii) *Grassland and tree management system in subtropical and temperate areas*
    In the hills of the South and North, steep slopes under private ownership are usually maintained under grass and tree production system. These are locally called *ghasini* in Himachal Pradesh and Uttar Pradesh. Such areas are also under the control of government in revenue and forest departments. While the *ghasinis* under private ownership are well managed and protected, those under the control of government are often degraded and overgrazed.
    The grass species growing in subtropical areas include: *Cymbopogon* spp., *Thameda triandra*, *Brachiaria mutica*, *Cenchrus ciliaris*, *Bothriochloa* spp., *Panicum antidotale*, *Chrysopogon* spp., *Apluda mutica*, *Bromus* spp., *Pennisetum* spp., etc. In temperate areas, species such as *Phalaris tuberosa*, *Agropyron* spp., *Lolium* spp., *Poa* spp., etc. predominate. The tree species include: *Pinus roxburghii*, *Albizia chinensis*, *Bauhinia variegata*, *Coraria nepalansis*, *Celtis australis*, *Grewia optiva*, *Ficus* spp., and several others. Some species, e.g. *Quercus leucotrichofora*, *Aesculus* spp., *Ficus* spp., *Alnus* spp., etc., occur in sheltered areas and along water courses. In temperate areas, species such as *Quercus himalayana*, *Q. semecarpifolia*, *Abies pindrow*, *Cedrus deodara* and *Picea smithiana* may predominate.
    The areas under the control of forest and revenue departments of governments are in poor condition due to overgrazing. The areas under private ownership are protected and grasses are usually cut or a limited number of cattle are allowed to graze. More recently, the forest department has taken up the management of such lands which includes providing protection by stone-walling and plantation of fuelwood and fodder species with the co-operation of local villagers. Once the area is protected, it starts giving a

good yield of grass. The cutting of grasses is free but grazing is usually restricted up to the first 4–5 years before plantation gets established.

### (iv) *Grassland and tree management in alluvial regions*

In the alluvial regions of Gujarat, Madhya Pradesh, Uttar Pradesh and elsewhere, closure of the area against grazing has brought about improvement in grass composition and yield. It is reported that in the ravine area of Gujarat, as a result of closure to grazing, the dominant grasses, e.g., *Aristida paniculata* and *Themada triandra* were first replaced by annual grasses, e.g., *Apluda mutica* and then by perennials such as *Eremopogon faveolotus*, *Heteropogon contortus*, *Dichanthium annulatum* and *Cenchrus* spp. (Tejwani *et al.*, 1960). The closure also results in improvement of the tree flora and substantial reduction in soil erosion. Similar results have been reported in the ravines of Chambal and Yamuna (Singh, 1971; Singh and Verma, 1971). Under Dehra Dun conditions, which has relatively humid and subtropical conditions an annual yield of 10.55 tonnes/ha of *Chrysopogon fulvus* and other grasses was obtained under *Dalbergia sissoo* (about 100 trees/ha) in a bouldery river terraced area (Mathur and Joshi, 1972). At Dehra Dun, *Chrysopogon fulvus* has been tried under *Eucalyptus* hybrid and the results indicate that there is a gradual decrease in the yield of grasses as closing of the tree canopy progresses (Puri and Joshi, 1979).

There is a vast scope for increasing fodder production by adopting a proper pastural silviculture system. However, there is immediate need to work out the proper association of tree species and fodder grasses and legumes, proper management systems, carrying capacity of grasslands, etc., in order to make the system more effective.

### Silvipasture System

In a silvipasture system, the silviculture aspect of land use dominates and grazing or fodder production is subsidiary. The system involves grazing under the tree storey and lopping of fodder trees in forests and plantations. One of the reasons attributed to destruction and degradation of forests in India is *uncontrolled grazing* in the forests. While human and livestock populations were within reasonable limits, the forests were able to meet the needs without degradation. Except remotely located forests, almost all forest lands are degraded now due to overgrazing. Interpretation of recent satellite imagery has indicated that only about 38 million ha forests have a density of more than 40 per cent (FSI, 1989). *Dense forests* having a density of 75 per cent and above are fewer. Generally, the people enjoy the right for grazing their animals in adjoining forests. The practice differs from state to state. The present position in some of the states is as under (Anon., 1984C):

*Andhra Pradesh*

Free grazing is allowed in all the forests of the state except in the areas under regeneration. This concession includes grazing in the reserved forests as well. There is also no restriction on the number of cattle which graze. It is pointed out that relaxation in grazing was introduced due to the situation created by drought but relaxation has continued however, even though the drought is over. Migratory cattle are allowed to graze in the forests.

*Bihar*

Grazing has been recognised as a right to villagers in the protected forests and as a concession in the reserved forests of the state. No grazing fee is realised for grazing, nor is the number of cattle which graze restricted. However, the regeneration areas and plantations are closed for grazing usually for a period of 5 years. There is almost no provision for grazing in respect of migratory cattle.

*Himachal Pradesh*

Local population has the right of grazing. Any number of cattle can graze in the forest area adjoining villages. Grazing is also allowed to nomadic graziers of the adjoining states both in respect of sheep and buffaloes. The number of animals allowed to graze differs from area to area. The right-holders are also permitted to lop specified species of trees in different lopping regions both for fodder and bedding material for the cattle.

*Karnataka*

Free and unrestricted grazing is allowed in all forest areas including reserved forests. The regeneration areas and plantations are closed for a specified period. The incidence of migratory grazing is almost negligible.

*Madhya Pradesh*

Before 1979, grazing was permitted in the forests free or at concessional and commercial rates. The grazing fee was determined on the basis of number of cattle in the family. Up to 4 cattle units, free grazing was permitted. Up to 8 cattle units, cattle units in excess of 4 were allowed to graze at concessional rates. Above cattle units 8, the commercial rate of grazing was charged. These rights and concessions were applicable for cows, bullocks and buffaloes. Where these were maintained for commercial purposes, commercial rates of grazing were realised. Now the grazing policy in the forest has been revised and cows, bullocks and buffaloes are permitted free grazing throughout the year. A nominal grazing fee is charged for grazing of sheep, goats, camels, horses, ponies, elephants, etc. Though regeneration and plantation areas are supposed to be closed for grazing for a period of 5 years, the regulation is rarely implemented.

## Maharashtra

In Maharashtra, grazing is allowed free in the forests up to 8 essential cattle units. A grazing fee is charged above 8 essential cattle units and also for non-essential cattle. All animals up to the age of 6 months are exempted from fee. Though grazing incidence is prescribed subject to the carrying capacity of the forests, which is fixed on an *ad-hoc* basis as 1 cattle unit in 10 acres of protected forests and 3 acres of tree forests, control in the field is rarely exercised. A system of closure is also prescribed based on the silvicultural requirements. The period of closure prescribed ranges from 5 to 10 years.

## Meghalaya

Grazing in the forest is allowed free to the forest villagers and others living adjacent to forest areas. The forest villagers are allowed free grazing but cattle coming from outside the reserved forest are charged a grazing fee.

## Rajasthan

In Rajasthan, grazing concessions are allowed to cultivators and tribals. Usually cows and oxen are exempted from payment of a grazing fee. The tribals including professional graziers are allowed to graze their cattle, but not camels, free of charge in areas not closed for grazing. In hilly areas, the cultivators and tribals have the right to graze their cattle free of charge in the adjoining forest areas. Other animals, e.g., camels, sheep and goats, are charged a grazing fee. However, in the village and open forest, free grazing of sheep and goats is allowed. The professional nontribal graziers are usually charged a grazing fee. There is no control on the number of animals allowed for grazing.

## Tamil Nadu

Grazing in forests is allowed free of cost to animals of the tribals who live in and around forests. Others pay grazing fees, as fixed by the government from time to time. Grazing is also permitted in wildlife sanctuaries on payment of fees at twice the normal rates. There is no restriction on the number of cattle to be grazed.

## Uttar Pradesh

The cultivators living in and around the forests have the right and concession for grazing their animals in the forests. In most of the areas, the rightholders and concessionists are allowed to graze their cattle free of cost. Professional graziers are charged a separate grazing and lopping fee as fixed from time to time. There is a system of closure which includes regeneration and plantation areas.

*West Bengal*

No grazing rights have been admitted in the reserved and protected forests in the state. However, local people residing in and around forests are allowed a concession of grazing their cattle in open blocks. A grazing fee is usually charged.

SYSTEMS OF FOREST GRAZING

The system of forest grazing is organised on the basis of grazing passes or licences. These grazing passes specifically mention the area where grazing is allowed, the number of animals, grazing fee, etc. Foresters have long stressed the need for determination of the carrying capacity of the forests and that grazing should be limited to the carrying capacity. Several silvicultural conferences and committees have recommended that the number of animals admitted for grazing should be determined by the carrying capacity of the forests and that such carrying capacity should be scientifically investigated. Several state forest departments have adopted arbitrary figures of grazing incidence. However, controlled experiments have been few and sufficient information still needs to be collected. Furthermore, free grazing in the forests as practised in several states with no consideration for the carrying capacity of the forests has made such studies irrelevant.

The following systems of grazing in the forest are commonly practised:

  i) Free system
 ii) Permit system
iii) Village system
 iv) *Kancha* system
  v) Migratory grazing

(i) *Free system*

This grazing system is followed in several states where the number of animals allowed to be grazed is not restricted and grazing is free. The areas closed for grazing are usually publicised and the rest of the area open for grazing. Where grazing is free, there is no record of the number of animals which graze per unit area. In such areas, the grazing incidence is very high. The forests have been deteriorating continuously due to overgrazing in such areas.

(ii) *Permit system*

In the permit system, some fee is charged for allowing grazing in the forest. A permit may also be issued for those animals for which a concession is allowed. For example, in some states, grazing in the forest is free up to four cattle units and some fee is charged for more than four animals. Permits are also given for establishing *grazing camps* inside the forests. In Madhya Pradesh, where grazing is free, permission for establishing grazing camps is needed. Ordinarily the animals are taken to the forest and brought back in the evening after grazing. In *Grazing camps*, however, a large herd of animals belonging to the people located away from the forests, are kept. Such grazing camps are detrimental to forests.

### (iii) *Village system*

In some states, village panchayats are responsible for grazing in a forest area. The panchayats have the obligation of restricting the number of animals and protecting the area. A proportionate rebate is usually given if these obligations are fulfilled. In some states a particular panchayat is allotted a specific forest area for grazing but in practice it has been observed that usually there is a demand to increase this area. In some states a lump-sum grazing system is practised in which a lump-sum payment is made to the forest department for grazing all the animals of the village.

### (iv) *Kancha system*

This system of grazing is practised in some areas of Andhra Pradesh and Karnataka. In this system, the grazing area is divided into convenient blocks known as *kanchas* and foreign *kanchas*. Local *kanchas* are made to meet the needs of villages within 8 km of the forests. The remaining are dubbed foreign *kanchas*. If local kanchas are surplus to the requirement of the local population, they may be allocated to foreign cattle. Permits are issued by the forest department according to the estimated capacity. The *kanchas* are closed for grazing from July to September to conserve the edible grasses.

### (v) *Migratory grazing*

There are a number of tribes in Himachal Pradesh, Uttar Pradesh, Gujarat and Rajasthan who practise migratory grazing. The most popular tribes involved in migratory grazings are the Gaddis, Gujars, etc. The Gaddis are true shepherds and belong to Chamba and other districts of Himachal Pradesh. These graziers graze their animals on the higher hill areas during summer and autumn for about 8 months. They descend to lower hills during winter. The alpine pastures of Bara Bengal, Chamba, Lahaul etc. provide fine grazing facilities to the animals of the Gaddis. The Gaddis are required to pay a grazing fee to the forest department. The Gujjars are found in Himachal Pradesh, Uttar Pradesh, etc. They are also migratory graziers like the Gaddis. In summer and autumn, they graze their animals on higher hills and come down in the winter. Not all the Gujjars are nomads. The migratory Gujjars are called Ban-Gujjars. Some of these Gujjars have a traditional right for grazing their animals in forest areas falling in their route. There are also migratory graziers in Gujarat and Rajasthan who graze their animals in the forests of Madhya Pradesh, Maharashtra, Andhra Pradesh, Orissa, etc.

LOPPING

A large number of trees, shrubs, and climbers provide fodder leaves to the animals. These are usually lopped by the graziers for feeding their animals. In some of the working plans, there is specific provision for forest species to be lopped, the area and the season, etc. Some species are protected from lopping. The species which are lopped include: *Bauhinia vahlii, Garuga pinnata, Grewia optiva, Kydia calycina, Bauhinia variegata, B. racemosa, Aegle marmelos, Nectanthes arbortristis, Stereospermum suaveolens, Albizia lebbeck, Cordia dichotoma,* etc. Some protected species are: *Shorea robusta, Acacia catechu, Toona ciliata, Adina cordifolia, Terminalia alata, Anogeissus latifolia, Terminalia chebula, Dendrocalamus strictus,* etc. The other regulations regarding lopping usually include the following: (i) No lopping is permitted in areas liable to erosion or in areas which are protected. (ii) Saplings and poles are not to be lopped. Trees below 120 cm in girth are not recommended for lopping. (iii) The upper one-third of the crown of all trees is protected from lopping; branches over 8 cm in diameter in the lower two-thirds of the crown are not to be cut. (iv) The *lopping cycle* should be so designed that trees get complete rest for at least two growing seasons. (v) The lopping season is usually between November to March. Lopping is to be avoided in the growing season.

Fodder from trees and shrubs is usually considered scarcity fodder and is used only in seasons when grass is scarce or unavailable. The availability of grasses is restricted in dry periods in southern India and in winter in northern India. Leaf fodder in these periods provides a good alternative to grass fodder.

Lopping, however, affects the growth of trees adversely and sometimes is the cause of fungal infection. Continuous lopping deforms the crown of the tree. Limited information is available about proper lopping intensity, appropriate period and effects on growth in respect of different species in different edapho-climatic conditions.

EFFECTS OF GRAZING IN FORESTS

Heavy incidence of grazing in the forest is almost a common feature in nearly all the states. All the forest areas are not available for grazing because of serious soil erosion, invasion by bushes and weeds which are not browsed and closure of some of the areas for afforestation. The National Commission on Agriculture estimated that 88 per cent of the forest area is open to grazing and only about 12 per cent of the forest area remains closed to grazing for regeneration purposes (NCA, 1976). The total number of animals which graze in the forests increased from 35 million in 1956 to 54 million in 1972 and 90 million during 1982 (Anon., 1988b). The livestock population in all the states and union territories has been continuously increasing while the areas available for grazing have decreased. As a result most of the areas, particularly near habitations, are overgrazed. The figures presented in Table 55 show the intensity of grazing in important states.

**Table 55: Livestock population and grazing incidence in forests (Anon., 1984c)**

| State | Total number of livestock (000) | Percentage grazing in forest | Equivalent cow units (000) | Total forest area 1967—68 (000 ha) | Number of cow units per 100 ha of open areas |
|---|---|---|---|---|---|
| Andhra Pradesh | 33,060 | 9.0 | 2 328 | 6 405 | 44 |
| Assam | 8,210 | 1.3 | 129 | 4 573 | 3 |
| Bihar | 27,946 | 30.5 | 8 945 | 3 059 | 385 |
| Himachal Pradesh | 4,703 | 85.1 | 3 312 | 2 158 | 173 |
| Jammu & Kashmir | 4,285 | 13.6 | 353 | 2 108 | 19 |
| Madhya Pradesh | 39,989 | 15.8 | 10 131 | 17 169 | 65 |
| Maharashtra | 26,361 | 19.9 | 5 310 | 6 672 | 89 |
| Punjab | 9,295 | 3.2 | 384 | 197 | 492 |
| Rajasthan | 38,878 | 7.7 | 2 947 | 3 758 | 93 |
| Tamil Nadu | 23,979 | 5.8 | 991 | 2 214 | 52 |
| Tripura | 738 | — | 793 | 630 | 133 |
| Uttar Pradesh | 49,099 | 4.1 | 1 837 | 4 282 | 44 |
| Total | 2,66,543 | 13.0 | 37,460 | 53,225 | 81 |

Most forest areas of the country are overgrazed. The incidence of grazing in the forest is several times more than its carrying capacity. This is causing serious degradation in the forests. However, it is known that light to moderate grazing in the forest is advantageous, particularly in areas of heavy grass growth.

Regulation of forest grazing, limited to its carrying capacity, has not been implemented; overgrazing has been the common practice almost throughout the entire forested area. The following adverse affects of overgrazing are usually observed.

(i) Continuous overgrazing by large herds of animals destroys the seedlings and recruits of tree species due to: (i) grazing and brousing of young succulent plant parts and (ii) trampling. Most of the regeneration, which is in the form of seedlings, saplings and recruits, is destroyed. The forest floor is cleaned by the grazing animals and consists of only the older crop. Due to overgrazing the entire regeneration process gets disturbed or almost ceases, which adversely affects the forest crop. When a younger crop is not available to take the place of the older one, large blanks are created when the older trees are removed or die.

(ii) Overgrazing destroys the perennial and better grass species. In a semi-arid climate, in forests of *Prosopis* spp., *Acacia* spp., and *Capparis* spp., the grass community consists of perennial grasses, e.g., *Dichanthiam annulatum*, *Cenchrus ciliaris*, *C. setigerus*, *Bothriochloa pertusa*, *Heteropogon contortus*, etc., and annual grasses, e.g., *Eragrostis tremula*, *E. viscosa*, *E. ciliaris*, *Cenchrus biflorus*, *Aristida* spp., etc. *Dichanthium annulatum* and *Cenchrus ciliaris* are the most economic grasses. As grasslands are subject to heavy grazing, they undergo various stages of degradation through gradual disappearance of perennial species to sparsely populated annual types and ultimately to almost bare soils (Whyte, 1964). Similarly, in other areas almost in all edapho-climatic conditions, heavy incidence of grazing deteriorates the grass complex and reduces the carrying capacity. Heavy grazing is ecologically most undesirable as it retrogrades the grass complex and the whole vegetation.

(iii) Overgrazing not only disturbs the grass complex, but in certain circumstances leads to the appearance and development of unpalatable grazing-resistant herbs and shrubs, e.g., *Lantana camara*, *Partheneum* spp., *Eupatorium* spp., *Cassia tora*, *Xanthium strumarium*, *Holarrhena antidysenterica*, *Caseria tomentosa*, *Strychnos nuxvomica*, *Woodfordia fruiticosa*, *Dodonaea viscosa*, *Adhatoda vasica*, *Carissa spinarum*, etc. (Troup, 1926; Whyte, 1964). These shrubs and plants are avoided by cattle and their presence indicates excessive grazing and browsing. In the temperate regions of Jammu & Kashmir and Himachal Pradesh, the useless grass *Oryzopsis acquiglumis* has occupied several areas. Overgrazed grasslands in these areas have been occupied by *Careya herbacea* (Bor, 1942).

(iv) Uncontrolled grazing reduces the soil cover which increases surface run-off resulting in increased soil erosion. Increased soil erosion is caused due to reduction in tree density and ground flora.

(v) Overgrazing affects the soil properties adversely. It causes compaction in the soil and reduces soil porosity and soil aeration. It reduces permeability and water-holding capacity. It disturbs the soil structure.

(vi) Overgrazing reduces the organic matter content and fertility of the soils. Overgrazed soils are eroded and poor in nutrients.

(vii) Free and concessional grazing in forests has led to reckless increase in the number of livestock which further increases the pressure on forests. The animals are of poor productivity and low value.

(viii) Overgrazing in forests makes scientific forest management difficult. Selection of the management system, choice of species, etc., are dictated by the grazing problems.

(ix) Overgrazing results in lowering the productivity of forest lands in terms of grass, fodder and wood. It also reduces the nutritive value of fodder and quality of wood produced.

(x) Overgrazing in forests reduces species diversity of flora and fauna which is essential for preservation of the environment.

Grazing in the forests, particularly unlimited and uncontrolled, is incompatible with scientific forestry. This fact was recognised in the forest policy of 1952 wherein it was recommended that grazing in the forest may be allowed after due consideration of the carrying capacity. But the local population consider grazing in the forest a traditional right.

There is an urgent need to work out the carrying capacity of different forest types and to limit the number of animals accordingly. Plantation and regeneration areas should be closed to grazing for at least a period of 8 to 10 years. Stall feeding and cutting of grasses should be encouraged.

# CHAPTER 9

# Windbreaks and Shelterbelts

A *shelterbelt* is defined as a belt of trees and/or shrubs maintained for the purpose of shelter from wind, sun, snowdrift, etc. Shelterbelts are generally more extensive than windbreaks and cover areas larger than a single farm and sometimes a whole region on a planned pattern (Anon., 1966). Shelterbelts are also sometimes called *protection belts*. A windbreak is a narrow shelterbelt or other obstacle maintained against the wind. It is also called a *windbelt* and *wind breakage*, i.e., breaking the force of wind by trees or branches (Anon., 1966). A windbreak means any barrier erected to break down or slow down the effects of wind. Multirow windbreaks are often called shelterbelts. The purpose of windbreak and shelterbelt is to provide protection from the wind.

Shelterbelts and windbreaks consist of groups of trees and shrubs maintained in such a way that they work as a protective mechanism against wind, sun, snow, etc. Various parts of the tree, e.g., crown, leaves, stem and roots supported by other vegetative forms, e.g., shrubs, herbs, etc., play a significant role in making this protective mechanism efficient. The height of the trees in windbreaks and shelterbelts is important as it decides the area of protection. The crown and the leaves effectively retard wind action. In India, shelterbelts and windbreaks are required mainly for protection against wind and sun.

## Wind Conditions in India

Winds are an important climatic factor and influence the growth of all forms of vegetation. They influence evaporation, transpiration, pollination, fruit and seed dispersal, etc. Surface winds in the country are generally weak. Mean daily wind speeds are less than 10–15 km per hour except along the Saurashtra coast where winds are usually stronger. The wind speed differs according to season also. During the cold weather period, i.e., from December to February, the average velocity of wind ranges between 2.5 km/hour to 5 km/hour over the greater part of northern India. Over the greater part of the peninsula, it is 5 km/hour to 10 km/hour and above. Along the coast from Madras to Konkan, wind velocity exceeds 10 km/hour. In the hot weather period, i.e., during March to May, the wind movement increases almost throughout the entire country and averages about 10 to 12 km/hour. However, the increase is larger along the coast where it exceeds 12 km/hour. Occasional high winds in northern India, particularly in Rajasthan and adjoining areas, are common. During the monsoon period, i.e., June to September, the distribution of wind velocity in northern India is almost similar to that in hot weather. In Rajasthan and Gujarat, wind velocity averages about 15 km/hour. Much higher wind velocities have been recorded in these areas for several days. Along the coast, the wind movement is much stronger and averages about 18 km/hour. During the retreating monsoon period, i.e., during October to December, the average wind velocity decreases throughout the country.

However, maximum wind forces are several times much greater than the average wind speeds. A knowledge of the maximum wind force expected in any locality is useful in planning the windbreaks and shelterbelts. The northwesters of Bengal, the dust-storms and the thunder-storms of northern India, associated with sudden squalls, are several times greater in force and cause wide destruction of agricultural and forestry crops. Such thunder-storms are also common in peninsular India. Dust-storms of great force are a common occurrence in northern India, particularly in the plains and in Rajasthan, Gujarat and Madhya Pradesh.

Strong winds are usually associated with thunder-storms, dust-storms, depressions, cyclones, squally weather along the coast, etc. The *squalls* are momentary gusts of high winds while *cyclonic storms* are of much longer duration. *Dust-storms* are a common feature during summer in several parts of the country, particularly in the alluvial plains, Madhya Pradesh, Orissa, Maharashtra, etc. These winds are sometimes very strong and cause serious damage to crops and property. Strong winds, due to cyclonic storms along the coastal areas, are almost a regular annual feature causing severe damage to crops and property. During cyclonic storms, maximum wind speeds of 160 km to 200 km/hour or more have been recorded. The wind speeds during squalls far exceed 60 km/hour. In severe *thunder-squalls* wind speeds of 160 km/hour have also been recorded. Thunder-squalls are generally localised and gusts of very strong winds last only for a few minutes.

**Wind and Vegetation**

Normal wind movement is not harmful and does not create a problem. High winds cause: (i) excessive evapo-transpiration, (ii) uprooting and breaking of trees and lodging of agricultural crops, (iii) disturbance in the normal form of the stem of the plant, (iv) conditions favourable for frost when air is cold, (v) injury to plants near the sea coast due to spray of saltish water, (vi) injury to plants in temperate climate, (vii) wind erosion and (viii) damage to houses, communication lines, etc.

A high wind force accelerates the process of *transpiration* by removing the sheath of moist air which surrounds the leaf. This results in an increase in the diffusion gradient from the stomata or pores to the surrounding atmosphere. In a strong wind, bending of leaves and alternate contraction and expansion of the tissues expel more moisture. The *evaporation* of moisture from the wet surface, e.g. soil, plant parts, etc., is also increased considerably. The increased evaporation is due to increased diffusion from moist surfaces to the surrounding atmosphere. The hot and dry air causes rapid evaporation from lakes, reservoirs etc., reducing the level of water. The evaporation losses from soil and water bodies are serious in arid and semi-arid areas of the country. Strong winds cause uprooting and breaking of trees and lodging of agricultural crops. Isolated trees are subject to greater wind force than trees growing in compact blocks. Shrubs and other vegetation growing under trees are protected. Some trees can resist a strong wind force and do not break or become uprooted because of their anatomical features and root system. Most agricultural crops, such as paddy, sugar-cane, mustard, wheat, etc., are damaged due to lodging when wind speed exceeds 30 km/hour. Such damages are often localised as lodging depends upon the stage of growth of the crop. In coastal areas where strong winds are a common occurrence, lodging losses in agricultural crops are high.

Strong winds cause deformation of stems and bole forms of the trees. The trees tend to bend due to the force of the wind. Continuous shaking of leaves due to strong wind results in the death of the leaves. Small branches are also sometimes killed due to shaking and rubbing. There is greater loss of moisture by transpiration, if the leaves are constantly shaken and the roots are often not able to replenish quickly in spite of a sufficient amount of rainfall. This results in desiccation of buds, leaves, twigs, etc., which die as in drought. Windward sides suffer more and trees are often distorted. In northern India, during January and February winter disturbances occur across northern India. Strong cold winds blow in the plains of northern India, particularly in Uttar Pradesh, Haryana, Punjab, Rajasthan, Bihar, and some parts of Madhya Pradesh creating conditions for frost damage. Some valleys, such as the Doon valley, are seriously affected by frost due to the accumulation of cold winds.

Near the sea coast, salt-laden winds cause serious damage to vegetation because saltish water is sprayed on the leaves. The common symptoms of salt damage are scorching along leaf margins and veins and burnt patches on the leaves. The burnt patches on leaves are caused due to spray of concentrated saltish water, which dehydrates the tissues through osmotic pressure. Tender parts of the plants, e.g., flowers, new leaves, buds, etc., usually die.

In temperate areas, cold winds, spray of hail and snow damage the agricultural crops considerably. Damage is also due to the higher speed of wind, common in higher altitudes.

Due to strong winds, soil materials on the surface of the land are loosened, particularly in arid and semi-arid areas where the soil mass consists of sand with a small proportion of fine particles and organic matter. The sand particles are lifted and bounced along the surface due to turbulence and irregularity of the movement of wind. The tender seedlings are usually killed due to mechanical injury caused by moving soil particles. The seedlings and saplings are buried under the *shifting sand dunes* which are of common occurrence in arid areas.

### Functions of Windbreaks and Shelterbelts

Windbreaks and shelterbelts perform several functions. Some of the important functions are:

(1) Controlling the ravages of wind
(2) Improving environmental conditions
(3) Improving output from arable lands and grazing lands
(4) Controlling wind erosion and shifting sand dunes
(5) Countering the salt-laden winds along the coast
(6) Providing shelter to houses and other constructions
(7) Yielding firewood, fodder and timber
(8) Improving aesthetic value and generation of recreation area

1. CONTROL THE RAVAGES OF WIND

Winds of gentle and moderate speed cause no serious damage. However, winds more than 40 km/hour in speed cause uprooting of trees and lodging of agricultural crops. In arid areas, they cause movement of sand and other soil particles and cause serious soil erosion. Strong winds cause mechanical injury to plants and adversely affect their growth. The effect of windbreaks and shelterbelts depends upon the quality of the windbreaks and shelterbelts. Important factors determining the quality of windbreaks and shelterbelts are their height, width, penetrability, orientation, etc. Windbreaks obstruct the flow of wind and create a zone of shelter nearby, mainly on the leeward side and to a lesser extent on the windward side. The destructive play of strong winds is controlled in the sheltered areas. The sheltered area is usually over a distance forty times the height of the windbreak; three-quarters of the protected area lie on the leeward side and a quarter on the windward side (Caborn 1957; 1965). The efficiency of a windbreak in reducing the wind velocity and providing shelter largely depends upon the density, height and the crown spread of the trees comprising the windbreak. The ideal *shelterbelt should filter the wind* without any efforts to block it completely. When the wind encounters the obstruction of windbreaks.

the uniform wind current is disturbed and the turbulent motion exerts a brake on the forward wind velocity. On the windward side a *cushion* of slow moving air develops which accounts for shelter on the windward side. The major part of the horizontal air stream is deflected upward towards the top of the obstruction and the remaining part, which is much weaker, passes through the cushion and the shelterbelt (Fig. 9a). The dense and impenetrable windbreaks divert the whole force of the wind over their tops. This creates a higher wind pressure near the top of the belt which is sucked down within a short distance behind the windbreak resulting in a comparatively much smaller sheltered area. Sometimes this also causes *vigorous eddying* which causes lodging in agricultural crops immediately behind the windbreak (Fig. 9b). Moderately dense shelterbelts are preferred due to their greater sheltering effect and smaller eddying effect. In such cases, maximum wind abatement occurs not immediately behind the trees but between two and four times the height on the leeward side. However, it should be noted that *dense shelterbelts* afford better protection in smaller area immediately near the shelterbelt on the leeward side.

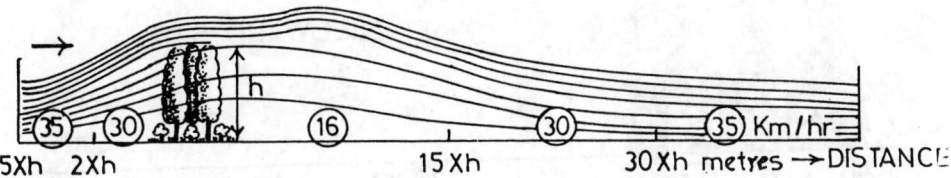

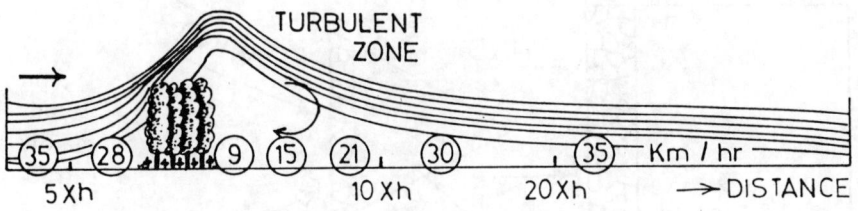

**Fig. 9: Effect of (a) Moderately penetrable and (b) Dense windbreaks on the flow of wind**

Though the effectiveness of an individual windbreak and shelterbelt in reducing the wind speed and providing shelter may vary considerably, it is often recognised that under average conditions (shelterbelt density 50 to

60 per cent or wind permeability about 40 to 50 per cent), the sheltered area may go up to 15 to 25 times the height of the windbreaks on the lee-ward side and 2 to 3 times on the windward side when the wind speed is about 40 km/hour (Caborn 1957; 1965; Kunkle, 1978; Shankarnaryan *et al.*, 1987). It is often recommended that the porosity may be evenly distributed throughout the entire height of the shelterbelt or it should be denser to-wards the ground and slightly open towards the top. The sheltering effects has been found to be limited to 10 to 15 times the height of the shelterbelt in the case of dense shelterbelts.

Maximum protection is provided when wind speed and temperature are higher. The requirement of protection, however, is much less when wind speed is low. Physiography also affects the protection provided by wind-breaks and shelterbelts. The ground which slopes away on the leeward side increases the range of protection, while a rise behind a shelterbelt shortens the area of protection. The maximum area is protected by windbreaks and shelterbelts when the direction of the wind flow is at right angles to the direction of the windbreak (Fig. 10).

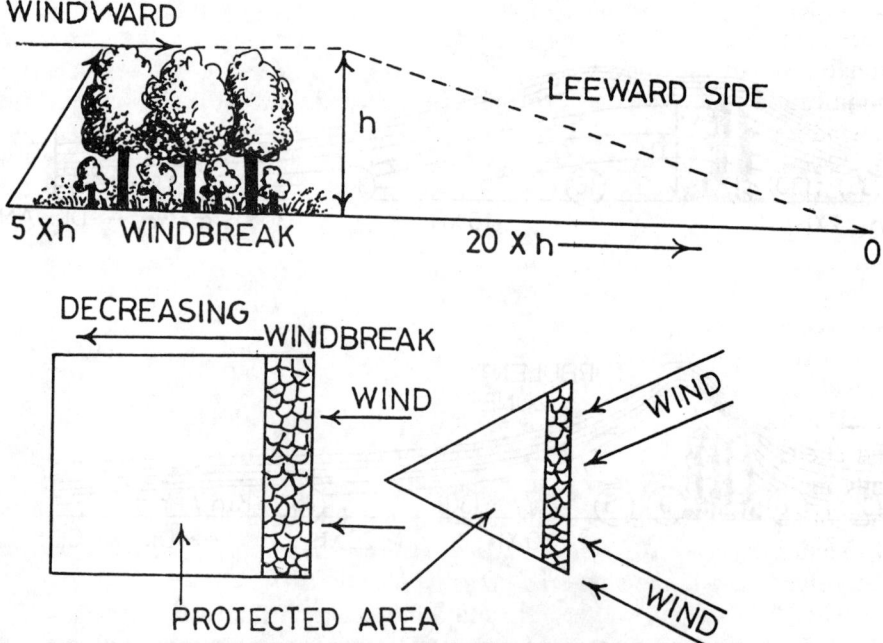

Fig. 10: **Windbreak and protected area**

The cumulative effects of a series of windbreaks and shelterbelts which check the windspread over a large area, have been the subject of much dis-cussion. It is generally believed that windbreaks and shelterbelts planted in parallel lines one after another across the direction of the wind should necessarily reduce the wind velocity progressively. However, studies conducted in this regard indicate that each windbreak acts independently and parallel windbreaks have no apparent cumulative effect on wind abatement. In fact, windbreaks and shelterbelts obstruct the air flow in the lowest layers of the atmosphere. Nevertheless, it is possible to plant windbreaks and shelterbelts sufficiently near together to ensure that their zones of shelter overlap and that free wind is not resumed at any point in between the shelterbelt (Caborn, 1965).

## 2. IMPROVE ENVIRONMENTAL CONDITIONS

Windbreaks and shelterbreaks grown continuously over a large area bring overall improvement in several environmental conditions. The environmental conditions are influenced due to the planting of trees and their impact due to tree crowns which intercept the sun's rays and precipitation, retard wind speed and loss of heat by radiation. A large quantity of leaf litter falls on the soil surface which protects and influences the soil properties and hydrological conditions. The interlocked root system of trees bind the soil together. The trees and other forms of vegetation planted as a windbreak and shelterbelt tend to bring about the same kind of influence on climate, soil and hydrology as other forms of forest with a certain degree of modification depending on the extent of shelterbelts and width and intervals of windbreaks.

Several climatic factors, such as temperature, precipitation, humidity, wind, snow, evapo-transpiration, frost, etc., are influenced to some extent by windbreaks and shelterbelts. Shelterbelts like other forests exercise a moderating influence on air temperature. Although there is a considerable variation in the data obtained regarding the effects of forests on temperature, the evidence is conclusive that forests reduce air temperature (Caborn, 1957; Kittredge, 1962; Dabral *et al.*, 1969; Warren, 1974). Observations recorded indicate that: (i) shelterbelts and windbreaks lower the daily mean temperature in the summer and raise it slightly in the winter and (ii) windbreaks and shelterbelts lower the daily maximum air temperature and raise the daily minimum. Thus, they reduce the daily range of variation. This influence of shelterbelts on air temperature is not limited in the close vicinity of the shelterbelts but goes far beyond. However, an appreciable influence is observed only in the sheltered areas.

The influence of shelterbelts and windbreaks in increasing precipitation has not been amply demonstrated. There is, however, some evidence which proves that forests increase precipitation (Hill, 1906; Nicholson, 1960; Rakhmanov, 1966; Warren, 1974). It is widely recognised that the *microclimate* created by windbreaks and shelterbelts extends a few hundred metres horizontally and vertically. When rain-bearing winds pass through this zone of microclimate, condensation takes place because of higher humidity and lower temperature. Also, the blocks of tall trees force the rain-bearing winds to move upwards which releases more moisture due to the presence of highly saturated air above the shelterbelts. In the absence of shelterbelts and windbreaks, these winds would otherwise pass without causing any local rain (Nicholson, 1960). However, such influences are quite difficult to measure. Dew and mist have been recorded in higher intensities in areas having enough trees. In the USSR, due to afforestation annual precipitation has increased by 12 per cent; also there has been 4 to 5 per cent more condensation (Rakhamnov, 1966). It has been concluded that a positive relationship exists between afforestation and precipitation which is close to a linear function. Experience in India indicates that forests increase summer and pre-monsoon rainfall and tend to reduce the heavy downpour (Rangnathan, 1949; Nicholson, 1960; Warren, 1974).

Windbreaks and shelterbelts influence the humidity in the atmosphere. Trees absorb a large quantity of water from the soil and only about one per cent of the water absorbed is utilised in the tree body for building up processes and the greater part is lost in the form of water vapour in the atmosphere by the process of *transpiration*. Release of a large quantity of water adjacent to the canopy of the trees increases the relative humidity. If windbreaks and shelterbelts are of sufficient coverage, the relative humidity will always be higher than in the open. Observations taken inside forests indicate that the relative humidity is about 5–8 per cent higher than in the open (Zon, 1927; Caborn, 1957; Warren, 1974; Champion and Seth, 1968; Ganguli and Kaul, 1969). Windbreaks and shelterbelts raised continuously over a large area would definitely increase the relative humidity of the area.

In temperate areas, shelterbelts affect snowfall and snow melting. Sheltered area contains more snow-water than unsheltered and treeless area and delay in the melting of snow is also observed.

Due to the moderating influence on solar radiation, temperature and wind, windbreaks and shelterbelts tend to reduce evaporation of moisture from surface soil. Dewfall in sheltered areas has been found to be 200 per cent greater than on the exposed ground; the difference was less in weather favourable for dewfall than in windy nights (Caborn, 1957). The heaviest dewfall occurs at a distance of 2–3 times the height on the leeward side of the windbreak. Evaporation in sheltered areas may be considerably less compared to unsheltered areas. Reduced evaporation helps to minimise the needs of irrigation in agricultural fields.

Another important role which shelterbelts perform is regulating the output of water. Trees affect the hydrological cycle considerably. The value of trees lies in prolonging the water-cycle from its inception as precipitation to the final disposal as run-off into streams and the ocean. The longer the water is retained on the land, the greater its usefulness in nurturing crops and trees, in maintaining a regular supply of water in streams and in preventing soil from washing (Stebbing, 1952). Trees, due to their crowns, intercept the rain and cushion its impact on the ground and curb its erosive power. They reduce surface run-off and increase infiltration which help in building a large underground reservoir.

Tree crowns *intercept* as much as 15 to 30 per cent of the total precipitation (Dabral and Rao, 1968; Ray, 1970; Ghosh *et al.*, 1980). Infiltration, which is the flow of water through the soil surface, is increased considerably due to penetration and decay of roots and improved soil conditions. Trees and other woody vegetation reduce *surface run-off* and increase *subsurface* run-off and *ground-water run-off* due to increased infiltration and water-holding capacity, maintaining partial or complete soil cover by leaf litter, creating mechanical obstruction in surface run-off and greater absorption and infiltration due to decayed plant roots and animal burrows. The precipitation which falls on bare ground is easily washed away in the form of surface run-off which carries soil particles along with it. In areas covered with trees, more precipitation goes inside the soil, remains absorbed and moves slowly in the form of subsurface run-off which tends to maintain the year-round flow of water in the streams with least damage to the soil (Fig. 11). The role of trees and forests in protecting the soil from *water erosion* is very well established (Bennet, 1955; Anon., 1975; Sinha, 1975; Singh, *et al.*, 1981). In high rainfall areas, shelterbelts can be effective tool in minimising flood peaks (Anon., 1975; Mathur *et al.*, 1976).

Shelterbelts and windbreaks planted over a large area in sufficient proportion may play an important role in improving the water yield and its quality. Experience in India indicates that trees and other woody vegetation help to maintain the flow of water in streams after the rainy season while in treeless tracts, such streams usually dry up soon after rains. Several perennial streams in the Himalayas become dry after the monsoon due to deforestation in their catchments. Martin (1944) observes that there were many perennial streams in Chittagong hills but due to deforestation, several perennial streams have dried up. Trees tend to reduce silt and other particulate and dissolved substances. Streams flowing in forested areas usually have clear water while those flowing through cultivated areas have turbid water. Plantation in the form of shelterbelts and windbreaks, therefore, would help to improve water yield and its quality.

Windbreaks and shelterbelts can play an important role in water and soil conservation. A large number of multipurpose reservoirs have been constructed at huge cost. Due to serious soil erosion problems in the catchments of these reservoirs, the actual rate of siltation is 5 to 10 times more than the anticipated rate (Anon., 1972). Nizam Sagar reservoir has lost 60 per cent of its storage capacity within 45 years. The reservoirs will give full service only when they are saved from siltation.

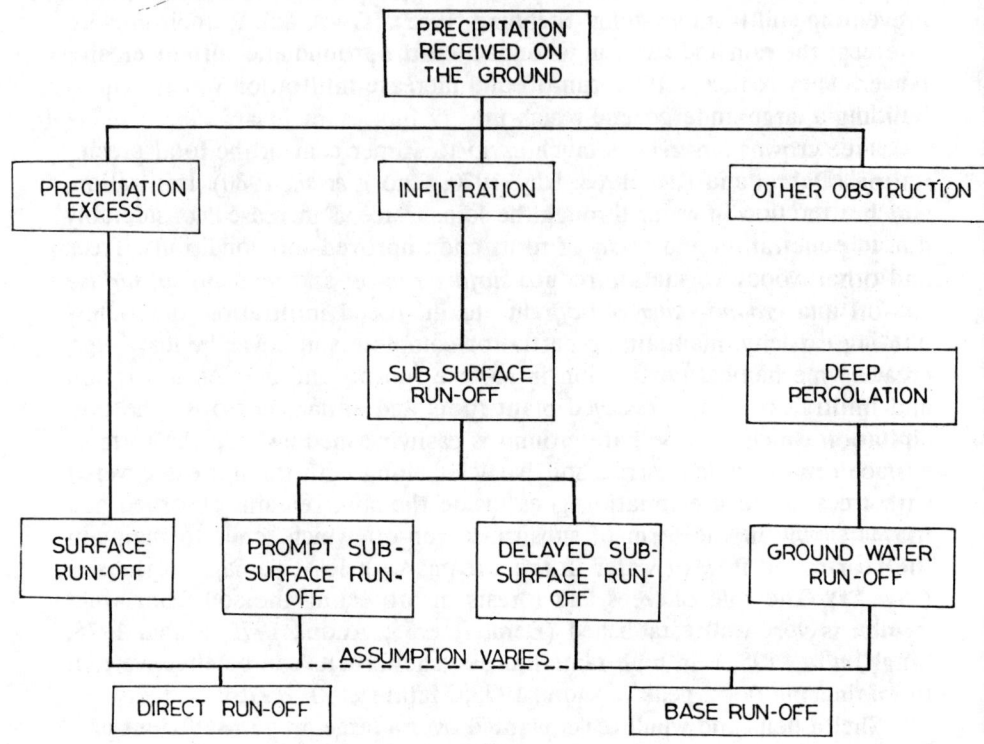

**Fig. 11: Forms of run-off**

Windbreaks and shelterbelts can be effective in controlling physical, chemical and noise pollution to a large extent. Dust and other particulate pollutants are filtered by trees. Windbreaks and shelterbelts increase dust precipitation due to reduction in wind velocity and increase in relative humidity. One hectare of shelterbelt can filter 50 to 100 tonnes of dust. Several chemical pollutants, e.g. gases, hydrocarbons, etc., are also removed from the atmosphere by the process of precipitation, scavanging chemical reaction, dry deposition, and absorption. Rows of trees have been found effective in reducing noise pollution. A 50-metre belt consisting of trees can reduce the traffic noise considerably. A few rows of trees along road and rail sides, are effective in reducing the noise pollution generated from traffic.

### 3. INCREASE YIELDS OF AGRICULTURAL AND PASTURE LANDS

As discussed earlier, windbreaks and shelterbelts help to conserve soil and water in the sheltered areas. Extremes of temperature, which adversely affect the plant growth, are moderated due to the influence of windbreaks and shelterbelts resulting in better temperature range for the growth of crops. Similarly, the positive influence of windbreaks and shelterbelts on the precipitation and relative humidity results in creating better growth conditions for the crops. In general, it can be concluded that windbreaks and shelterbelts create favourable climatic conditions for the growth and yields of crops. Soil water, which is very important in arid and semi-arid conditions, is conserved as *evaporation losses* are reduced considerably in the sheltered areas. Studies conducted in Czechoslovakia showed a decrease in evaporation over a distance equal to 33 to 46 times the height of the windbreaks. Similar investigations carried out in Japan indicated a 60 per cent reduction in evaporation up to a distance equal to height, 40 per cent reduction up to 5 times the height and 20 per cent reduction up to a distance of 10 times the height in comparison to open country under similar conditions (Ganguli and Kaul, 1969).

Windbreaks and shelterbelts have been found to improve the soil properties up to 5 to 7 times the height of the windbreaks. The leaf litter from trees improves the humus content in the soil. The soils near shelterbelts have been found to have higher organic matter, two to two-and-a-half times higher aggregate stability and to be richer in nutrients in comparison to open areas.

Favourable climatic and edaphic conditions created due to windbreaks and shelterbelts increase crop yields. Increased agricultural production due to windbreaks and shelterbelts has been reported from several countries (Bates, 1944; Caborn, 1957; 1965). In Germany, shelterbelts consisting of barriers (10 m wide and 6.1 m high) and spaced 199 m apart resulted in increased production of small grains ranging from 6 to 19 per cent, of hay by 8 per cent, of sugar-beet by 16 per cent, of potatoes by 12 to 16 per cent and of green beans by 57 per cent . In Italy, wheat crop protected by 4.8 to 7 m shelterbelts gave an increased yield of about 18 per cent in a belt extending from 6.1 m to 83.8 m on the leeward side of the windbreak. Similar increased yields of different crops have been reported in the USSR, the USA, and Italy (Stoeckeler, 1943, 1965; Bates, 1944, 1945). In the USSR, wheat yields are 20–30 per cent higher with shelterbelt protection and shelterbelts are most beneficial during extra dry years (FAO, 1969). In Guangdong province of China, *Casuarina equisetifolia* windbreaks established over 4034 ha, reduced the wind damage, stopped shifting sand dunes, reclaimed more than half of the land and crop yield increased from 1.4 tonne/ha to 4.8 tonne/ha (Ern, 1977). In one of the classic studies on the subject, C.G. Bates took crop measurements in Nebraska, Iowa, Minnesota and South Dakota in the USA and found that shelterbelts and windbreaks increased agricultural production; a maximum increase of yield of about 45 per cent above the field average was found at 4 to 5 times the heights from the shelterbelts.

The useful role of windbreaks and shelterbelts in improving agricultural yields has also been reported in India (Bhimaya *et al.*, 1958; Kaul, 1959;

Cornelius *et al.*, 1977; Rao, 1980; Reddy *et al.*, 1981). The studies indicated that increased agricultural yields are due to controlling evaporation losses of water and favourable climatic and edaphic conditions. Observations indicate that immediately near the windbreaks, the agricultural yields are somewhat adversely affected but beyond 10 m or so no adverse effects are visible and the positive role of windbreaks and shelterbelts becomes evident. The role of windbreaks and shelterbelts in improving agricultural production is more evident in arid and semi-arid areas. The windbreaks and shelterbelts are equally important in temperate areas for improving agricultural production.

Favourable effects of shelterbelts have been reported from Andhra Pradesh by Rao (1980). In Tirupati, the groundnut yield has been increased by about 75.6 per cent in the protected area as compared to the unprotected area. The red gram yield was also found to be more in the protected area. A significant increase in grains and straw yield of bajra by 63.6 per cent and 4.7 per cent respectively was observed in Rajendranagar (Hyderabad). Reddy *et al.* (1981) carried out studies at Tirupati and Rajendranagar (Andhra Pradesh) during 1978 and 1979 and found an increased yield in groundnut of 39.60 per cent, pigeon-pea 46.90 per cent and millets from 23.28 to 63.64 per cent.

Orchards seem to be especially responsive to the protection afforded by windbreaks and shelterbelts (Ganguli and Kaul, 1969). Windbreaks and shelterbelts improve pollination and fruit-setting and reduce dropping and rubbing of fruits. Mango is particularly responsive to shelterbelts and windbreaks where higher yields are assured due to higher setting and reduction in droppings during the early period of growth. It is due to this reason that mango orchards in Uttar Pradesh and other areas are generally protected all around by windbreaks. Pasture lands have also been found responsive to windbreaks and shelterbelts and they give increased yields, particularly in arid and semi-arid areas.

The influence of windbreaks and shelterbelts on crop production is mainly attributed to: (i) protection from hot and dry winds during critical stages of the growing season, (ii) reduction in evaporation losses due to lower wind velocities, (iii) reduction in crop damage due to storms and strong winds, (iv) reduction of fallen ears in crops infested with borers, (v) reduction in injury to young seedlings by blown soil through uprooting and (vi) creating favourable edaphic and climatic conditions for the growth of crops. The increase in yields is higher in tropical arid and temperate areas where strong hot dry and cold winds respectively are common. Similar increased yields in more favourable areas have not been confirmed.

### 4. CONTROL WIND EROSION AND SHIFTING SAND DUNES

Approximately 32 million ha area in India are subject to wind erosion which includes 23.5 million ha of desert and 6.5 million ha of coastal sandy areas. Due to the action of high-velocity winds, soil material mostly sand, from the soil surface is loosened, rolled, lifted, slid or bounced along the surface of the ground by the turbulence and irregularity of the wind movement. In soil erosion caused by wind, three processes take place, namely: (i) loosening and initiation of movement of soil particles, (ii) movement of soil particles and (iii) deposition of soil particles. Loosening of soil particles and initiation of the movement start when the wind speed is about 12 km/hour. The soil particles move by three kinds of movement, namely, saltation, suspension and surface creep. The soil moved by *saltation* is chiefly composed of fine particles ranging in diameter from 0.1 mm to 0.5 mm. In *suspension*, only very fine particles which are less than 0.1 mm are lifted. In suspension, the particles are lifted to great heights and soil particles are blown over long distances. In saltation and surface creep, the soil particles are not blown over long distances. The initiation of movement of soil particles and their transportation is the function of wind velocity. Therefore, the prevention and control of wind erosion would have to be based on the slowing down of the wind velocity at the ground surface.

Windbreaks and shelterbelts reduce the wind velocity and create a favourable microclimate which helps in controlling wind erosion. As discussed earlier, the wind speed is reduced up to a distance of 5 times the height on the windward side and up to a distance of 20 times the height on the leeward side. The wind erosion is controlled therefore, in the sheltered areas. The action of wind which causes loosening of the soil particles from the surface soil, initiation of movement of soil particles and transportation of the soil particles to other areas is considerably reduced, resulting in minimum wind action which does not cause erosion. On the other hand, when wind carrying soil particles is obstructed by windbreaks and shelterbelts, its velocity is reduced, resulting in deposition of soil particles which are in suspension. The effectiveness of windbreaks and shelterbelts depends upon their width, density and height. The microclimatic effect, e.g., increase in humidity and decrease in temperature and evaporation etc., helps in dust precipitation and deposition in the sheltered areas.

Stabilisation of shifting sand dunes is one of the primary concerns in controlling wind erosion. In order to successfully control the sand drift and stabilise the sand dunes, it is necessary that primarily loosening and blowing of sand from its source must be stopped. Stabilisation of sand dunes is based on the principle of reducing the threshold velocity of wind by providing obstruction to the wind. The efficacy of windbreaks and shelterbelts in controlling the sand drift and checking the sand dunes is well established. However, their efficiency depends upon the degree of aridity, wind velocity, duration of high wind period and quality and extent of obstruction offered by windbreaks and shelterbelts.

## 5. COUNTER SALT-LADEN WINDS ALONG THE COAST

Along the coastal regions, generally in about a 10-km wide belt, salt-laden winds cause considerable damage to all plants, including agricultural crops, home gardens, orchards, etc. During storms, the damage occurs in a much wider belt. The important symptoms of salt damage are scorching along leaf margins and veins and burnt patches on leaves. The burnt patches are caused by the spray of concentrated saltish water which dehydrates the tissues through osmotic pressure. Tender parts, e.g., flowers, buds, leaves, etc., usually die and injured leaves die prematurely. Plants with hairy or thick leathery leaves often escape the damage. But salt spray is not the only cause of these damages; the velocity and nature of the wind off the sea and its temperature also play a role. Shelterbelts created along the coast moderate the wind velocity, reduce the salt content of the winds and thus protect the agricultural crops, other plants and atmosphere in the protected side.

## 6. PROVIDE SHELTER TO HOUSES, GARDENS, ROADS AND OTHER CONSTRUCTIONS

Windbreaks and shelterbelts provide protection and shelter to houses, gardens, roads, rail lines, parks, etc., against several adverse climatic factors. Protection is provided against hot sun during summer and cold winds during winter. It is also provided against strong winds, hailstorms and avalanches in the mountains. They also provide protection against physical, chemical and noise pollution. Planting of trees and shrubs around houses, parks, garden, roads, and other installations is a common practice in India. These trees moderate the air temperature during summer. They protect against hot winds, locally called the *loo*. The temperature under the trees and forests is 3° to 8° C less than in the open (Caborn, 1957; Dabral *et al.*, 1969; Warren, 1974). Trees also protect from cold winds. Strong cold winds blow during winter, particularly in the higher altitudes of Himalayas. The main fear in planting windbreaks near houses in colder regions is that they may restrict sunlight which is essential during winter. Some adjustment in the distance of the windbreak and its height is needed to allow sufficient light required in the house, particularly in the temperate areas.

Strong winds cause serious damage to houses, communication networks and other installations. Most houses in rural areas are *kachcha* and strong winds cause serious damage to these houses. Though trees are damaged and uprooted during strong winds, they absorb the impact and provide protection to houses and other installations. In the areas susceptible to very strong winds, the trees selected for windbreaks should have a strong root system. The stems and branches of the trees should be able to withstand the impact of strong winds. The windbreaks and shelterbelts provide protection against dust and other particulate and gaseous chemicals. Noise pollution is also checked. In Rajasthan, where strong winds carrying dust and sand particles are common, windbreaks and shelterbelts near the houses provide a great comfort to the inhabitants.

Shelterbelts and windbreaks are useful for gardens and parks where a large number of tender plants and flowers grow. The windbreaks and shelterbelts afford protection and create a favourable microclimate for the growth of these plants. Shelterbelts and windbreaks moderate temperature which sometimes in summer goes very high. Higher humidity, protection against hot and cold winds and moderation of temperature are some of the important types of protection offered by windbreaks and shelterbelts.

Windbreaks and shelterbelts also provide protection to the roads and railways. The multirow plantations along roads provide shade to *pedestrians* and a journey through roads protected by trees is more comfortable than one performed in the open. In arid areas, these windbreaks can be effective tool for protecting the roads, railways, etc., from drifting sands.

The roots of trees bind the soil together and check landslides and landslips in steep terrain in the hill areas. Enough evidence exists to indicate that removal of trees has caused serious soil erosion and problems of landslides and landslips and afforestation has resulted in their stabilisation. The shelterbelts planted along the contour can be effective in checking landslips and landslides and protecting the houses, roads, soils, etc., from destruction.

## 7. SUPPLEMENT TIMBER AND FIREWOOD

Windbreaks and shelterbelts are generally planted for providing protection to crops, houses, roads, etc. and bringing about improvement in the environmental conditions due to their microclimatic effects. However, these windbreaks and shelterbelts can necessarily be a good source to yield timber, firewood, fodder, fruits, etc. A single-row windbreak planted along field bunds can yield about 2.5 $m^3$/ha/annum (Dwivedi and Sharma, 1990). Multirow windbreaks and blocks of shelterbelts would produce enough timber, firewood and other produce to meet the demands of the local population. The plantations along roadsides, canal sides and railway lines in some states have started giving a handsome yield of timber and firewood when harvested.

Selection of species can be made in such a way so that local demand for timber, firewood, fodder, etc. is effectively met. The yield of these plantations would depend upon the area, species, extent of windbreak and shelterbelt, etc. In order to achieve quick results, it is desirable to grow fast-growing species so that desirable benefits can be achieved within a reasonable period.

## 8. ENHANCE AESTHETIC VALUE (SCENIC BEAUTY) OF AN AREA

Windbreaks and shelterbelts improve the recreational value of an area due to the presence of a variety of trees, shrubs, herbs, flowers, etc. Some spots in the existing shelterbelts and windbreaks have already been developed into recreational areas by the introduction of flowering plants, ornamental shrubs and climbers, making enclosures for wild animals and birds and with additional arrangements for rest and parking. Such recreational areas have become popular along roadsides in several states.

**Designing Windbreaks and Shelterbelts**

The basic concept behind designing a windbreak or a series of windbreaks to afford protection depends upon the nature of the wind, the physiography, the area and the object to be protected. Usually, normal winds possess two important components: (i) The horizontal component which flows more or less parallel to the ground and (ii) the turbulent component which gives the gusts and lulls. In the plains, where there is no disturbance, most winds flow as a horizontal component. However, under disturbed conditions, the winds become more turbulent. The turbulence in the winds may be caused by physiographic factors or any obstruction in the wind direction. The designing of windbreaks will also depend upon the area and object to be protected. For example, if the area to be protected is small or large, or the object is a house, park, garden or agricultural fields, the design will vary. The area of protection in any single windbreak system depends upon several factors, e.g., orientation, interval, continuity, height, density, form of species and the width of the barriers in the pattern (Ganguli and Kaul, 1969).

ORIENTATION

It is generally accepted that the maximum protection is obtained when windbreaks are at a right angle to the direction of the wind and the protection effect obtained for a direction of wind flow at a right angle to the windbreak does not appreciably change up to a deviation of approximately 45 degrees. However, if the angle between the windbreak and the wind direction is further reduced, the protection effect reduces rapidly. In most areas, the wind direction is not constant and it is almost always changing. Under such conditions, isolated windbreaks may not be effective at all. In order to provide protection in the fields where wind direction is always changing, a series of windbreaks with suitable pattern is essential. Planning of windbreaks in the plains poses no serious problem. However, in undulating terrain orientation and establishment of windbreaks are difficult. In areas with more than 5 per cent slope, windbreaks should be planted, as far as possible, along the contours. *Cross windbreaks* are essential to obtain maximum protection.

INTERVAL

The planting of a series of windbreaks would reduce the area of arable land for cultivation of agricultural crops. Therefore, this point deserves careful attention. The interval between two windbreaks largely depends upon the maximum height of the windbreak. The interval of the windbreaks should ordinarily not exceed 20 times the maximum height of the windbreaks attainable. In sloping areas, the interval should be slightly less. Since trees may take 10 to 15 years to attain their maximum height, temporary windbreaks involving crops and grasses may be grown. For temporary windbreaks, crops such as arhar (*Cajanus cajan*), dhaicha (*Sesbania* spp.), etc., which grow tall could be used. Some grasses, e.g., napier grass (*Pennisetum purpurium*) and munj grass (*Saccharum munja*) could also be tried.

## CONTINUITY

The windbreaks need to be continuous. Any gap in the windbreak will give rise to a funnel action where wind velocity through these gaps is usually high. Similarly, the wind velocity at the ends of the windbreaks is also high. It is, therefore, necessary to see that no gaps are created in the windbreaks. Any passage or roads should be avoided through windbreaks. When essential, they should cross the barrier at an angle so that prevailing winds not blow straight. Cross windbreaks should be provided wherever necessary. Around corners curved rows are reported to provide better continuity.

## HEIGHT

The protected area created by a windbreak depends almost entirely upon its height and the protected area on the leeward side is proportional to the height of the windbreak. When the direction of the wind is at a right angle to the direction of the windbreak, the wind velocity is considerably reduced up to a distance of about 25 times the average height on the leeward side. On the windward side the protected area extends up to 2 to 5 times the height of the windbreaks. Trees which are capable of growing 20 to 30 m within a short period should be preferred. The trees should be strong enough so that they are not damaged. The windbreaks should be able to reduce the wind velocity to less than 20–25 km/hour which is the threshold velocity for wind erosion or lodging in agricultural crops.

## WIDTH AND DENSITY

The width of a windbreak is important as it decides its density. As already discussed, the ideal windbreak should filter the wind without any effort to block it completely. The width of the windbreaks decides the degree of density required at different levels. Moderately dense barriers are considered better as they filter the wind but do not block it. A density of 50 to 60 per cent is considered the best for field windbreaks. In order to maintain the required density up to its height, the windbreaks have to be wide enough to accommodate large trees, small trees, shrubs, etc.

FORM OF SPECIES AND CROSS-SECTIONAL AREA

Studies have indicated that more protection is provided if the windbreak is more vertical. Round shapes have comparatively less influence in comparison to vertical windbreaks. It is, therefore, necessary that only such trees be selected for windbreaks which are tall and possess a cylindrical or conical crown.

SELECTION OF SPECIES

Windbreaks may be classified on the basis of vegetation. A windbreak for a farm may consist of a single-row, three-row, five-row or multiple-row belt. It may comprise tall trees, medium-height trees, shrubs, etc., to provide uniform density throughout the height. Some trees may be evergreen and others may be deciduous in nature. The windbreak for a house or park may consist of only a hedge of tall shrubs.

Windbreaks may be: (i) permanent windbreaks, (ii) intermediate windbreaks and (iii) temporary windbreaks. While permanent windbreaks are provided near or around reservoirs, along roads, rails or canals and around large farms, the other two types are meant for small holdings or for farm villages and individual holdings. The interval between two successive windbreaks depends upon the height, density, orientation and speed and direction of the wind. Usually the tallest trees are arranged in the middle row and small trees and shrubs are kept in the outer rows. It is difficult to achieve vertical sides in a windbreak and therefore, a conical shape is generally aimed for. Tall trees and shrubs are also intermingled in order to ensure the desired density of the windbreaks. Mixing can be done at random or in lines or patches.

Trees growing as windbreaks are constantly subjected to usually high stresses of wind, temperature, etc. The species selected should first be adapted to the site. The other necessary characteristics may include wind firmness, moderate to dense crowns, strong stem, low branching, uniform fast growth, good height growth, etc. On the whole, these characteristics along with timber value of the trees determine the suitability of species for windbreaks.

Under Indian conditions, Table 56 gives a list of suitable species for different regions (Ganguli & Kaul, 1969).

**Types of Windbreaks**

Generally three types of windbreaks or shelterbelts are maintained. Different types of windbreaks along with species used under each are as given below:

A. *5-Row Type*: This type consists of central flank and side rows numbering five. The species for each row may be as under:

**Table 56: List of suitable species for windbreaks**

| Region (1) | Trees (2) | Small trees/shrubs (3) | Grasses (4) |
|---|---|---|---|
| **1. Arid region** | *Acacia nilotica, A. tortilis, A. leucophloea, Eucalyptus camaldulensis, Albizia lebbeck, Prosopis* spp. | *Balanites aegyptica, Calligonum polygonoides, Capparis decidua* | *Saccharum munja, Penicum turgidum* |
| **2. Northern alluvial plains** | *Acacia nilotica, Albizia lebbeck, Dalbergia sissoo, Syzygium cumini, Eucalyptus hybrid, Melia azedarach,* etc. | *Parkinsonia aculeata, Jatropha curcus, Vitex negandu, Capparis decidua, Euphorbia royaleana, Agave* spp. | *Saccharum benghalense, S.munja* |
| **3. Central region** | *Acacia nilotica, A. catechu, Cassia siamea, Hardwickia binata, Tamarindus indica, Madhuca latifolia, Pongamia pinnata, Dalbergia sissoo,* etc. | *Leucaena leucocephala, Gliricidia maculata, Jatropha curcus, Pithecellobium dulce, Agave* spp. | -do- |
| **4. Southern region** | *Acacia nilotica, A. auriculiformis, Albizia lebbeck, Bambusa* spp., *Casuarina equisetifolia, Anacardium occidentale, Tamarindus indica, Borassus flabellifer, Eucalyptus* spp., *Grevillea robusta, Cocos nucifera,* etc. | *Tephrosia candida Puraria javanica, Gliricidia muculata, Jatropha curcus, Sesbania sesban, Agave* spp. | *Saccharum benghalense, S. munja* |
| **5. Eastern region** | *Dalbergia sissoo, Albizia lebbeck, Syzygium cumini, Casuarina equisetifolia, Borassus flabellifer, Acacia auriculiformis, Pongamia glabra, Grevillea robusta, Bambusa* spp., *Artocarpus heterophyllus, Ficus* spp., etc. | *Thespesia populnea Tephrosia candida, Vetex negundo, Agave* spp. etc. | *Imperata cylinderica, Spinifex littoreus* |

(a) For Central Rows: *Acacia nilotica, Albizia lebbeck, Azadirachta indica, Dalbergia sissoo, Eucalyptus* spp., *Ficus* spp., *Holoptelea integrifolia, Kigelia pinnata, Prosopis cineraria, Tamarindus indica, Syzygium cumini,* etc.

(b) For Flank Rows: *Acacia senegal, A. leucophloea, Anacardium occidentale, Casuarina equisetifolia, Cassia siamea, Inga dulce, Moringa oleifera, Prosopis chilensis, Tamarix articulata,* etc.

(c) For Side of Outer Rows: *Acacia jacquemontii, Agave* spp., *Capparis aphyla, Calligonum polygonoides, Cassia auriculata, Dodonea viscosa, Glyricida maculata, Parkinsonia aculeata, Saccharum munja, Salvadora persica, S. oleoides, Vitex negundo, Ziziphus* spp., etc.

B. *3-Row Type*: This type of windbreak may consist of outer and central rows of the species discussed above.

C. *Single-Row Type*: Such types of windbreaks are also called buffer strips. *Prosopis* spp., *Acacia nilotica, Tamarix articulata* are the most suitable species for this system, since these species start branching from near the ground level and provide a partially penetrable barrier throughout their height. These species can be easily grown on saline and sandy soils of low rainfall area. The other species for a single-row windbreak system are *Casuarina equisetifolia, Parkinsonia aculeata, Ziziphus* spp. and *Prosopis chilensis*, etc.

These species are easy to grow in arid areas. *Tamarix articulata* can be grown by branch cuttings. *Acacia nilotica, A. tortilis, Ziziphus* spp. and *Prosopis* spp. can be grown by seeds which are easily available. *Parkinsonia aculeata* can also be raised by root-shoot cuttings.

Spacing

The spacement of trees has to be close enough to provide a moderately dense barrier against the wind. At the same time, it has to be wide enough so that growth is not adversely affected. Tall trees are to be planted in the centre row and smaller trees and shrubs in the outer rows. Shrubs should also be intermingled between the trees. The spacing between row to row may be kept 3 to 5 metres depending on species and area. The spacing between plant to plant in the row should be kept as under:

  (i) *Shrubs*—0.6 to 1.0 m
 (ii) *Small trees*—1.5 to 2.0 m
(iii) *Tall evergreen trees*—2.5 to 3.0 m

In areas where climatic conditions are favourable or where windbreaks could be irrigated, the spacement may be kept a little wider.

**Windbreaks and Shelterbelts for Coastal Regions**

Coastal regions face complex problems of aridity, salinity, high winds, tides, etc. Along the seashore, a large quantity of sand is often deposited during high tides. High sand dunes are usually formed along the coast. When a strong wind blows, the sand is often blown and deposited in agricultural fields and causes extensive damage. In order to control coastal wind erosion and to protect agricultural land, establishment of shelterbelts or windbreaks along the coast is essential. The species selected for windbreaks should include trees, shrubs, grasses, etc. A number of tree and shrub species, e.g.,

*Prosopis cineraria, P. chilensis, Casuarina equisetifolia, Borassus flabellifer, Acacia planifrons, A. auriculiformis, Salvadora persica, Capparis decidua,* etc., can be grown. Some species of grasses, e.g., *Erianthus munja, Atriplex nu-mularia* can also be grown.

Where strong winds are common, a littoral dune is erected along the sea face to act as a barrier to the prevailing wind to stop the sand blowing inland. This artificial sand dune is an effective barrier for providing protection against blowing sand. The artificial sand dune should be made at a right angle to the prevailing wind. In order to protect the dune from high water, it should be established sufficiently far inland to avoid damage. It is recommended that the barrier be put up about 120 m from the high-water mark (Ganguli and Kaul, 1969). Wooden planks, e.g., rejected railway sleepers, would form a suitable barrier for creating a littoral dune. Sand is deposited against the planks. As soon as the sand reaches the upper end of the planks they are pulled up about 1 m to form a fresh barrier over the previous barrier. The operation is continued until the dune reaches the desired height. The next work is to carry out afforestation on the sand dune. The shifting dune is established by providing brushwood barriers. *Brick planting* has been recommended as it has given better results. Several species including *Casuarina equisetifolia, C. jhunjhuniana, Acacia cyanophylla, Salvadora persica, Simmondisia chinensis, Tamarix,* etc. can be tried. Grasses, e.g., *Ammophila arenaria, Atriplex nummularia, A. rhagodioides, Juncus rigidus,* etc. are also useful and can be grown easily in these areas.

# CHAPTER 10

# Social Forestry

## Historical Genesis of the Concept of Social Forestry

Forestry practices which aim at providing goods and benefits to a nearby society are usually referred to as social forestry (Fig. 12). In India, forests have always played an important role in bettering the lives of the local population. In ancient times, when forests were plentiful, people lived in complete harmony with nature. People met their needs of timber, firewood, fodder and minor forest produce easily. There were no serious droughts and floods, duststorms, soil erosion and desertification. These problems arose when man and his livestock increased and started to destroy forests. It is recorded that during Ashoka's period, there were large forest areas identified for the use of the public besides closed areas for the king and eminent Brahmins. Successive invasions, increase in population and foreign domination contributed to rapid destruction of forests in India. In the early days of British administration in India, forests were ruthlessly exploited. Large forest areas were cleared for extension of agriculture. During the middle of the nineteenth century, difficulties began to be experienced in north-west India. Construction of railway lines necessitated clearance of forests for sleepers. The Indian Forest Act of 1865 was, perhaps, the first attempt towards forest management which gave powers for the constitution of 'reserved' and 'protected' forests.

The first national forest policy resolution issued in 1894, provided that the sole object for which state forests were to be managed was the public benefit. The policy, however, considered it necessary in the interest of the preservation of the forests, that the rights and privileges enjoyed by the inhabitants of the neighbourhood, be regulated and restricted. Four categories of forests were recognised, namely: (i) forests, the preservation of which is essential on climatic or physical grounds; (ii) forests, which afford a supply of valuable timbers; (iii) minor forests and (iv) pasture lands. The first category of forests was to be preserved for protective functions. The second category of forests was to be managed for the production of valuable timbers for commercial purposes. However, it was provided that every

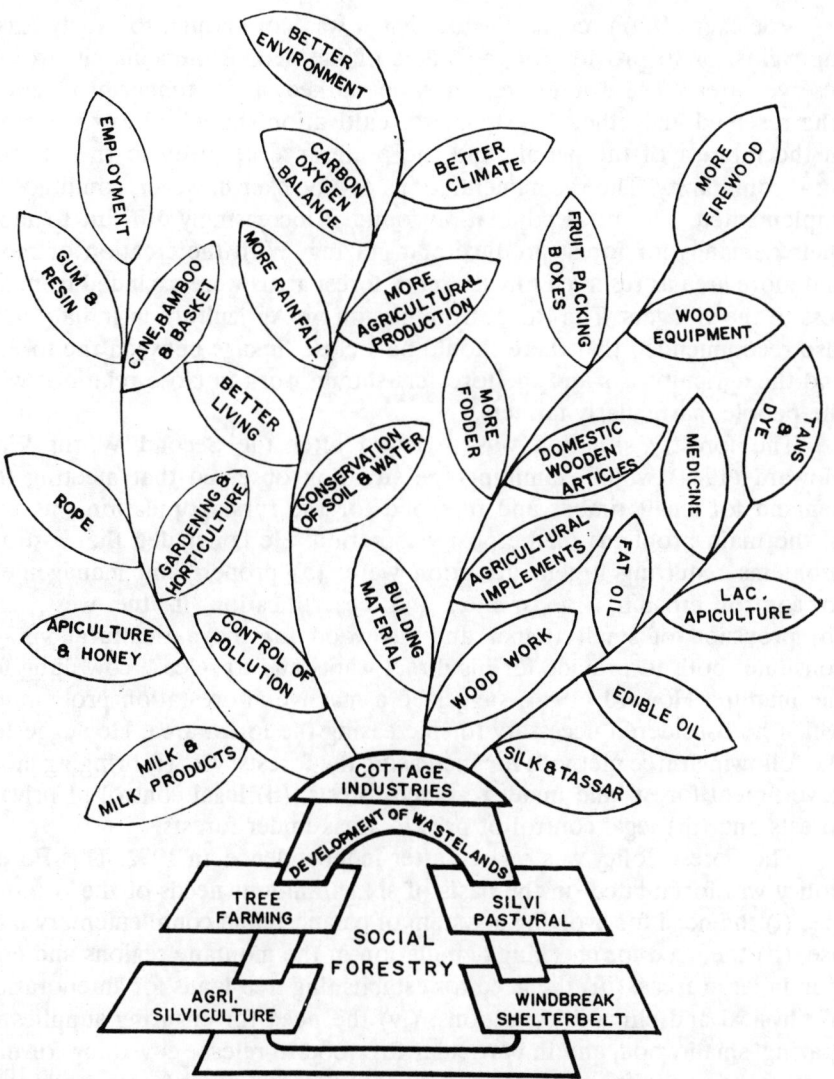

**Fig. 12: Components of social forestry**

reasonable facility was to be provided to the people living near such forests for easy satisfaction of their needs in respect of small timber, fuelwood, leaves for manure and fodder, thorns for fencing, grass and grazing for their cattle and edible products for their own use. The third category of forests was meant for supply of fuelwood, fodder or grazing and minor forest produce. The pasture lands were meant for grazing of the livestock. Keeping in view the importance of agriculture, it was recognised that the claims of agriculture are stronger than forest preservation in land use.

Voelcker (1897) recommended that it was not enough to satisfy existing rights, or to provide for the wants of the people immediately around reserved areas. The distant areas may be marked off for timber production. The reserved and other forests nearer cultivation should be worked more in the interest of the people and the preference be given to the agricultural community. The recommendations of Voelcker however, could not be implemented. The rural population found it increasingly difficult to meet their demands for forest produce and grazing. With the creation of more and more area as reserved forests, more forest areas were excluded from access to the villagers. The Royal Commission on Agriculture in India (1928) also recommended that there should be a close linkage between the forests and the agriculturists and the foresters should work in close relations with the people, particularly farmers.

The forestry situation was reviewed after the Second World War. Howard (1944) while examining the situation observed that meeting the demand for small timber and firewood for the rural population was one of the main problems in the post-war period. He concluded that forestry problems requiring urgent attention were: (a) proper land management to control erosion, floods, etc., and desertification in the west; and (b) provision of small timber and fuelwood for the agricultural village consumer both to provide for his direct wants and to release cow dung for the manure. Howard (1944) suggested a massive afforestation programme which he considered necessary for increasing the forest area. He suggested the following three methods for increasing the forests, viz., (i) bringing more government forest land under reserved forests, (ii) legal control of private forests and (iii) legal control of private areas under forests.

The Forest Policy was revised after independence, in 1952. This Forest Policy was formulated on the basis of six paramount needs of the country, viz., (i) the need for evolving a system of balanced and complementary land use; (ii) the need for checking denudation in the montane regions and erosion in large areas; (iii) the need for establishing tree-lands for amelioration of physical and climatic conditions; (iv) the need for ensuring supplies of grazing, small wood, and in particular firewood to release cow dung for manure; (v) the need for sustained supply of timber and other forest produce to meet national needs; and (vi) the need for realisation of maximum annual revenue consistent with the fulfilment of the above needs (Anon., 1952). The Forest Policy recognised the need for meeting the requirements of the villagers for forest produce. The forests were classified as: (i) protection forests; (ii) national forests; (iii) village forests and (iv) tree lands. The protection forests were to be preserved for physical and climatic considerations. The national forests were to be managed primarily to meet the demands of timber for defence, communications, industry and other purposes of public importance. The village forests were to provide small timber for house construction and agricultural implements, firewood, fodder and other forest products for local requirements, besides providing grazing to the livestock. The tree lands, which were outside the scope of the regular forest management, were meant essentially for the amelioration of the physical and climatic conditions of the country.

The Forest Policy of 1952 broadly provided for meeting the requirements of the people in the villages neighbouring the forest area, but it provided that the use of forest produce for the satisfaction of needs of neighbouring communities should not be permitted at the cost of national interest and the mere location of a village close to the forests, should not prejudice the right of the country as a whole to receive the benefit from the forests. It was stated that while the need of the local population must be met to a reasonable extent, the national interests should not be sacrificed. The importance of forests in protecting the montane areas was realised and it was recommended that about 60 per cent of the land in the montane tracts liable to erosion should be kept under forests and in the plains about 20 per cent of the area should be under forests.

In 1950, the then Union Agriculture Minister, K.M. Munshi, started '*van mahotsav*', a festival of planting trees. The objective was to popularise tree planting among the masses so that more area could become forested. The idea was to involve the masses in the tree-planting programme. Since 1950, however, '*van mahotsav*' is celebrated every year but it has not been able to generate the kind of enthusiasm in the plantation programme with which it was started. The programme remained almost a formality to be performed by government departments.

Jack Westoby (1968) in his inaugural address to the IX Commonwealth Forestry Conference used the term 'social forestry' for the first time and observed: (i) A distinction can be drawn conceptually between production forestry—forestry which aims at producing wood for industrial or household use and social forestry—forestry which aims at producing a flow of protection and recreation benefits for the community. (ii) In principle, production forestry should pay. (iii) The goals of social forestry should be determined by the amount of investment the community is prepared to allocate to secure the desired social benefits. (iv) The fact that wood and the physical protection and social benefits are frequently joint products does not rule out this approach. Social forestry should aim at creating protection and recreation benefits for the community.

The National Commission on Agriculture (1972) in its interim report on production forestry discussed the idea of social forestry and observed that social forestry should also include the activities concerned with growing and meeting the firewood needs of the community. Some amount of plantation of protection forests in areas liable to soil erosion must be accepted as a social investment (NCA, 1972).

The National Commission on Agriculture in its final report reviewed the forestry situation in the country and recommended that by taking up the programme of raising trees, grasses and fodder in the farmers' own lands, village commons, wastelands and degraded forests close to habitations, it would be possible to meet the requirements of fuelwood, fodder, small timber for small housing and agricultural implements, thorns for fencing, etc., (NCA, 1976). The objective of the social forestry programme was to meet the basic and economic needs of the community aimed at bettering the conditions of living by ensuring: (i) fuelwood supply to the rural areas for replacement of cow dung, (ii) small timber supply, (iii) fodder supply, (iv) protection of agricultural fields against winds and (v) recreational needs. The scope of social forestry included: (i) farm forestry, (ii) extension forestry, (iii) reforestation of degraded forests and (iv) recreation forestry. The farm forestry practices covered growing of trees on agricultural lands integrated with other farm operations. The basic component in farm forestry is to organise a substantial programme of tree planting on bunds and boundaries of fields to be taken up by the farmers themselves. The programme of *van mahotsav* was proposed to be strengthened to motivate farmers to take up tree planting. Necessary organisation and infrastructure for this was recommended by NCA. Extension forestry, envisaged as part of social forestry, covered: (i) mixed forestry on wastelands, panchayat lands, village commons, (ii) raising of shelterbelts in dry and arid areas and (iii) raising of plantations of different quick-growing species on lands on the sides of roads, canals and rails. The *degraded forest* areas were to be afforested with species which could supply fuelwood and small timber to villagers. *Recreation forestry*, as a part of social forestry, was envisaged near urban centres to meet the recreational needs of urban society. Recreation forestry aimed to contribute both towards social development and education in environmental matters.

Social forestry programmes were strengthened almost everywhere in the country along the lines of the recommendations of NCA. The programme has received sufficient budgetary allocations since 6th Five-Year Plan. Social forestry programmes have been further strengthened by procuring assistance from different international donor agencies, such as the World Bank, Swedish International Development Agency (SIDA), United States Agency for International Development (USAID), Canadian International Development Agency (CIDA), etc.

**Progress of Social Forestry**

Social forestry programmes began in 1950 in the form of *van mahotsav* (festival of tree planting) with a view to increasing public awareness about trees and forests. This programme was reinforced in 1952 when the National

Forest Policy recommended plantation of trees outside forests to increase the forest area in the country. Over the next few years, the achievements from these initiatives were not substantial. The *van mahotsav* programme continued to be observed every year but it could not generate such public awareness as would lead to a tree-planting movement and remained almost a governmental ritual (Dalvi, 1983). The number of seedlings distributed to farmers and others remained minimal. Since then, several kinds of programmes have been taken up in different states with a varying degree of success.

Several states, viz., Gujarat, Uttar Pradesh, Punjab, Haryana, Tamil Nadu, Karnataka, etc., started various social forestry programmes during the Second and Third Five-Year Plan periods. In Tamil Nadu and Uttar Pradesh, forest departments started raising plantations along canal and river banks as early as in 1956 for preventing erosion and increasing the supply of timber and firewood. In 1960, a programme of planting village wastelands was started in Tamil Nadu, Uttar Pradesh, Gujarat and some other states. Under this scheme, village common lands, government wastelands, etc. were planted with trees yielding fuel and fodder. In Uttar Pradesh, Haryana, Punjab and Gujarat, *Eucalypts* was grown on these lands. In Tamil Nadu, tank foreshores were planted with such species as *Acacia nilotica*.

Gujarat was perhaps the first state to create a separate social forestry wing in the forest department to give organisational support to the programme. Since then other states have also created social forestry wings in state forest departments. Substantial emphasis on social forestry programmes was given in the recommendations of the National Commission on Agriculture in its interim report of 1972 and the final report of 1976.

Social forestry programmes usually have three components, namely (i) farm forestry, (ii) community forestry and (iii) rehabilitation of degraded forest area. *Farm forestry* is plantation and management of trees on private lands, purely by private efforts, with or without government support. *Community forestry* is a practice of forestry on community lands, panchayat lands, government wastelands, etc., to be managed by the community. *Rehabilitation of degraded forest area* is afforestation of degraded forests with a view to meeting the requirements of fuelwood and fodder of the local population. Here the management usually remains with the forest department.

The afforestation programmes executed earlier, involved species which were not important for their fuelwood and fodder values. Invariably, the species were such which were required by the industry. A quantum jump can be seen in the afforestation activity from the Sixth Five-Year Plan onwards (Table 57).

It can be seen from the Table 57 that, although increasing emphasis was given to afforestation during the First to the Fourth Five-Year Plans

**Table 57: Progress of afforestation in Five-Year Plans (Anon., 1989)**

| Period | Afforestation in '000 ha | Expenditure in millions Rs. |
|---|---|---|
| First Plan (1951–56) | 52 | 12.80 |
| Second Plan (1956–61) | 311. | 68.60 |
| Third Plan (1961–66) | 583 | 211.30 |
| Annual Plans (1966–69) | 453 | 230.20 |
| Fourth Plan (1969–74) | 714 | 443.40 |
| Fifth Plan (1974–79) | 1221 | 1072.80 |
| Annual Plan (1979–80) | 222 | 371.00 |
| Total (1951–80) | 3556 | 2410.10 |
| Sixth Plan (1980–85) | 4650 | 9260.10 |

(1951–74), schemes which could meet the requirements of social forestry received little attention. The NCA's compelling plea for change in the traditional approach and to generate forest resources for local people outside reserved and protected forests through well-organised social forestry organisation and community participation considerably influenced the afforestation programmes. In the Fifth Five-Year Plan (1974–79), social forestry components were allocated about 50 per cent in the total forestry planting programme. There was a substantial step-up in outlays and targets during the Sixth Five-Year Plan (1980–85) and the total afforestation targets for the Sixth Plan far exceeded the total planting targets of the First to the Fifth Plans (1951–80) put together. It was during the Sixth Plan period that social forestry received considerable emphasis and social forestry components comprised almost 78 per cent of the total forestry planting targets. Block plantations on common land, strip plantation on the sides of roads, canals and railways, farm forestry and agroforestry on private marginal lands were carried out. Rehabilitation of degraded forest area and planting of fuel and fodder trees were given priority. Funds continued to be made available through the state forest departments and were further supplemented with externally aided social forestry projects in 14 states (NWDB, 1989b). Efforts were made to popularise social forestry programmes in rural areas particularly in the poorest section of the community by distribution of seedlings free of cost.

The beginning of the Seventh Five-Year Plan (1985–90) saw special emphasis on social forestry and wasteland development programmes with the creation of a National Wasteland Development Board (NWDB) during 1985 with the principal aim of reclaiming wastelands in the country through a massive programme of afforestation of fuel and fodder species with the active participation of the people. Some policy changes were brought out to make afforestation a people's movement, especially for production of fuelwood and fodder. A large number of small decentralised nurseries were set up in the private sector in order to involve small and marginal farmers, schools, women groups, etc. The raising of saplings by individuals was made commercially viable by a buy-back arrangement with the forest department. Voluntary agencies were encouraged to take up the work of social forestry and to spread its massage among the rural masses. Tree growers' cooperatives were encouraged to provide technical and material inputs on tree planting and to make arrangements for harvesting and marketing of produce. At the initiative of the NWDB, the Rural Development Department stepped up allocations for social forestry and earmarked 20 to 25 per cent of allocations for this purpose. The institutional arrangement for providing assistance to individuals for tree planting was initiated with the National Bank for Agriculture and Rural Development (NABARD) and Land Development Bank (LDB). The physical and financial targets were considerably increased during the Seventh Five-Year Plan (Table 58).

**Table 58: Year-wise afforestation programme in 7th Five-Year Plan (Anon., 1989)**

| Year | Afforestation (million ha) | Financial outlays (million Rs) |
|------|------|------|
| 1985–86 | 1.51 | 3746.0 |
| 1986–87 | 1.76 | 4917.0 |
| 1987–88 | 1.77 | 5334.0 |
| 1988–89 | 2.00 | 6205.0 |
| 1989–90 | 2.70* | 8589.0* |

* Provisional

Social forestry projects are under implementation in 14 states with financial assistance from international donor agencies. The World Bank is a major donor agency involving social forestry projects in nine states, namely, Uttar Pradesh, Gujarat, Rajasthan, Himachal Pradesh, Haryana, Jammu and Kashmir, Karnataka, Kerala and West Bengal. SIDA is assisting projects in Bihar, Orissa and Tamil Nadu. CIDA help is involved in Andhra Pradesh and Maharashtra. The total outlay involved under these projects for the five-year period is Rs. 9940 million involving plantation over 2.0 million ha including distribution of 747.5 million seedlings. The main objective of these projects is the production of fuelwood, fodder, small timber and other products to meet the needs of the rural communities by raising plantations on wastelands, including community lands, sides of roads, rails, and canals, degraded forests, other government wastelands and private marginal lands.

Important schemes being implemented for social forestry are as under (NWDB, 1989b).

## 1. RURAL FUELWOOD PLANTATION SCHEME

The Rural Fuelwood Plantation Scheme was introduced in the 6th Five-Year Plan to take up plantation of fuelwood in 157 districts of the country which were considered deficient in fuelwood. It involves raising of fast-growing species which may yield fuelwood, timber and fodder. In the Sixth Five-Year Plan, about 0.31 million ha have been planted. The objective of this scheme is to augment fuelwood production and make it available to the rural poor near their habitation. The scheme has been continued in the Seventh Five-Year Plan with increased targets.

## 2. OPERATION 'SOILWATCH'

The operation 'soilwatch' scheme was introduced in 1977–78 for treatment of selected catchments in microwatersheds in the Himalayan states. The objective of the scheme is to treat the microwatersheds with soil and water conservation measures. Afforestation and pasture development form essential components of the scheme. In the Sixth Plan, about 0.11 million ha have been covered with an expenditure of Rs. 264.2 million. During the first three years of the 7th Plan about 0.12 million ha have been covered. The treatment given under this scheme has helped in reducing soil erosion.

## 3. EXTERNALLY AIDED PROJECTS

In 14 states, externally aided social forestry projects are in progress. The total cost of these projects involves about Rs. 9940.44 million with a plantation target of 2,068,483 ha and other services (Table 59).

## 4. CATEGORISATION AND IDENTIFICATION OF WASTELANDS

Wastelands had earlier neither been systematically defined nor identified. A national wastelands identification project has been taken up in 146 districts of the country. District-wise maps showing different categories of wastelands have been prepared with the help of remote-sensing techniques with the assistance of the Department of Space and Survey of India. The identified wastelands will be superimposed on Survey of India topographical maps showing village and district boundaries.

## 5. GRANTS-IN-AID TO VOLUNTARY AGENCIES

This scheme was initiated in the Seventh Plan with a view to involving non-governmental organisations in the process of wasteland development with the people's participation. Voluntary agencies can approach people and can function in a flexible manner for obtaining people's participation in afforestation programmes. Grant-in-aid is provided under the scheme to the voluntary agencies, tree growers' co-operatives and non-governmental organisations involved in nursery, plantation, training of rural people, awareness enhancement, etc. In all, 289 projects had been funded up to April, 1989, involving a financial commitment of Rs.225 million (NWDB, 1989b).

6. DECENTRALISED NURSERIES PROGRAMMES

The scheme of decentralised nurseries was introduced in 1986–87 with a view to involving people in production of nursery stock and afforestation programmes. The cost of production of seedlings @ Rs. 0.45 per seedling is paid in instalments in cash and kind to the farmer by the forest department. The object is to produce the maximum possible seedlings in the private sec-tor. The farmer is free to sell the seedlings in the market. The remainder seedlings are to be purchased by the forest department. Such nurseries could be established by the small farmers, schools, women groups, voluntary agen-cies and co-operatives. The seedlings produced during 1986–87, 1987–88 and 1988–89 were 307 million, 288 million and 280 million respectively (NWDB, 1989b).

7. SILVIPASTURE SCHEME

The silvipasture scheme was started in 1986–87 to augment the pro-duction of grass and fodder. The objectives of the scheme are to increase grass production and to improve the degraded lands. The scheme envis-ages raising of fodder trees, shrubs, legumes and grasses on degraded and marginal lands of the farmers. About Rs. 1250 per ha was provided as central assistance. The assistance is available to non-governmental organisations, co-operatives, milk unions, etc. The scheme is also being implemented in government wastelands and degraded forest areas.

8. MARGIN MONEY ASSISTANCE SCHEME

The margin money assistance scheme has been taken up with the ob-jective of raising fuelwood, fodder, small timber and commercial timber plantations with the help of institutional finance. The scheme is for private lands and lands belonging to public sector undertakings. Assistance up to 25 per cent of the total project cost is available, provided at least 50 per cent of the project cost is met from loans from financial institutions. The scheme will increase the flow of institutional finance for wasteland develop-ment.

9. TREE PATTA SCHEME

The tree patta scheme envisages to give usufruct rights in trees planted and grown on earmarked land. The rights would include collection of dead-wood, lopping, tree produce, such as fruits, flowers, etc. The patta would be for a period equivalent to the normal silvicultural life of trees. The patta would be granted on government land to beneficiaries selected from the landless poor with emphasis on scheduled caste and scheduled tribe. The average size of patta land is not generally more than 1 ha.

**Table 59: Physical and financial components of externally aided social forestry projects**

Financial: Rs. in million.   Physical: In hectares

| Sl. No. | Name of state project | Project period | Project cost | Project components | | | | Total |
|---|---|---|---|---|---|---|---|---|
| | | | | Farm forestry | Village woodlots | Strip plantations | Rehabilitation of degraded forests | |
| 1 | 2 | 3 | 4 | 5 | 6 | 7 | 8 | 9 |
| 1. | Uttar Pradesh Social Forestry Project | 5 years 1985–86 to 1989–90 | 1611.60 | 147,210 | 14,000 | 740 | – | 161,950 |
| 2. | Gujarat Social Forestry Project | 5 years 1985–86 to 1989–90 | 1296.50 | 230,500 | 35,000 | 17,500 | 30,400 | 313,400 |
| 3. | Himachal Social Forestry Project | 5 years 1985–86 to 1989–90 | 572.90 | 66,838 | 41,000 | – | 5,000 | 112,823 |
| 4. | Rajasthan Social Forestry Project | 5 years 1985–86 to 1989–90 | 391.90 | 91,500 | 5,000 | 4,300 | 20,000 | 120,800 |
| 5. | Haryana Social Forestry Project | 1982–83 to 1986–87 extended to 1989–90 | 333.25 | 30,000 | 12,000 | 9,500 | 15,500 | 67,000 |
| 6. | Jammu & Kashmir Social Forestry Project | 1982–83 to 1986–87 extended to 1989–90 | 237.40 | 19,000 | 5,000 | 3,000 | 17,000 | 44,000 |
| 7. | Karnataka Social Forestry Project | 1983–84 to 1987–88 | 552.30 | 120,500 | 25,000 | 4,000 | – | 149,500 |

| No. | Project | Period | | | | | | |
|---|---|---|---|---|---|---|---|---|
| 8. | Kerala Social Forestry Project | 1984–85 to 1989–90 | 599.11 | 69,200 | 14,100 | 2,000 | — | 35,300 |
| 9. | West Bengal Social Forestry Project | 1981–82 to 1986–87 extended to 1989–90 | 348.65 | 52,000 | 6,000 | 20,000 | 15,000 | 93,000 |
| 10. | Bihar Social Forestry Project | 1985–86 to 1990–91 | 538.57 | 71,750 | 30,750 | 1,200 | 64,500 | 168,200 |
| 11. | Orissa Social Forestry Project: Phase-I | 1983–84 to 1987–88 | 281.70 | 26,500 | 21,700 | — | 35,300 | 134,400 |
| | Phase-II | 1988–89 to 1992–93 | 783.40 | 62,000 | 52,500 | 650 | 19,250 | 83,500 |
| 12. | Tamil Nadu Social Forestry Project: Phase-I | 1981–82 to 1985–86 extended to 1987–88 | 591.38 | 85,165 | 131,405 | 7,925 | — | 224,495 |
| | Phase-II | 1988–89 to 1992–93 | 854.00 | 18,000 | 56,300 | — | 4,080 | 78,380 |
| 13. | Andhra Pradesh Social Forestry Project | 1983–84 to 1987–88 extended to 1989–90 | 383.78 | 108,100 | 25,000 | 3,785 | 13,840 | 150,725 |
| 14. | Maharashtra Social Forestry Project | 1982–83 to 1989–90 | 564.00 | 44,035 | 33,975 | 2,990 | — | 81,000 |
| | Total | | 9940.44 | 1,242,298 | 508,730 | 77,590 | 239,870 | 2,068,483 |

## 10. PROMOTION OF CO-OPERATIVES

Three major types of co-operatives have been promoted with a view to increasing people's participation in social forestry programme. *Tree Growers' Co-operatives* on the Anand pattern have been created in five states, viz., Gujarat, Andhra Pradesh, Rajasthan, Orissa and Karnataka. The village wastelands are taken on a long lease and trees and fodder crops are planted by the members of co-operatives. About 120 such co-operatives have been set up up to March 1989 (NWDB, 1989b). *Farm Forestry Co-operatives* aim at establishing integrated farming systems including not only agroforestry, but also fodder, poultry, piggery, etc. The main objectives of these co-operatives are to develop farm forestry on a large scale on wastelands to bring ecological balance and to conserve soil and moisture. Under the Operation Flood Programme, *Dairy Co-operatives* are functioning in many areas. The National Dairy Development Board's concern is to produce enough fodder for milk cattle. These co-operatives therefore are helped to produce fodder on community wastelands.

## 11. AERIAL SEEDING

Aerial seeding has been taken up in eight states covering an area of about 0.8 million ha. Aerial seeding has been done in ravines and other degraded areas in these states.

## 12. AWARDS

Several awards have been instituted in recognition of exceptional contribution in the field of afforestation and other fields of wasteland development. The important award is Indira Priyadarshini Vrikshamitra Award. Certificates are awarded to persons/institutions with outstanding performance. From 1988 onwards, a cash award of Rs. 50,000 has also been given.

## 13. AREA-ORIENTED FUELWOOD AND FODDER SCHEME

The objective of this scheme is to create fuelwood and fodder resources on the basis of watershed. The idea is to take up watershed-based projects for integrated development.

## 14. SCHEME ON MINOR FOREST PRODUCE

A scheme on minor forest produce has been introduced with the object of taking plantation of fruit trees, bamboos, canes, tendu, harra, lemon grass and medicinal plants.

## 15. GREENING OF HIMALAYAS

The Himalayas have been recognised as a national asset providing ecological balance to the whole subcontinent and means of livelihood to the people living in the hills. Development of wastelands and degraded forest areas in the Himalayas has been taken up in different forms, e.g., watershed development projects, afforestation by eco-task force, etc.

The Government of India has announced a National Mission on wastelands development with a view to checking land degradation, putting wastelands to sustainable use, increasing biomass availability, especially fuelwood/fodder and restoring ecological balance. The strategies to be adopted for achieving these goals are proposed to involve people's participation at all stages, integrated land use planning on a watershed basis, village level action plans, conservation and natural regeneration, fuelwood, fodder and timber production, and technology extension (NWDB, 1989). To implement the strategy, the Mission proposes to have six Mini-Missions, namely, (i) planning and policies, (ii) people's participation, (iii) technology extension, (iv) regeneration of degraded forests, (v) greening of public lands and (vi) farm forestry. About 17 million ha area is proposed to be covered under this Mission (NWDB, 1989).

**People's Participation in Social Forestry Programmes**

It was realised that social forestry programmes must have the participation of the people at large and the rural poor in particular, in order to make this programme a real people's programme. It was envisaged that a social forestry programme cannot succeed in isolation. It has to be a people's programme. Participation of the local population at every stage, viz., planning, executing, selection of area, choice of species, management, harvesting, distribution of produce and benefits, etc. was considered essential so that control of these resources would be in the hands of the poor. In fact, the social forestry programme aims to be a programme of the people, for the people and by the people.

Efforts have been made to obtain people's participation in several ways. These efforts may be grouped into: (i) process participation, (ii) cognitive participation, (iii) interactive participation and (iv) material participation.

The *process participation* involves the people's participation in the processes leading to better decision making. Such participation of the people is usually obtained by the following methods: (i) By holding regular meetings with the local people to involve them in the process of planning. (ii) Local leaders are involved in the process of planning in several ways. In several states, there are village level, block level, district level and state level committees in which local leaders are members. (iii) Organising contests for deciding plantation plan, layout of nursery, etc. (iv) Involving local school/college and voluntary organisations and using their surveys and reports for decision making.

In *cognitive participation* one involves oneself with the concept, idea or task without participating in it physically. Efforts have been made to utilise the radio and television for spreading the message of social forestry among the people. Similarly, other agencies, e.g., advertising agencies, calendar manufacturers, publishers, newspapers and magazines, etc. have been working for popularising social forestry programmes. These agencies have been instrumental in finalising and spreading the message through slogans, pamphlets, popular articles, books, magazines and publicity materials. Other agencies such as the railways, the pulp-and paper-mills, match-box units and other industries were also involved in the publicity campaign. In some states, some industries such as WIMCO, etc. have their own publicity infrastructure to motivate farmers for raising tree plantations.

In social forestry programmes, efforts have been made for obtaining *interactive participation*. Interactive participation involves active physical participation in educating, motivating, organising, guiding and preparing people for social forestry programmes. It also involves active publicity carried out with the help of various agencies such as, films and newsreels, exhibition and publicity campaign, central and state publicity wings, etc. For obtaining maximum participation by the people, several states have created a special social forestry organisation on the lines of an agriculture extension organisation. The function of this organisation is to obtain the people's participation in all programmes of social forestry.

A more successful method for obtaining people's involvement is *material participation*, which involves active participation of individuals and organisations by way of sharing material inputs. Some examples of material participation include: providing fencing material, seedlings, fertilisers, etc., either free or on subsidised rates, providing technical and material help in establishment of nurseries and farm forestry programmes. Some of the social forestry schemes, such as decentralised nurseries, tree patta scheme, *hitgrahi yojana* (beneficiary scheme), etc. are typical examples of obtaining people's participation.

### Constraints in Obtaining People's Participation

In social forestry programmes, people's participation is the central thrust. In several schemes, the desired level of people's participation was not obtained. Therefore, some of the programmes in some states became programmes of the government department with little or no participation by the people. It is generally agreed that despite impressive achievements in area coverage, increased wood production, generation of employment opportunities and similar other benefits, there have been deviations and distortions (Banerjee, 1989). In some parts of the country, substantial success has been achieved in social forestry programmes primarily through resource-rich farmers taking advantage of the subsidised inputs and incentives. Some of the early adopters and innovators in Punjab, Haryana and Gujarat have reportedly reaped rich benefits. However, the situation is gradually changing and resource-poor farmers have also started participating in the programmes. Some of the major constraints in people's participation in social forestry programmes are as under:

### (i) *Most people show future-ignoring behaviour*

Most forestry programmes tend to take a considerably long period to give results. The people in general tend to ignore the long-term effects of the programme. Most people act for short-term considerations and they are greatly concerned for fulfilling immediate needs. The immediate needs of grazing may be felt more than the need of forest protection, which may yield dividends after a few years. Planting fast-growing multipurpose tree species is likely to generate interest among the people. Similarly, introduction of better strains of fodder grasses might improve the fodder yield in plantation, protecting both short- and long-term needs.

### (ii) *Social forestry generates common property resource*

Most programmes of social forestry generate a common property resource. There are several problems associated with a common property resource. One such problem is the tendency of people to pluck apples green without waiting for them to ripen, on the assumption that someone else will pluck them anyway. This is called 'the green apple picking phenomenon'. The immature plantation may be cut by people on the apprehension that someone else will cut it. The other problem is generation of concern of all sections of society. For example, the needs of the poor sections are less voiced and social forestry programmes have offered very little to the landless and similar other people. The poor are less able to buy their supplies and least able to produce it (Banerjee, 1989). Unless all sections of society have interest in social forestry, its success can not be ensured.

The common property resource faces another problem, usually referred to as *collective action dilemma*. This can be explained by the behaviour which villagers exhibit in protecting a plantation from grazing. If grazing is not done, the plantation is capable of yielding enough grass which could be cut and carried for stall-feeding. But the first person to refrain from grazing his cattle will be at a loss. For example, in a crowded show, almost all spectators tend to stand in order to get a better view, even though if everyone sat, there would be an unobstructed view for all. However, no one sits down, as the first person to sit will be at a disadvantage and will not be able to see anything. This type of behaviour, applicable with a common property resource creates problems in protection, management and distribution of benefits of social forestry plantations, particularly those raised on government wasteland, degraded forest areas and community lands.

Another important problem associated with social forestry plantations is social cost-ignoring behaviour of the people. Most of the social forestry plantations and wasteland development programmes involve a huge cost which is borne by society. However, since individual money is not involved, there is no realisation by the individuals of the expenditure incurred. The plantations and other wasteland development works are not protected solely due to non-realisation of the cost and consequent damages involved.

### (iii) *Lack of suitable organisation*

Lack of suitable organisation at different levels is one of the serious constraints in social forestry programmes. At the village level, the panchayats, at the block level, the block samiti and at the district level, the janpad, etc. are not adequately geared to take up social forestry works. In fact, the programme of social forestry has to be a programme *of the people and by the people*; consequently, then *for the people*. The social forestry organisation in the forest departments is also not adequate and lacks trained staff. The forest staff traditionally engaged in forestry activity is good for technical decisions but lacks extension acumen. Therefore, in several areas, social forestry programmes have resulted in good plantations but no or very little people's participation. Chaudhari (1989) observed: "Social forestry required a people-centred approach and special sensitivities in dealing with the rural poor. The forest officials, entrusted with social forestry, were expected to work collaboratively with village communities when their whole tradition and approach was to keep the people away from their forests. It was therefore not surprising that the afforestation schemes of roadside plantations, village wood lots, nurseries, etc. became essentially departmental programmes."

The organisation of the forest department is said to be too bureaucratic and people's participation in such structures is difficult to achieve (Chaudhari, 1989).

The role of the forest department in social forestry has to be one of the promoter and facilitator. The local people should be organised into an effective system to interact with the government department. They have to undertake forestry activity in which they become the principal actors and primary beneficiaries. In West Bengal, social forestry has been more successful because the village panchayats were actively involved in identifying land and beneficiaries. A group scheme of social forestry emerged in these villages for landless people who were distributed surplus land. Fuelwood, fodder and also cash incomes increased (Chaudhari, 1989).

### (iv) *Lack of appropriate policies and public awareness*

In many areas, people are not aware of the existing provisions of laws relating to the use of land, growing of timber and fuelwood trees, management and marketing, etc. The two main reasons for the situation are:

(i) low level of literacy among the rural population and (ii) failure of various extension organisations involved in the programme. People are not fully informed about what they can legally collect from forests and what is prohibited. A government evaluation of a tribal district revealed that of the 767 tribals interviewed, only 145 said they had the right of collection of timber from the forest for making agricultural implements, 222 said that they could send their cattle for grazing, 143 said that they had no right in forests and the remaining did not know and were not sure about anything (Anon., 1987). In social forestry programmes, rights over trees and procedure for marketing are not clearly defined. Where clear policies with regard to these different issues have been framed, these are not publicised. A government evaluation of the Orissa social forestry project indicated that about 82 per cent of the villagers did not know how the produce from village wood-lots would be distributed and most of them did not expect any share from the final produce (Anon., 1987).

### (v) *Lack of appropriate technology*

The traditional forestry practices are characterised by an overwhelming concern for meeting commercial and industrial needs. The silvicultural practices developed have centred around generating single-tiered, single species, single produce and with culture of elite trees over a large area. In social forestry, the practices should aim at developing multitiered, multispecies, and yielding multiple products to meet the variety of needs of the farmers, i.e., firewood, timber, fodder, minor forest produce, income, etc. In social forestry programmes, a variety of wastelands and marginal lands are to be utilised for production of goods and services. It requires development of site-specific and situation-specific technology which should be cheap and easy. Failure in afforestation of certain categories of wastelands in some areas is due to lack of appropriate technology (NWDB, 1989a). The programmes of social forestry were initiated with little research support. Even now, the desired research support, particularly in the field of socioeconomics, is not available, which restricts participation of the people.

### (vi) *Small size of holdings and land ownership*

In India, most of the holdings are marginal and small. The distribution of holdings in India indicates that of the 89.4 million holdings, 50.5 million holdings (56.49 per cent) are classed as marginal with an average area of 0.39 ha per family. About 16.1 million holdings (18.01 per cent) are categorised as small holdings with an average area of holding of 1.43 ha per family. The small and marginal farmers, who are in the majority, cannot earn enough to make their living. In the absence of job opportunities in the villages, a large number of them live below the poverty line. The first preference of these farmers is to produce foodgrain for sustenance.

In several areas, these holdings are in several fragments located in different areas and cause a serious problem in management for optimum production. These problems are further aggravated by erratic monsoon and other calamities, e.g., drought, floods, cyclones, frost, etc. The individual size of holdings being small, planting of trees is not common as trees tend to cause some adverse effect on agricultural crops in their near vicinity. In order to encourage poor farmers to plant trees, seedlings were distributed free of cost under the farm forestry programme in several states. It is alleged that the programme was elected by the bigger and resource-rich farmers and farm forestry became almost a subsidised programme for the bigger farmers (Chaudhari, 1989; Banerjee, 1989). The evaluation of this programme in Gujarat indicates that small and marginal farmers also actively participated as the number of seedlings lifted was in the proportion of 63 by marginal farmers, 112 by small farmers and 196 by big farmers (Anon., 1990). It appears that it may be true that initially only big farmers planted trees on their holdings but looking to the benefits, more and more small and marginal farmers are gradually taking up tree planting on their farm lands.

The social forestry programmes taken up on the village commons, panchayat lands and degraded forests were largely considered as plantations of the forest department due to the absence of a clear understanding among the villagers and the absence of appropriate village institutions for the purpose. The Forest Conservation Act, 1980 may also have come in the way of handing over the forest management, especially of degraded forest areas, to panchayats or individuals. The schemes authorising usufruct rights, such as tree patta scheme, *hitgrahi yojana*, etc., in which forest lands are utilised have not become very popular as ownership of the land rests with the government.

(vii) *Credit schemes, market support measures and tax rules*

In several areas, people are willing to plant trees but their inability to obtain credit may be an important factor in preventing them from doing so. If tree growing is to be taken as a land use on a significant scale, it will require investment, labour and time. Temporarily, it would mean diversion of agricultural or pasture land for forestry purposes. The present credit schemes are not attractive in view of the long gestation period of forestry.

For continuous tree cropping, it is necessary to have a stable commercial market for tree products. Uncertainty of market conditions is one of the serious constraints in agroforestry programmes. For example, farmers in Haryana, Punjab, western Uttar Pradesh and Gujarat harvested rich dividends from eucalypts planting about 5 to 10 years ago, but recently the price of eucalypts wood has fallen due to market saturation. This has adversely affected the entire tree planting programme in general and eucalypts planting in particular. In fact, some of the farmers are reportedly uprooting eucalypts saplings which they had planted recently. Most of the farmers feel that pure agriculture may be more profitable. The poor farmers are not able to protect themselves as there is a complete absence of any government support. Establishing new wood-based industries may create competition for the raw material and the farmer may get some benefit from this. Price support for forest products may help poor farmers. Reduction in marketing cost is also necessary as there is a large gap between the producer and consumer price. In some states, income from trees is tax free. Such rules tend to help only big farmers.

#### (viii) *Antipathy to trees*

In some areas, people have a genuine antipathy to trees due to cultural traditions. Farmers object to planting trees as they reportedly compete with agricultural crops, resulting in reduction in yield in the zone of competition. Some trees are believed to be inauspicious. In some parts, bamboos are not planted near the house as they are believed to bring destruction. In some areas, tamarind, pipal, bar, etc., growing close to a village are believed to house evil spirits. In several areas, tree groves are not planted near habitation as people regard them as a potential hiding place for thieves and robbers (Foley and Barnard, 1984). In some areas, tree planting is discouraged because trees increase wildlife population, particularly monkeys, blue bull, etc. which cause serious damage to crops. Birds also cause serious damage to crops in some areas. Stricter wildlife laws do not permit killing of wildlife and birds. Therefore, in certain areas, people avoid planting trees on their agricultural lands.

#### (ix) *Other constraints*

Several other constraints, such as lack of sufficient incentives, legal complications, long gestation period of forestry plantations, poverty, absence of developed marketing infrastructure, etc., have also inhibited people's participation in social forestry programmes. Incentives have their own role in establishing new programmes. There are several examples of governments offering grants, subsidies and other facilities for industries, development of agriculture, etc. However, such liberal incentives have not been given in social forestry programmes. Even in the programme of decentralised nurseries, the incentive provided is inadequate.

The present legal position is perhaps one of the serious constraints in people's participation. Even private forests or private lands brought under forest plantations are reported to be under the Forest Conservation Act, 1980. The Forest Conservation Act, 1980 is reportedly creating problems in Haryana, West Bengal, Himachal Pradesh, etc. where the schemes involve management of forest land through village societies and the future of these schemes is in jeopardy (Saxena, 1989).

Other provisions which restrict people's participation include: nationalisation of forest produce and provisions in the transit rules. In Madhya Pradesh, trade of teak, bamboo, etc., is controlled by the state government. Similarly, in Karnataka, trade of sandalwood is under the control of state government. The private farmers and others who grow such produce are forced to sell through the forest departments. Such restrictive trade policies often discourage planting of such species. *Transit rules* provide obtaining transit permit from the local forest department for transporting wood from one place to another. Such restrictive rules often discourage planting of trees on private ownership. In some states, restriction of transport of eucalypts, babul wood, etc., has been liberalised and needs no transit permit. In general, however, such restrictions discourage people's participation.

Poverty in rural areas also restricts people's participation. Most of the poor people have small holdings and cultivation of agricultural crops has the first preference. Growing of trees is limited in such holdings, although this activity might be more economical and sustainable in the long run. It has already been established that a tree-based land-use system in the hills and marginal lands of arid and semi-arid areas is more economical and has a greater human carrying capacity than agriculture and other land-use systems (Gupta and Mohan, 1982; Martins and Nautiyal, 1986). However, for poor people immediate living is of greater significance than long-term sustenance.

In several areas, people took to tree farming because of financial gains. In the Punjab, Haryana, western Uttar Pradesh and Gujarat, farmers have planted eucalypts on a large scale during the last 10 to 15 years. But the price of eucalypts wood has come down due to over supply. The farmers who planted eucalypts with great hopes are now facing problems in disposal of their produce. Such conditions work as serious constraints in farm forestry programmes.

**Criticism of Social Forestry Programmes**

Social forestry programmes were initiated with a view to improving fuelwood and fodder production, reducing pressure on forest land, regenerating wastelands, involving the community in forest management and helping the rural poor through improved access to forestry products and improved income through assets creation. Implementation of the programmes over the last 20—25 years has helped in revegetating a large area of wastelands and created significant wood resources in the country. In some states, such as the Punjab, Haryana, Gujarat, Uttar Pradesh, Karnataka, etc., the programmes succeeded in generating a people's movement in tree planting. Haryana, once a totally deficit state in wood is surplus in wood today, mainly due to success in agroforestry. The state is now exporting wood and charcoal to ine adjoining states of Himachal Pradesh, Jammu and Kashmir, etc. It is reported that during 1967–68, Haryana was short of about 1 million m³ of wood but had a surplus of about 0.2 million m³ in 1988–89; it is expected that in 2005, the surplus will be more than 10 million m³ (Mallik, 1989). Similarly, in Punjab, about 1.5 to 2.0 million m³ of surplus eucalypts wood is available every year (Kapur, 1989). Similar success is reported in Gujarat, Karnataka and western Uttar Pradesh. In farm forestry, significant success has been achieved in several states. However, the community forestry programmes have not shown similar successes. Village communities have not always come forward to protect and manage the forests. In Uttar Pradesh, Karnataka and some other states, *gramsabhas* are reported to be reluctant in taking over the management of community forests and village wood lots (Raizada, 1989; Gaonkar, 1989).

Some of the common criticisms of the social forestry programmes are as under (Foley and Barnard, 1984; Saxena, 1989; Chaudhari, 1989): (i) In order to encourage poor farmers to take up afforestation, seedlings were distributed free. The benefits from farm forestry tend to favour rich farmers. The programme has not been able to generate similar interest among poor farmers. Therefore, farm forestry became a subsidised programme of bigger farmers. (ii) Fuelwood and fodder species were not always planted. Several times market-oriented species were planted which did not improve consumption within the village. (iii) As the demands of the poor for fuelwood and fodder are not met, pressure on forest land continues unabated. The process of forest degradation shows no decrease. (iv) A large area of wastelands is encroached upon and not available for revegetation. These lands are in scattered blocks of small sizes and require a large investment. (v) Social forestry programmes could not become a people's movement. Communities were generally not given funds. The plantations raised by the forest department on community lands were seen as government trees. Sharing of goods and benefits emanating from these plantations among different sections of the community remains unresolved in many areas. (vi) Participation of small farmers in farm forestry remained poor. Diversion of land from agriculture to forestry reduced employment opportunities. The tribals whose dependence on forest is acute, were generally ignored. (vii) Social forestry programmes have failed to provide adequate social and environmental benefits. In farm forestry, farmers grow trees for maximising financial returns. Individual farmers do not bother about social and environmental considerations. This, however, does not mean that trees planted in farm forestry do not have beneficial environmental impacts. The criticism is that the wood from farm forestry goes in the market and is not consumed in the same area where it is grown.

These criticisms of social forestry programmes are not always well founded. Experience indicates that during the initial years some mistakes did creep into the implementation of the programmes, but these mistakes were corrected in due course. Evaluation of social forestry projects in some of the states confirms that the programmes have achieved their goal to a large extent. It is true that the farm forestry programme was initially adopted by only well-to-do farmers. But looking to the benefits, the programme became popular among poor farmers as well. Since poor farmers have smaller holdings, the number of trees per ha may be slightly less in poor farms in comparison to rich farms. The number of seedlings distributed indicates that almost 50 per cent seedlings were lifted by small and marginal farmers in Gujarat, Karnataka, West Bengal and Haryana (Gaonkar 1989; Anon., 1990).

It is wrong to criticise the programme by saying that appropriate species were not planted. In Karnataka, the study indicates that 72 per cent seedlings planted under social forestry were those of fuelwood and fodder species (Gaonkar, 1989). In Punjab, Haryana and Gujarat, eucalypts was the dominant species under farm forestry. However in community forestry programmes, several fuel and fodder species were given preference. Eucalypts was not usually planted in forest lands. In farm forestry programme, eucalypts, poplars, subabul, babul, etc. were planted by farmers because these are fast growing, have a ready market, can be easily grown with agricultural crops, and give a good financial return. It has been said that farmers wanted to plant such trees which would give them money. They were not very concerned for their firewood and fodder needs. In fact, the firewood need was easily looked after because enough lops and tops were available, even though the trees were harvested for timber.

Experience indicates that there were several weaknesses in the community forestry programmes. One serious constraint was lack of effective organisation at the village level which could look after the interests of all sections of society. However, there are some examples where effective community participation was obtained. Much publicised examples are Sukhumajhari in Haryana, Arabari in West Bengal, Ransinghee in Maharashtra, Dhari in Madhya Pradesh, etc. But these examples have not been multiplicated on a large scale. In spite of the best efforts, the Sukhumajhari type of community participation has not been achieved in other places in Haryana, to say nothing of other parts of the country. In fact, these models of people's participation are the consequence of an evolutionary process to avoid the tragedy of the common property resource. To replicate these models for resource management in other areas widely differing in people, in socioeconomic conditions, in cultural traditions etc., some adjustments are necessary; some unforeseen problems will emerge which can only be solved through learning and adjustments (Chopra et al., 1989). Several non-governmental organisations have started playing the role of catalyst in propagating community participation in management.

One can see a large number of trees on farm lands and on sides of roads, rails and canals. But still degradation of the natural forest continues. It has been observed that in some areas farmers planted trees on their farm lands, maintained them and sold them in the market but still collected fodder, firewood and timber from the adjoining natural forest to meet their needs. However, this tendency will also change someday.

CHAPTER 11

# Development of Wastelands

## Wastelands

Wastelands may be defined as lands where the production of biomass is less than its optimum productivity. In other words, any land which is not being used to its optimum productivity is wasteland (NWDB, 1989). In land management system, the basic principle is that every piece of land must be put to use in accordance with its capability in such a way that it produces the maximum without deterioration. Lands which have some limiting factors are also included in the category of wastelands. Wastelands are also de- scribed as degraded lands which can be brought under vegetative cover with reasonable effort and which are currently underutilised and lands which are deteriorating for lack of appropriate water and soil management or because of natural causes. Wastelands can result from inherent or imposed disabilities, such as location, environment, chemical and physical properties of the soil or financial or management constraints (NWDB, 1987). The technical task group of experts set up by the Planning Commission at the instance of the National Wasteland Development Board classified wastelands into two categories: (i) *Cultivable Wastelands*: lands whose level of productivity of vegetative cover can be increased with or without inputs and include: (a) gullied and ravined lands, (b) salt affected lands, (c) surface waterlogged and marshy lands, (d) degraded lands, (e) deserts, (f) fallows and other lands lying unutilised. (ii) *Uncultivable Wastelands*: wastelands on which vegetative cover cannot be developed. Such lands include: (a) barren rocky areas and (b) snow covered or glacier area.

The Interministerial Working Group classified wastelands as under (NWDB, 1987): (1) Physical or chemical features which generally characterise wastelands as follows: (i) gullied and/or ravined land, (ii) upland with or without scrub, (iii) waterlogged and marshy land, (iv) land affected by salinity and alkalinity — coastal/inland, (v) shifting cultivation area, (vi) sands — desertic/coastal and (vii) wasteland arising out of mining and industrial activity. (2) Land with one or more of the foregoing features or otherwise underutilised or misutilised may be within the notified forest area. Such lands may also be used now or might have been used in the past for agriculture or grazing purposes.

These lands are ecologically most unstable and suffer due to various problems. These problems may include: (i) low nutrient status, (ii) top soil completely washed away due to serious soil erosion, (iii) difficult surface due to formation of ravines, gullies, landslides, etc., (iv) lack of moisture due to aridity, (v) development of toxicity in the soil in the root zone due to salinity, alkalinity and acidity, (vi) development of waterlogging, and (vii) presence of rock outcrops, boulders, etc.

In the absence of any detailed land-use survey, several estimates have been made about the extent of wastelands in the country. Some of the estimates of wasteland indicate that about 175 million ha land falls in this category (Anon., 1976; Singh *et al.*, 1981). The other estimate by the Society for the Promotion of Wastelands Development puts it at 129 million ha --93 million ha in non-forest area and 36 million ha in forest area (Anon., 1984a). NWDB (1989) estimates wastelands to the extent of 120 million ha, with 40 million ha in forest lands and 80 million ha in agricultural lands.

There are several causative factors for the creation of such wastelands. Faulty land use, defective tillage practices, excessive land use due to increased human and livestock population, improper water use, defective drainage conditions, adverse climatic factors, etc. have led to the formation of these wastelands.

## Technology for Development of Wastelands

The use of these wastelands for agriculture requires large inputs and has a very limited scope. However, these lands could easily be developed through afforesting them with suitable species. For successful afforestation of these areas, the operations which need maximum attention are proper soil-working and plantation techniques and selection of suitable species. Soil-working techniques should aim at: (a) proper conservation of soil and moisture, (b) drainage of excess moisture, (c) removing obstruction for root growth and (d) removing toxicity at the root zone. The species should be such that: (i) it is adapted to the site, (ii) shows high establishment rate, (iii) has good root system, (iv) has faster rate of growth, (v) shows good recovery from damage and (vi) produces sufficient regeneration. Plantation techniques should be such that planted seedlings strike the root system, show high establishment and a faster rate of growth. The technology for afforestation of some wastelands is discussed.

### 1. HOT DESERT AND SHIFTING SAND DUNES
The total area of hot desert in India is approximately 31.3 million ha.

Of this, 12.6 million ha is occupied by moving and semi-established sand dunes of different degrees and extent, mainly in Rajasthan.

Desert soils are purely mineral soils obtained by mechanical disintegration of rocks. These soils are characterised by: (i) very low organic matter, (ii) high percentage of soluble salts, (iii) low in nutrient status, particularly nitrogen, (iv) high pH and calcium carbonate, (v) structureless and coarse texture, (vi) very poor water-holding capacity and (vii) absolute deficiency of soil moisture. Climate of the area is highly unfavourable. Rainfall is very low. Temperature is very high during May and June and reaches 49°C. During winter, it goes below the freezing point. The wind velocity is generally high.

Seeds fail to germinate in the unstabilised windblown sand and often seedlings get buried by drifting sand. Sand dunes keep moving from one place to another. In areas where sand has become stabilised, soil may be fertile but is usually associated with accumulation of lime nodules. The height, length and width of sand dunes may vary; they may be small to very big and are generally classified into three categories: (a) small sand dunes —less than 1.5 m in height; (b) medium —1.5 m to 3 m in height and (c) big —more than 3 m in height.

Afforestation of these areas is associated with fixation of sand dunes. The smaller the dune, the more active it is. Fixation of moving sand dunes in Israel has been done by planting cuttings and potted stock, generally 60—80 cm deep, as reported by Kolar *et al.* (1966). Rooted cuttings of 60—80 cm length are set into the sand with the aid of an iron bar, at least 50—70 cm deep. Deep planting protects the planting stock against deflection and secures root development in a comparatively moist sand layer. Use of bitumen emulsion to fix shifting sand has been done in Middle Asia and the Ukraine in the USSR since 1943. This method has been tried in Libya, Tunisia, and some Middle East countries. The method has also been tried in India in collaboration with Esso Oil Company in various areas. Wilson (1965) tried to work out the practicability of such a method over a large area and reported that it would prove costly.

Any programme of afforestation in these areas calls for rapid stabilisation of the drifting sand dunes, maximum moisture conservation and proper utilisation, selection of suitable species and efficient planting technique. For fixing sand, palisading with brushwood of locally growing bushes such as *Crotalaria burhia*, *Ziziphus* spp., *Calligonum polygonoides* and *Calotropis* spp. buried in the sand and running along and parallel to the crest of dunes has proved successful in Rajasthan and reduced evaporation loss from the surface and increased the humus content (Bhimaya, 1960). Protection from cattle is also an important factor (Shrivastava and Qureshi, 1966). Live hedge should be preferred because it protects against cattle menace and also provides protection against wind action. *Euphorbia tirucalli* and *Jatropha curcus* are often handy for live hedges.

Plantation of shelterbelts should also form a part of desert afforestation. They reduce the wind velocity and thus check the shifting of dunes on the leeward side. Windbreaks of *Dalbergia sissoo, Albizia lebbeck, Acacia nilotica, Ziziphus mauritiana, Tamarix* spp., *Azadirachta indica, Parkinsonia aculeata,* etc. have been successful. Planting of trees should be done by raising container plants. As far as possible, tall plants should be used for planting. According to researches conducted by the Central Arid Zone Research Institute Jodhpur (Anon., 1965), it has been found that in Barmer (Rajasthan) where rainfall is only 200 mm, *Acacia tortilis* and *Calligonum polygonoides* have proved successful and at Gadra Road (rainfall, 150 mm) *Acacia seyal, A. aneura, Eucalyptus camaldulensis, E. occidentale* (from Israel), *E. olcosa* (Perth, N. Australia) have given encouraging results. Among grasses, *Saccharum munja, Panicum antidotale, P. turgidum, Cenchrus ciliaris, C. latharicus, Eleusine* spp., *Elionurus hirsutus* and *Pennisetum* spp. have been found successful in these areas.

Different species which have been successful and may be recommended for planting are as under (Anon., 1965; Ghosh, 1977; Muthana, 1986):

(a) Rainfall 150 mm to 300 mm: *Acacia tortilis, A. senegal, Prosopis chilensis, P. cineraria, Tecomella undulata,* etc.
(b) Rainfall 300 mm to 400 mm: *Prosopis chilensis, P. cineraria, A. tortilis, A. senegal, A. nilotica, Tecomella undulata, Eucalyptus camaldulensis, Parkinsonia aculeata, Ziziphus* spp., etc.
(c) Rainfall 400 mm and above: *Prosopis chilensis, P. cineraria, Acacia tortilis, A. senegal, A. nilotica, Parkinsonia* spp., *Albizia lebbeck, Ailanthus excelsa, Azadirachta indica,* etc. have been successful. Some other exotic species, e.g., *Acacia aneura, A. albida, A. salichina, Colophospermum mopane, Dichrostachis glomerata, Eucalyptus terminalis, E. goolabah,* etc. have also given good results (Das *et al.,* 1986; Muthana, 1986).

Stabilisation of sand dunes is based on the principle of reducing the threshold velocity of wind at the dune surface by providing obstruction, such as a bunch of shrubs and grasses, etc. A uniform method of reclamation for all sand drift areas cannot be suggested, as conditions vary in respect of degree of aridity, wind velocity, and availability of different materials.

Important factors in afforestation of sand dunes and desert areas are: (i) protection of area against biotic factors, (ii) treating the dunes by fixing barriers in parallel strips using locally available shrubs and (iii) planting the area by nursery-grown seedlings.

Ten-to twelve-month-old seedlings should be planted in pits of 45 cm$^3$ or 60 cm$^3$ filled with weathered soil. After planting, a saucer-shaped depression (1 m diameter) may be made around the plant to harvest rain-water.

The plantation needs effective protection against biotic factors for at least 10 years.

## 2. SALINE/ALKALI AREAS

Saline and alkali soils cover nearly 7 million ha occurring in Uttar Pradesh (1.3 million ha), Gujarat (1.2 million ha), West Bengal (0.8 million ha), Rajasthan (0.7 million ha), Punjab (0.7 million ha), Haryana (0.5 million ha), Maharashtra (0.5 million ha), Karnataka and Orissa (each 0.4 million ha) and smaller areas in Andhra Pradesh, Madhya pradesh and other states. Various factors which inhibit the growth of plants in these areas are: (a) high salt concentration, (b) poor moisture availability, (c) poor permeability, (d) presence of hard kankar pan, (e) high pH, (f) nutritional disorders and (g) toxic effect of sodium and other salts.

Several factors contribute to the development of salinity and alkalinity in soils. The important factors are: (a) arid and semi-arid climate, (b) impervious hard subsoil due to clay or kankar pan, (c) basin-shaped topography, (d) high water-table, (e) impeded drainage, (f) salt-bearing substrata, (g) excessive canal irrigation, (h) use of saline and brackish water for irrigation and (i) flooding by sea water. The poor physical condition of the soil, caused by high osmotic pressure due to an excessive concentration of soluble salts, is the chief inhibiting factor for plant growth. Decreased availability of soil moisture and air caused by impermeability and dispersed condition of the soil, toxicity of sodium salts and poor availability of certain essential nutrients further deteriorate the unfavourable conditions for plant growth.

From the management point of view, these soils can be classified into two categories; namely (a) sodic or alkali soils and (b) saline soils. Sodic soils are characterised by high exchangeable Na (more than 15%) and pH more than 7.5 and usually exceeding 10.0. The dominant soluble salts are carbonates of sodium. An excessive amount of exchangeable sodium disturbs various physical, chemical and biological properties of soil. These are usually associated with a hard clay layer or kankar pan in lower depths of the soil. Saline soils are distinguished from sodic soils by the dominance of neutral salts, such as chlorides and sulphates. These soils have a low proportion of exchangeable Na (less than 15%) and pH usually below 8.5. These soils are injurious to plant growth mainly because of higher salt concentration.

Good soil working is very important in afforestation of saline alkali soils. The requirements of good soil working are: (a) production of loose soil suitable for rapid root development, (b) breaking the kankar pan or clay pan for deeper penetration of roots, (c) production of soil mass which may facilitate leaching of soluble salts, retention of soil moisture, draining excess water, (d) replacement of exchangeable sodium by addition of amendment and (e) maintenance of fertility through addition of manures and fertilisers. Several methods of soil working, e.g., pits and trenches of various sizes are used in different states. It is reported that the trench-ridge method of soil working is better than pits (Yadav and Singh, 1970; Ghosh, 1977). It has been observed that in the pit method, salts accumulate in the pit and their higher concentration results in death of the planted seedlings. Pits or trenches of 45 cm or 60 cm deep have been found to be better. The addition of 3.5 kg of gypsum and 2.5 kg of farmyard manure has been found helpful in improving the soil conditions and was reflected in better survival and growth of several species (Ghosh, 1977; Yadav, 1980). The addition of fertilisers helps the growth of seedlings. Digging of deeper pits up to 90 to 120 cm, breaking the kankar pan and replacement of soil have been found to give better results (Ghosh, 1977; Yadav, 1980). Some soil-working methods are shown in Fig. 13.

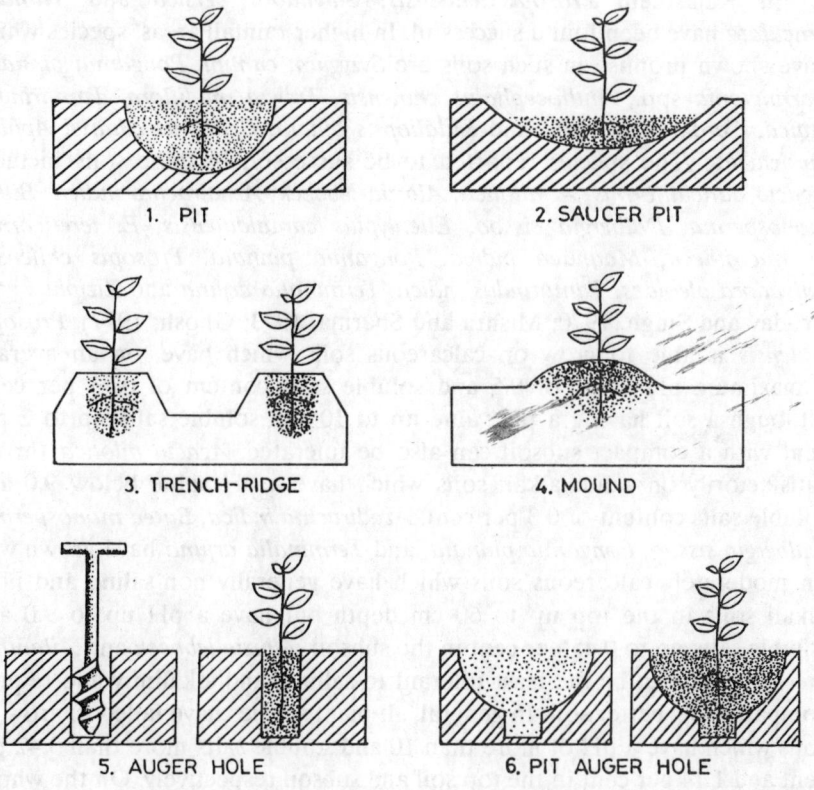

Fig. 13: Methods of soil working in saline-alkaline soils

More recently, tractor-mounted augers have been used for breaking the kankar pan and digging deeper pits. Usually 120 cm to 150 cm deep and 25 cm diameter pits are dug and refilled by the amended soil before planting the seedling. It was found that mixing of 2 to 3 kg of gypsum and 7 to 8 kg of farmyard manure with alkali soils used for refilling gave better results (ICAR, 1985; Chhabra and Kumar, 1989).

Species with a tolerance for high salts, having soil-binding characteristics and fast growing should be selected for the afforestation of saline/alkali soils. The species used in these soils will also depend on the climate of the area. Pandey (1961) reported that *Acacia nilotica*, *A. catechu*, *Albizia procera*, *Azadirachta indica*, *Butea monosperma*, *Crataeva religiosa*, *Dalbergia sissoo*, *Madhuca latifolia*, *Prosopis chilensis*, *Tamarindus indica* and *Terminalia bellerica* were tried in usar land near Lucknow by planting these species in pits of 0.6 × 0.6 m and 0.6 × 1.2 m, filled with non-usar soil; of these species, *Acacia nilotica*, *Prosopis chilensis*, *Azadirachta indica* and *Albizia procera* showed comparatively better results.

In Rajasthan, *Prosopis chilensis*, *Salvadora persica* and *Tamarix articulata* have been found successful. In higher rainfall areas, species which have shown promise in such soils are *Syzygium cumini*, *Pongamia pinnata*, *Barringtonia* spp., *Anthocephalus chinensis*, *Trewia mudiflora*, *Tamarindus indica*, *Cordia dichotoma* and *Eulaliopsis binata*, *Cenchrus ciliaris*, *Apluda mutica*, etc. The species reported to be successful in these soils include: *Acacia auriculiformis*, *A. nilotica*, *Albizia lebbeck*, *Azadirachta indica*, *Butea monosperma*, *Dalbergia sissoo*, *Eucalyptus camaldulensis*, *E. tereticornis*, *E. microtheca*, *Madhuca indica*, *Pongamia pinnata*, *Prosopis chilensis*, *Salvadora oleoides*, *Tamarindus indica*, *Terminalia arjuna* and *Ziziphus* spp. (Yadav and Singh, 1970; Mishra and Sharma, 1973; Ghosh; 1977). *Prosopis chilensis* is able to grow on calcareous soils which have, on an average a maximum pH value of 9.5 and soluble salts content of 0.53 per cent, although a soil having a pH value up to 10 and soluble salts up to 2 per cent with a compact subsoil can also be tolerated. *Acacia nilotica* thrives satisfactorily on saline/alkali soils which have a pH value below 9.0 and soluble salts content of 0.3 per cent. *Azadirachta indica*, *Butea monosperma*, *Dalbergia sissoo*, *Pongamia pinnata*, and *Terminalia arjuna* have grown well on moderately calcareous soils which have generally non-saline and non-alkali soils in the top up to 60 cm depth but have a pH up to 9.0 and soluble salts up to 0.45 per cent in the subsoil. *Albizia lebbeck* and *Ailanthus excelsa* appear to be the least tolerant to salinity and alkalinity and require comparatively better soil. In general, all plant species have failed to grow on soils which have a pH of more than 10 and soluble salts more than 3.42 per cent and 1.14 per cent in the top soil and subsoil respectively. On the whole, various species can be graded in descending order of their tolerance of alkalinity as *P. chilensis*, *A. nilotica*, *Butea monosperma*, *Azadirachta indica*, *Dalbergia sissoo*, *Pongamia pinnata*, *Terminalia arjuna*, *Albizia lebbeck*, and *Ailanthus excelsa*. Several grasses, e.g., Karnal grass (*Diplachue fusca*),

*Cynodon dactylon, Chloris gayana, Brachia mutica*, etc., could also be grown (Chhabra and Kumar, 1989).

## 3. RAVINE LANDS

Ravines are a network of deep gullies. The whole land mass is cut by a number of deep gullies. These lands occupy about 4 million ha mainly on the banks of the rivers Yamuna, Chambal, Mahi and Betwa, which lie in the States of Uttar Pradesh, Rajasthan, Gujarat, and Madhya Pradesh. The soil is generally calcareous and compact with wide variations in texture. The average rainfall in ravine areas is about 750 mm to 900 mm, mostly during the monsoon. The summer temperature may go up to 47°C during May and June and the winter temperature may touch the freezing point. Severe winter frosts may also occur.

The ravines are characterised by the absence of vegetation of any type. The run-off is maximum, slopes are steep and gullies are of varying depth from a few metres to 100 m or more. Ravine afforestation has to be carried out on a catchment basis with other operations of soil and water conservation. In order to achieve better results, the afforestation of ravines may be accompanied by proper management of agricultural lands.

It has been observed that the prevention of erosion by easing of the slopes of gullies and by diverting the surface flow is an essential prerequisite for any successful afforestation programme. The gullies have to be plugged with suitable mechanical measures to conserve soil and moisture. The slopy areas of the ravines have to be clothed with grasses, trees, etc., to avoid further deterioration.

Various soil working and moisture conservation measures, such as contour furrowing, contour ridging, contour trenching, check damming, gully plugging, etc. have been found useful in checking soil erosion. Contour furrows 15—25 cm deep are dug at regular intervals of 2 to 3 m along the contour. A contour ridge is like a contour bund and is made to check the flow of water and impound it so that it is soaked in the soil. Staggered contour trenches of 3 m × 45 cm × 30 cm are prescribed at suitable spacing, depending on the degree of slope, soil type, etc. Check dams are useful to store water and improve water regime in the area. Plugging of new gullies is essential. Ringing the gully head by a continuous trench to divert run-off water from the active gully is important in preventing its extension. In the case of a comparatively easier slope, pits of 30 cm³ or 45 cm³ have given good results.

The species which have been grown successfully in India are *Eucalyptus* spp., *Dalbergia sissoo, Albizia lebbeck, Pongamia glabra, Acacia catechu, Dendrocalamus strictus, Acacia nilotica, Ailanthus excelsa, Prosopis chilensis,* etc. Afforestation by nursery raised seedlings is the most successful method but it needs more time, expenditure, great care and skill. Direct seeding is also recommended as it is a quicker and labour-saving method of af-forestation. Experience in ravine afforestation in Uttar Pradesh, Gujarat and Madhya Pradesh has shown that the selection of species may be done as follows (Sharma, 1973; Ghosh, 1977): (i) Area abutting: cultivated fields and flat tops with loamy soils—*Dalbergia sissoo, Albizia lebbeck, Prosopis chilensis*; (ii) Eroded domes with kankar accumulation: *Prosopis chilensis*; (iii) Upper eroded slopes: *Prosopis chilensis, Albizia lebbeck, Acacia nilotica*, etc.; (iv) Valley bottoms: *Dalbergia sissoo, Albizia lebbeck, Dendrocalamus strictus*, etc.

Deep planting is found to be better in several species (Singh, 1972). Protecting the area from grazing by providing effective fencing is absolutely essential. Verma *et al.* (1969) reported that grasses such as *Apluda mutica, Aristida royleana, Atylosia scaraboides, Cenchrus ciliaris, Dichanthium annulatum, Eulaliopsis binata, Eremopogon feveolatus, Heteropogon contortus, Panicum antidotale*, etc. have been found successful and economical in stabilising peripheral and contour bunds and bench terrace faces in gullied land. Afforestation trials carried out in ravine lands of Gujarat indicated that *Dendrocalamus strictus, Gmelina arborea, Eucalyptus camaldulensis*, etc. are suited for planting in gully beds, *Dalbergia sissoo* for gully slopes and *Acacia catechu, A. benthamii, A. tortilis* and *A. nilotica* for gully humps and marginal lands (Singh, 1972).

*Aerial seeding*

Aerial seeding of forests is a recognised technique of forest regeneration in Australia, New Zealand, Canada, the USA, etc. (Riley, 1980). Aerial seeding has been tried in India in the States of Maharashtra, Rajashthan, Uttar Pradesh and Madhya Pradesh for ravine afforestation. In Madhya Pradesh and Uttar Pradesh, approximately 50,000 ha ravine area was aerially seeded between 1980 to 1986. In Madhya Pradesh, about 40,000 ha were treated through aerial seeding during this period. Seeds of *Prosopis chilensis, Acacia nilotica, A. catechu, Dendrocalamus strictus, Albizia lebbeck* and *Dalbergia sissoo*, etc., at the rate of 20–40 kg/ha were sown using Beaver and Dakota aircrafts. Preliminary results indicate that in aerial seeding in Chambal and Yamuna ravines only *P. chilensis* and *A. nilotica* are partially successful with only 200–300 seedlings and other species, e.g., *Dalbergia sissoo, Holoptelea integrifolia*, and *Ailanthus excelsa* are almost a complete failure (Sharma, 1984; Chaturvedi, 1986; Ram Prasad, 1988). The seedlings have yet to establish and more deaths are expected. In aerial seeding, the desired results have not been achieved because seeds are either washed away or fall at places where they cannot strike root in the ground. For better results, proper seedbed and soil conservation measures appear necessary.

4. COLD DESERT

A large area of Leh and Kargil in Jammu and Kashmir and some area of Spiti valley in Himachal Pradesh fall in this category. The dryness is of two types here, one caused by the low temperature (below 0°C) inhibiting the absorption of water by plant roots and the other caused by the dryness of the atmosphere. The total precipitation usually does not exceed 400 mm. The soils vary from sand and sandy loam to loams in these regions. They are neutral or slightly alkaline in reaction.

The land is bereft of all vegetation except for isolated, scattered and overgrazed herbaceous shrubs. The growing season is limited to 3 to 4 months falling during the hot period. Therefore, any attempt at growing trees would depend on the availability of moisture. The growth of plants has to be slow because of the short growing season. Wind erosion is also common. Soil working is done during March to April followed by planting at April end. Soil working will depend upon the type of land. Kaul (1965) reported that two systems of soil working are common in this tract of Jammu and Kashmir. These are: (i) trench-cum-pit type and (ii) irrigation-cum-drainage. (i) *Trench-cum-pit type*: This technique is being extensively used in gently sloping areas with a stony surface. It consists of digging staggered contour trenches spaced 10 m × 10 m, each being 4.3 m long, 0.60 m wide at the bottom and 1.0 wide at the top. In the interspaces, pits 60 cm long, 60 cm wide and 100 cm deep, spaced at 2 m × 2 m are dug. Both trenches and pits are filled flush to the ground surface with a 50:50 mixture of dug soil and fine textured silt transported from nearby nala beds. (ii) *Irrigation-cum-drainage type*: This type of soil working is largely practised in low-lying areas, namely river banks and marshy lands with a high salt concentration. In order to facilitate leaching of salts from the root zone, planting is done in 45 cm–50 cm deep crowbar holes. Mounds 45 cm high are made around the plants by piling the excavated soil from two irrigation-cum-drainage ditches. Each ditch is 100 cm wide on the top, 60 cm wide at the bottom and 45 cm deep and each is placed 12.5 cm away on either side of the planted row. The depth and width of ditches are regularly maintained and excess water drained from these ditches into a permanent depression or through an artificially constructed channel.

In areas having slopes and river banks, species such as *Salix alba, S. fragilis, Populus alba* and *P. ciliata* are successful. In marshy areas with strong alkalinity, pH up to 9.5, pure *Salix* plantations are successful. Kaul (1965) reported that extensive field trials may be necessary for some species of *Populus* and *Salix*. Some species, e.g., *P. trichocarpa, P. tremuloides, P. tremula, P. alba, P. nigra, P. ciliata, P. euphratica, P. candicuas*, etc. can be cultivated. Among willows, *Salix fragilis, S. flabellaris, S. pycnostachya* and *S. angustifolia* are under cultivation. *S. hastata, S. wallichiana, S. daphnoides*, etc. grow naturally. Other species, e.g., *Juniperus wallichiana, J. communis*, etc., can also be grown. Planting is done by cuttings of *Salix* and *Populus*. Three sizes of cuttings are used: (i) Small size stem cutting, 0.5 to 0.6 m long and 3.5 to 7.5 cm in girth; (ii) standard size cuttings, 2 to 2.5 m in length and 3.5 to 7.5 cm in girth and (iii) medium size cutting, 1.3 m to 2 m long and 6.25 cm to 7.5 cm in girth. Rooted stem cuttings are better. Pollards do not develop adequate root system. Planting is done at the spacing of 2 m × 3 m or 3 m × 3 m.

## 5. LATERITE AND LATERITIC SOILS

Laterite and lateritic soils are characterised by the complete decomposition of parent rock, leaching of silicon from the soil surface, accumulation of sesquioxides of iron and aluminium and absence of humus. The soil is distinctively reddish because of the presence of sesquioxides of iron. The soil is poor in nutrients and the growth of vegetation is also slow.

Laterite and lateritic soils cover an area of 12 million ha in the states of eastern, central and southern regions of India. These soils are found in plateaus, high ridges, and plains in both high and low rainfall zones. The low fertility status, hard vesicular structure and deficient moisture supply render many of these soils unsuitable for growth of tree species (Shrivastava and Qureshi, 1966). Laterite and lateritic soils occur in arid, semi-arid and humid regions. The annual precipitation may range between 750 mm to 3750 mm. Due to poor vegetative growth, the rain-water moves very fast. The soil has a very poor water-holding capacity.

In preparation of the site, special care has to be taken for conservation of moisture. Of the various soil moisture conservation techniques tried in such areas, saucer-pit and rectangular ridge methods on the plateau area of Andhra Pradesh and continuous contour trenches in West Bengal have been found to give better results. Transplanting of seedlings has been more successful than sowing of seeds. Prasad (1965) reported that contour trenching and contour bunding are practised in Bihar. Contour bunding is done to minimise the soil waste and to conserve water. Planting is done in pits. The soil being poor in nutrients, the addition of inorganic fertilisers and manures has given better results. A fertiliser mixture of 2.0 oz of single superphosphate and 2.0 oz of ammonium sulphate added to the soil in the pit before it is filled has helped the growth of planted seedlings, especially of *Denarocalamus* spp. and *Eucalyptus* spp.

Ghosh (1961) reported that teak (*Tectona grandis*) introduced in West Bengal in the laterite area has shown encouraging results. Teak is planted in 'thalis'. Soil working consists of preparation of box-shaped trenches 0.45 cm × 0.45 cm at the interval of about 10 m roughly along the contour. Trenches are interrupted by retaining uncut wedges, 8 m apart, as this prevents lateral movement of water and also allows uniform accumulation of water along the entire length of the trench. A ridge is built up with the earth excavated from the trench along the lower side, leaving a berm of about 15 cm from the edge of the trench. The ridges are also interrupted with 'thalis' 0.6 m at the top, 0.45 cm at the bottom and 45 cm deep and a crescent-shaped bund is made at the lower side. A short contour trench of 0.6 m × 0.3 m is made on the upper side of the *thalis*.

Among many species, *Acacia auriculiformis, Xylia xylocarpa, Eucalyptus tereticornis, Grevillea robusta, Anacardium occidentale*, etc., have given good, results in West Bengal. *Anacardium occidentale* grown in 0.3 m × 0.3 m × 0.3 m pits has done well in Orissa and Andhra Pradesh. Banerjee (1972) reported that two species of *Agave* namely *Agave sisalana* and *A. cantala* were tried in lateritic areas of West Bengal and were found successful. *Agave cantala* gave a better performance than *A. sisalana*. Other species which are successful in these areas are *A. nilotica, Dalbergia sissoo, A. catechu, Ailanthus excelsa, Albizia lebbeck, Azadirachta indica, Cassia fistula, Hardwickia binata*, etc. In Madhya Pradesh, several species, e.g., *Emblica officinalis, Acacia auriculiformis, Albizia* spp., *Dendrocalamus strictus* and *Dalbergia sissoo* have been the most successful (Singh *et al.*, 1983).

6. COASTAL SANDS

Coastal lands occupy an area of about 8.4 million ha in India in the form of a narrow strip along the eastern and western coasts of the country. The soil is generally alkaline in reaction and the salt content is high. The soil is poor in nutrients and low in water-holding capacity. The water-table is generally high and, at places, the condition is waterlogged. The texture of the soil varies; mostly it is sandy but at places it is clayey. Sea water causes a high concentration of sodium salts. The soil in the Rann of Kutch is stiff and clayey (clay more than 60%). Lime deposition is also common at many places. The average annual rainfall varies from 500 mm to 3500 mm; the lowest occurs along the west coast of Gujarat. The humidity is generally high.

The urgency of afforestation of coastal areas has been felt due to the continuous occurrence of cyclones. The occurrence of cyclones has become almost a constant feature. The role of a forest as a moderator of the effect of a cyclone is well known. It has been suggested that a one-to two-kilometre wide forest belt is necessary for moderating the effects of cyclones all along the coastal areas of the country.

Shrivastava and Qureshi (1966) have reported that in areas with high salinity, it is necessary to leach the salts and plant the trees on ridges while in other areas planting in sunken pits is likely to give better results, as it would ensure protection from wind and salt spray in early stages of establishment. Plantations of *Casuarina equisetifolia* have been raised successfully on a large scale in coastal sandy soils. Where the water-table is high, plantations are done on raised mounds and where it is low, in pits. In the first two years, irrigation is necessary to save the plants from physiological drought *Cocos nucifera* and *Borassus flabellifer* have also been successfully planted in certain areas of Orissa.

In coastal areas, salinity is another problem. Organic matter is practically absent. The needles of *Casuarina equisetifolia* take a long period for decomposition. Other species which have been successful in Orissa and Gujarat are *Pongamia glabra, Nipa fruiticans, Avicennia officinalis*, etc. Planting of *Prosopis chilensis* on ridges about 3 feet high has given successful results where the soil is stiff clayey loam and highly saline. Other species tried in coastal belts are *Casuarina jhunjhuniana, Acacia auriculiformis, Eucalyptus tereticornis*, etc. In Gujarat, live hedges of *Agave* and *Aloe* on the boundary and *Ipomoea biloba* as a sand cover in plantations have been successful. Iyengar (1988) reported that Gujarat coastal areas present mainly two problems, namely, saline soils and dune sand. Several important species, e.g., *Salvadora persica, S. oleoides, Atriplex nummularia, Juncus regidus*, etc., have shown promise in saline areas and *Atriplex nummularia, Simmondsia chinensis* (Jojoba), *Simaruba* spp. and *Prosopis* spp. in sand dune areas.

## 7. DRY AND ROCKY AREAS

Dry, rocky and murramy soils are problematic areas and are generally termed skeletal soils. The total area occupied by such soils may be around 3 million ha in the country. These soils are very shallow, coarse, poor, eroded and degraded. Due to shallow soils, the vegetation is xerophytic in nature. These soils occur both in low and high rainfall areas. In most of the regions, rainfall varies from 650–1150 mm, largely during June to September. The temperature ranges from 21°C to 45°C.

Seth (1960) has suggested that soil working in these areas should aim at conserving soil and moisture and increasing soil depth. The soil-working technique may differ from area to area. In general, the soil being hard, the surface is to be pulverised by one method or another. Contour trenches or staggered trenches are made in order to conserve the maximum amount of moisture. In low rainfall areas, planting is done in furrows as in Maharashtra. In rocky areas, where soil materials are completely absent, digging of deep pits and filling them with imported soil brought from outside and planting of a seedling in each pit has been successful in the Bhopal hills in Madhya Pradesh and the Chaumandi hills in Karnataka. In some rocky areas, 100-cm deep pits have been dug to ensure the success of plantation. Where digging of pits is not possible, the pits are made by blasting the rocks. Blasting, though expensive, is the only answer in the case of rocky areas.

The selection of species depends on the rainfall and other climatic factors of the area. In low to medium rainfall areas, *Eucalyptus tereticornis, Dendrocalamus strictus, Melia azedarach, Albizia lebbeck, Acacia catechu,*

*Ailanthus excelsa, Hardwickia binata, Cassia siamea, Prosopis chilensis, Dalbergia sissoo,* etc., have been successful. Addition of fertilisers in small doses has given good response in the plantations of high rainfall areas of such species as *Eucalyptus globulus* and *Acacia mearnsii.* In higher elevations, up to about 2000 m, *Robinia pseudacacia, Populus ciliata, Pinus roxburghii* etc., have been successful. On rocky hill slopes at an elevation of 2300 m in Kulu, *Cupressus torulosa, Cupressus arizonica* and *Pinus ponderosa* have given good results and *Cedrus deodara* succeeded on better soils.

## 8. WET LANDS

Wetlands may be defined as lands where soil is badly or imperfectly drained. These soils are generally found in high rainfall areas. These wet soils may be classified into the following categories: (a) waterlogged areas caused by stagnant or impeded water, (b) waterlogged areas caused by impeded but mobile water, (c) swampy areas and (d) marshy areas and saline muddy areas.

These waterlogging conditions are produced because of many factors. Large areas in the Punjab and Uttar Pradesh have become waterlogged because of canal irrigation. Littoral salty marshes and swamps are confined to the coastal areas mainly in the estuaries of rivers, in creeks and lagoons and along the lower reaches of tidal streams. Certain areas in high rainfall zones where drainage is poor, become waterlogged generally in depressions and along river and nala beds. Such soils present a serious problem for afforestation owing to waterlogging, physiological drought, defective aeration, excessive salt concentration and poor nutrient availability.

The total area of all kinds of wetlands in the country is about 20 million ha. The areas occur under a great variation in annual rainfall. The operations which help in achieving successful afforestation are: (i) affecting drainage to remove excess water, (ii) reducing salinity and (iii) planting with suitable species.

Many unproductive waterlogged areas have supported good forest stands after affecting drainage in countries such as Finland, Sweden, the Netherlands, Germany and the United Kingdom. Lines and Neustein (1966) reported that improving the drainage facilities by deepening the drains increased the growth of sitka spruce by about 20 per cent in southern Scotland. Excess saline patches may be removed by proper soil working. The ridge and furrow method and open trenches of soil working have given good results in India (Shrivastava and Qureshi, 1966). The most common practice of soil working in Britain is to plough for temporary drainage and provision of planting turfs. The furrows are then traversed by a deep drainage plough (pulled by crawler tractors). The depth of cross-drains varies from 45—90 cm depending on the peat depth. The spacing varies from 20 to 50 m depending on the site and alignment varies with the catchment of the area and soil type. Tall planting has given good results. Jack (1966) reported that 3-year-old seedlings of sitka spruce have given better results than 2-year-old seedlings when planted in the blanket-bog area. Addition of chemical fertilisers increases the survival per cent.

In Uganda and East Africa, swamps have been afforested with *Eucalyptus robusta* in the wettest parts (where it is able to withstand submersion for as long as a fortnight during the rains) and with *Eucalyptus saligna* in the drained swamps in replacement of *Eucalyptus robusta* after the first fuel rotation of 8 to 10 years. In waterlogged areas of India, some species which have yielded good results are *Eucalyptus robusta*, *Syzygium cumuni*, *Salix* spp., *Populus nigra*, *Terminalia arjuna* and *Acacia nilotica*. In marshy areas species such as *Baringtonia* spp., *Bischofia javanica*, *Eucalyptus robusta*, *Eucalyptus rudis*, *Lagerstroemia speciosa*, *Casuarina equisetifolia* have been successful. *Diospyros embryopteris*, *Pterospermum acerifolium*, *Bischofia javanica* and canes are generally grown in fresh-water swamps. *Arundo donax*, *Sapium sebiferum*, etc. have been successful in swampy areas in many parts of the country.

## 9. DENUDED AND ERODED HILL SLOPES

Denuded and eroded hill slopes are generally found in subtropical and temperate regions of the country. The southern aspect in the Himalayas is generally denuded. These denuded hill slopes are heavily grazed and frequently burnt and are thus liable to serious erosion. Denuded hill slopes are more common in the western Himalayas than in the eastern Himalayas because the latter is more moist. In southern India, denuded hill slopes are generally covered with dense grass and, therefore, the destructive effects of erosion are less obvious. The denuded and eroded hill slopes are less common in tropical areas except in dry areas. Constant severe biotic pressure on forest lands has rendered a large forest area in several states, e.g., Jammu and Kashmir, Himachal Pradesh, Uttar Pradesh, Madhya Pradesh, etc., completely barren and denuded.

The hills have been denuded by unrestricted fellings associated with excessive grazing and frequent fires. In some areas due to growth of grasses, the erosion has been less damaging. But in most areas the surface soil has been eroded. In many areas, even subsoil has disappeared, leaving practically no soil material. The soil is generally poor in moisture and nutrients. These adverse conditions make the afforestation operation difficult.

When soil is present in the area, contour trenching may be done as it helps in soil and water conservation. When the slopes are steeper, digging of trenches may not be possible and, in such cases, preparation of pits for planting or preparation of 'thalis' for sowing may be adopted. Maintenance of regular spacing may not be possible in many areas, particularly in steeper slopes and badly eroded sites. In such areas, sowing or planting is generally carried out on the best available locations.

Before starting afforestation, the area should be closed effectively for grazing. If possible, the area should be fenced. Planting may be done in contour trenches which are generally 0.50 cm x 0.50 cm in cross-section. In steeper slopes, pits of 30 cm x 30 cm x 30 cm or 45 cm x 45 cm x 45 cm are dug and sowing or planting is done in them.

Only very hardy species are likely to succeed on exposed and eroded areas. Species which are propagated by vegetative parts, e.g., stem cuttings, are also likely to succeed. *Pinus roxburghii* and *Pinus wallichiana*, which are quite hardy, have been successfully used for afforestation of these areas in subtropical and temperate zones. The species tried in afforestation of denuded and eroded hills in temperate regions in different parts of the country with their common method of stocking, are given below (FAO, 1959; Ghosh, 1977; Tiwari, 1983):

(i) Direct sowing —*Pinus roxburghii, Pinus wallichiana, Cedrus deodara, Acacia dealbata, Ailanthus altissima, Cupressus torulosa, C. arizonica,* etc.

(ii) Transplants and —*Cedrus deodara, Cupressus toruloss, Morus* spp.
potted plants  *Pinus roxburghii, P. wallichiana, Morus serrata, Juglans regia, Robinia pseudacacia, Acacia dealbata,* etc.

(iii) Branch cuttings —*Acer negundo, Salix* spp., *Plantanus orientalis, Populus* spp.

(iv) Root suckers —*Ailanthus altissima*

In the dry areas of north-west India, species such as *Acacia modesta, A. catechu, Prosopis chilensis* etc., have been successful. In the peninsular region, *Eucalyptus tereticornis, Anacardium occidentale, Acacia auriculiformis, Cassia siamea, Albizia lebbeck,* etc., have been successful.

In subtropical and temperate areas of the Himalayas, tree planting has not been very successful. Efforts are being made to try multipurpose shrubs and small trees in greening of the Himalayas. Some of the multipurpose shrubs on which work has been taken up include *Grewia optiva, Debregeacia hypoleuca, Ficus roxburghii, Punica granatum, Desmodium* spp., *Indigofera* spp., *Coriaria nepalensis, Cotoneaster* spp., etc. (FRI, 1989; 1990).

## 10. Land Slips and Screes

Landslips are a common occurrence in the hills. Screes are generally found in dry and cold montane places and rarely form a good area for afforestation. The stabilisation of landslips and screes and prevention of their further extension is an important operation which the agroforester sometimes has to do. The area under land screes is not much but the extent of damage which they may cause is comparatively large. Landslips may cause a barrier on the road and sometimes may even destroy a road.

Landslips generally provide good sites with fresh mineral soil, moist but properly drained. Screes are composed of unweathered rock and are difficult to afforest. Landslips and screes do not provide a stable base for tree growth because the soil is largely exposed and unstable. Proper stabilisation of landslips and screes consists of a combination of engineering works and afforestation. The former deals with the construction of retaining walls, etc. The operation of stabilisation should be started from the top. The head should be rounded off and given an easy slope and a diversion channel should be made all along its upper part to prevent further extension.

The face of the landslip is sometimes paved with masonry but more often it is pegged with fascines made of 1.5-meter long tree posts of, e.g., *Salix*, *Poplar*, etc. These are buried about 0.5 to 0.65 m deep in the soil at 2 to 3.5 m apart, with wooden railings and packed behind with twigs and branches to hold up the loose material. Sometimes branch cuttings of *Populus* spp., *Salix* spp., *Ailanthus altissima*, *Alnus* spp. etc. are planted as they provide a temporary barrier and strike root quickly and stabilise the slip eventually.

As soon as the unstable places are fixed, these are planted with grasses and shrubs capable of growing quickly on the exposed areas. The grasses which form a massive root system may be useful. Planting of napier grass, *Saccharum* spp., etc., may be done. Sowing and planting of species, e.g., *Pinus roxburghii*, *P. wallichiana*, *Cupressus torulosa*, *Acacia dealbata*, *Rubinia pseudacacia*, *Populus* spp., *Salix* spp. may be undertaken immediately. No soil working is required. The cuttings and transplants may be planted in holes. The local shrubs, e.g., *Cotoneaster microphylla*, *Coriaria nepalensis*, *Viburnum cotinifolium*, *Berberis* spp., *Indigofera heterantha*, *Desmodium tiliaefolium*, *Lantana camara*, *Dodonea viscosa*, *Agave* spp., etc., may be used for stabilisation of landslips. Effective protection against grazing, browsing and fire is essential for success. As far as possible, the area must be fenced and the entry of animals must be strictly prohibited.

## 11. Shallow Black Cotton Soils

The total area under shallow black cotton soils is estimated to be about 5.8 million ha in the country, mostly in the Deccan plateau. These soils are generally heavy clays which are calcareous, rich in bases, high in water-holding capacity and poor in permeability. These soils form deep and wide cracks during summer on the evaporation of soil moisture. The formation of cracks exposes the roots of the plants and interferes with the establishment of the seedlings. The black cotton soils in certain areas contain lime nodules and form a hard pan at some depths. The areas in which such soils occur generally receive an annual rainfall of 600 mm to 1000 mm. The temperature variation may be from 7°C (average lowest) during winter to 43°C (average highest) during summer.

Soil working should aim at improving soil structure and permeability by breaking the hard pan through deep ploughing, conservation of soil moisture and improving soil fertility. The soil-working technique prevalent in different areas may differ slightly, depending on the climate of the area. In flat and gently sloping areas, contour furrows are prepared at an interval of 5—6 m. Generally, a 1.25-m wide area is ploughed and a bund is prepared having a base of about 45—60 cm and height of 25—30 cm. On hill slopes, contour trenches of 3 m × 0.6 m × 0.3 m or 0.45 m × 0.45 m × 0.45 m are dug and sowing and planting is generally done in these pits.

Species which have been successful in these areas are *Acacia nilotica, A. catechu, Azadirachta indica, Acacia leucophloea, Tamarindus indica, Prosopis chilensis, Eucalyptus tereticornis, Albizia lebbeck, Madhuca latifolia, Melia azedarach, Gliricidia maculata, Acacia modesta, Pongamia glabra, Cassia siamea, Chloroxylon swietenia, Hardwickia binata, Cleistanthus collinus*, etc. It has also been found that if the soil depth is less than 7 cm, afforestation might be difficult. Such areas may be maintained for grasses only. Effective closure against grazing may be necessary. Many grasses have been found suitable for these areas; a suitable silvipasture system needs to be evolved for these areas.

## 12. MINED AREAS

Open-cast mining for several minerals, e.g., coals, bauxite, limestone, slate, etc. is quite common. A large area of Madhya Pradesh, Bihar and other states are worked under open-cast mining. In this process, various heavy machines are used to excavate the earth. As a result of surface mining, all forms of vegetation are destroyed and the entire soil profile is disturbed. The area left, after surface mining of minerals is full of boulders, lateritic heaps, stones, and other inert substances which give a very desolate look. Afforestation of such mined areas is a challenging task.

Some attempts in afforesting these mined areas have been made in Madhya Pradesh, Bihar, Karnataka and other states (Sharma, 1980; Chaturvedi, 1983a; Ram Prasad and Shukla, 1986). The site conditions are very difficult for the growth of plants. The topography is highly undulating. The surface is usually covered with boulders and lateritic heaps. Organic matter is completely absent. Biotic interference is also high.

Before undertaking plantation, it is necessary to somewhat level the area with the help of dozers as it is easier to make pits in levelled areas. Pits of 60 cm³ are usually dug at a spacing of 2 m × 2 m. The pits should be filled with fertile soil from nearby forests. Addition of farmyard manure at the rate of 0.5 kg/pit has been found useful (Ram Prasad and Shukla, 1986). Addition of forest soil is necessary for mycorrhizal inoculation, required in pines and other broad-leaved species. Container raised seedlings 9 to 12 months old have been found to be good.

Several species have been tried in afforestation of these sites. The species which have shown success are: *Acacia auriculiformis, Dalbergia sissoo, Eucalyptus camaldulensis, Grevillea robusta, G. pteridifolia, Pinus caribaea, Albizia lebbeck, Prosopis chilensis* and *Cassia siamea* (Mathur, 1978; Sharma, 1980; Chaturvedi, 1983a; Ram Prasad and Shukla, 1986). Of these the species which have shown a faster rate of growth are: *Grevillea pteridifolia, Acacia auriculiformis, Pinus caribaea* and *Eucalyptus camaldulensis*.

# CHAPTER 12

# Interaction in Crops

In agroforestry systems, there are a variety of crop plants growing close to each other. These crop plants include large trees, small trees, shrubs, annuals, grasses, etc. Sometimes, animals are also included as a component in the agroforestry systems and their effect on crop plants is also important. The crop plants are grouped into: trees, agriculture, pasture, etc. From an interaction point of view, the following relations are important:

1. Effect of trees on agricultural crops
2. Effect of agricultural crops on trees
3. Effect of animals on trees
4. Effect of trees on animals
5. Other interactions

**Effects of Trees on Agricultural Crops**

The trees are woody perennials having a large size with a large root system, extensive crown and a huge mass. A single tree requires a large area. Agricultural crops are small in size with a smaller root system and crown spread. Some of the important characteristics of these two types of crop plants are given in Table 60.

When trees and agricultural crops are grown together, the following effects of trees on the agricultural crops are important:

1. Competition
2. Microclimate modification
3. Nutrient cycling

1. COMPETITION

Trees and agricultural crops compete with each other for moisture, nutrients, light and space. This competition could be grouped into two main classes: (a) above-ground competition which takes place for light and space and (b) underground competition which takes place for moisture, nutrients and space. The above-ground competition is usually limited in areas of physical interaction. The area which is under physical possession

**Table 60: Characteristics of an individual tree and agricultural crop**

| Characteristics | Tree | Agricultural crop |
|---|---|---|
| 1. Size | Large | Small |
| Height | 10 m to 40 m | less than 1 m |
| Root depth | 200 to 500 cm | 20–50 cm |
| Root spread | 50–150 m$^2$ | 0.25–1.0 m$^2$ |
| Crown size | 50–100 m$^2$ | 0.25–1.0 m$^2$ |
| 2. Mass | 500–5000 kg | 0.25–2.5 kg |
| 3. Life | Long, usually 25–150 years | Short, usually less than one year |
| 4. Density | 250–750 trees/ha | 10,000 to 150,000 plants/ha |
| 5. Utilisation | Wood, sometimes leaves, fruits, flowers, etc. | Grains, sometimes leaves, stem, etc. |

of the stem, root buttress, or clumps etc. is not available to any other plant. The tree crown is responsible for light competition.

### (i) *Light competition*

In considering the potential productivity of multiple cropping systems, especially intercropping systems involving trees, shrubs and agricultural crops, it is essential to be able to estimate the photosynthetically active radiation intercepted by each of the component crops at any given time and to integrate this over the time they occupy space. The productivity of a crop must ultimately depend upon its capacity to utilise solar energy. This capacity in the agricultural crop can be assessed by comparing responses of photosynthetic activity and grain growth of different cultivars to the variation in light intensity.

Low light intensity is one of the important constraints for higher yield. Tanaka *et al.* (1964) reported lower dry matter accumulation and decreased photosynthesis under shaded conditions. Stansel *et al.* (1965) stated that sterility increased, taller plants were produced and yield reduced under low light intensity conditions in rice cultivars. The yields were reduced from primordial initiation onwards under shade (Stansel *et al.*, 1965; CRRI, 1972). Wattal and Asana (1974), working with different varieties of wheat, observed that the tiller number increased with light intensity up to that of full sunlight and grain weight depressed more than grain number by decrease in light intensity. Naik *et al.* (1978) studied the effect of reduced light intensities, i.e., 35, 50 and 70 per cent of normal sunlight on leaf chlorophyll content, photosynthesis and photorespiration of two rice cultivars, i.e., Vijay (shade tolerant) and IR-8 (shade susceptible). The results indicated that Vijay was efficient under low light because of higher chlorophyll contents.

Venkateswarlu and Srinivasan (1978) studied the shading effect (40–50 per cent of natural light) on two rice crops at different phases with two different densities under field conditions. It was observed that the yield losses were maximum at reproductive and ripening phases. Janardhan & Murty (1980) observed that morphological changes, i.e., dry matter, photosynthetic rate, relative growth rate, net assimilation rate and specific leaf weight were reduced whereas height, leaf area, leaf area ratio and relative leaf growth were increased under low light as compared to normal light conditions.

Janardhan and Murty (1980) observed that grain yield of tall and semi-dwarf rice cultivars reduced under low light, due to low grain number per panicle and small grain size. Due to better mobilization capacity, the tall varieties, particularly the late-maturing types, recorded higher yield than the semi-dwarf ones under low light. Rao and Singh (1980) reported that a cut in the photosynthetic surface area either by shading or defoliation at the time of anthesis in *Pennisetum typhoides*, causes reduction in grain number per ear, grain size, rate of grain development and the final grain yield almost in proportion to the cut in leaf area.

Knight (1935) observed a two-third reduction in American cotton yield in Sudan in artificial shade under field conditions. Singh (1986a) reported that reduction in the light intensity decreased the rate of photosynthesis as indicated by reduced net assimilation rate, and leaf area efficiency, but leaf area ratio and specific leaf area increased in the four cotton genotypes. Dry matter of all the genotypes increased significantly with a 35 per cent reduction in light intensity.

*Light environment*: Of all the factors in a forest environment, light is one of the most elusive to measure. Not only does it vary with latitude, topography, weather, season and time of day, but at any given time in a wooded land the quantity and quality of light may vary in spatial distribution. For each additional degree of latitude from the equator, the angle of the sun above the horizontal is reduced by one degree. The intensity of light rays varies with the angle at which the rays meet the earth's surface and with the mass of air penetrated. Solar radiation decreases with the increase in distance from the equator because of the increased mass of air traversed. Some workers have sampled light only at mid-day (Wellner, 1948), others from 11.00 a.m. to 1.00 p.m. (Gatherum, 1961), or at 10.00 a.m. to 2.00 p.m (Ovington and Madgwick, 1955). Often, the method of sampling light throughout the stand is not described and one concludes that time and space variations were confounded.

*Quality of light*: The influence of light quality upon plants differs from one species to another. Atkins (1932) found the quality of light in coniferous forests is affected by the screening action of the foliage whereas beneath deciduous trees, the proportion of red light is higher and that of blue and violet is lower than in full sunlight. Toumey and Korstion

(1947) suggested that the total amount of light beneath a natural forest canopy with numerous openings probably affects the plants on the forest floor more favourably than the same quantity of light beneath a continuous canopy without openings. Nageli (1940) considered sun flecks of negligible importance but diffuse sunlight and skylight to be chiefly responsible for growth beneath a coniferous canopy. The importance of sun flecks on plant growth probably depends on the relationship between size of fleck and size of plant, the intensity of light in fleck and shadow and the duration of time that the plant is bathed in light, from the fleck.

A major advantage will accrue in growing a ground cover crop in addition to a tree crop only where the latter fails to intercept a high proportion of the available light. Any potential advantage will be small where the trees intercept most of the light. Many tree crops are inefficient in interception of radiant energy mainly in two main ways. First, they take many years to attain full canopy and second, this full canopy may still be incomplete for management reasons or as in the case of deciduous trees, be very inefficient in light interception. In the U.K., for example, strawberries are often planted with apples. The apple trees are grown in rows with the minimum distance between the rows being such, as to allow the movement of tractors and machines for spraying. The minimum row spacing is about 3 m, the more usual spacing in England being about 4.5 m. At this spacing, the trees intercept little of the available light in their early years. Using more vigorous trees, planted at 4.3 × 2.9 m, intercepted only 11 per cent of light at full leaf in the second orchard year, while a 5-year-old orchard intercepted only 30 per cent of the available photosynthetic active radiation (Jackson, 1980). The intercrop generally uses a greater part of the available radiation. The dominant factor in the choice of intercrop is then its actual value over a relatively few years. Strawberries often fulfil this role in apple orchards, as do field vegetables in some temperate fruit-growing areas. Coconut, oil palm and young rubber trees in Malaysia are often intercropped in their early years with annual crops, e.g., cassava, soyabean, peanuts, maize, etc., (Blencowe 1969). Upland rice and maize are grown in young mango and coconut plantations in the Philippines (Hardwood and Price, 1976). The intercrops in the open tree canopy situation do not need to be particularly shade tolerant or even to have strategies enabling them to grow at the time or season when the trees have least leaves.

Some tree crops show systematic changes in leaf area index over the years. This is so in an extreme form in deciduous temperate trees which may not attain full canopy until mid-summer and thus may have little or no foliage at times of high incident energy in the early summer and late autumn (Jackson and Landsberg, 1972). The undercrop should be so rapid growing that it can exploit the relatively short period during which radiation penetration through the tree canopy is maximum. Bunting (1976) cites cucurbits, with their very effective production of leaf area per unit of dry matter, as being outstanding performers in this respect.

The role of upper canopy of apple plantations in mixed cropping systems under different environmental systems has been studied in North America and Europe and it has been found that shade depresses the fruit quality in several important apple cultivars (Bassi *et al.*, 1980, Jackson, 1970, 1980). In the relatively high light intensity conditions of British Columbia, Heinick (1966) concluded that the best fruit colour was obtained when the fruits were exposed to more than 70 per cent of full sunlight, and inadequate for sale under 40 per cent. In addition, fruit size and soluble solids content were also reduced. Shade has also been noted to affect adversely fruit bud formation in South Africa, where a degree of shade otherwise benefits the cultivars grown by reducing the risk of sunscald. From such work, it would appear that there is little to be gained by having any apple canopy which is not irradiated to at least 50 per cent of the above canopy level. Citrus is usually grown in much denser plantations than apple. Although vegetative growth is vigorous under shaded conditions, a relatively high light intensity is required for maximum fruiting to the extent that thinning the trees in an overcrowded grove can increase yields by improving the light conditions within the canopy (Monselise, 1951). Where the understorey crop is grown only in the first few years after planting the trees, there is little interference in light interception. But this becomes more important as the trees utilise more of the available energy later. More often, $C_4$ plants have higher photosynthetic rates than $C_3$ plants at high photosynthetic active radiation densities (Hesketh, 1963), so the latter should be less reduced in performance when grown as understorey crops. Leaves of maize do not become light saturated in full sunlight whereas leaves of $C_3$ crops, such as sugarbeet and soyabean, may saturate at about half of full sunlight. However, Sale (1976) observed some reduction in potato yield at a shade level under which light was not considered limiting for photosynthesis. Barua (1968) reported that the reduction of light intensity (60 per cent of full sunshine) appeared to divert a larger fraction of assimilates towards economic yield (leaf) as opposed to total net production in tea plants.

Shade trees were introduced into tea plantations of north-east India towards the end of the first century (Watt and Mann, 1903). The first shade tree to be tried was *Albizia chinensis* and this was reported to be highly beneficial (Buckingham, 1885). Other leguminous trees, such as *A. odoratissima*, *Dalbergia assamica*, *Erythrina indica*, etc. were also tried with almost equal success (Watt and Mann, 1903).

The effect of trees on agricultural crops is not well researched but holm oak in south-west Spain has reduced per ha crop yield of crops by about 20 per cent and of pasture by 9 per cent. There is a loss in agricultural production because of the shade covering the agricultural crops and competition of nutrients and soil moisture. The presence of trees changes the microclimate near the ground level by reducing the wind velocity, intercepting light and heat radiation and also moisture. Reduced wind velocity reduces evapotranspiration from understorey plants. Interception of light radiation during daylight causes reduced photosynthesis and because of lower temperature, reduced evaporation.

The light reaching the agricultural crops is determined by site location (aspect, latitude, climate), time of year, cloud cover and interception by the tree canopy. The tree canopies change naturally with time but may also be changed within limits by pruning and/or thinning in many situations. Plant species differ in their adaptation to reduced light. Some may be shade dependent, others vary in degree of shade tolerance and others are extremely shade intolerant. It has been reported that vegetative yield of *Trifolium subterranecum* in winter was reduced by about 30 per cent with 50 per cent light reduction. In the field, light reduction is intermittent and depends on tree distribution and canopy density. Direct sunlight may be available to many plants for a part of each day.

Information on the productivity of agricultural crops grown under tree plantations in India is very limited and there is a high demand for the products of both agriculture and forestry. The natural forests remain less productive, so that any land for timber or wood production must come from land already in use for other purposes, predominantly agricultural. The potential advantage of growing trees around agricultural fields is mainly due to economic benefits to the farmer. The trend of growing trees around agricultural land has been adopted by the farmers. It would be worthwhile to work out an integrated approach with the help of agricultural and forestry scientists, depending on the suitability of the crops and locations. Therefore, there is a great need to identify the suitable agricultural and horticultural crops, which can grow well along with the tree plantations with limited solar energy available. Some of the crop plants which can grow well under shade include the following:

(i)   Vegetable          —Ginger, turmeric, potato, cucurbit, etc.

(ii)  Agricultural crops —Oats, maize, fodder crops and some varieties
                          of soyabean, beans and groundnut

(iii) Ornamental         —Ferns, orchids, cactus, *Hicotiana, Cineraria,
                          Adiantum, Seleginalla, Phyllocactus, Calisillaria,
                          Cyclamen, Streplocarpus*

(iv)  Grasses            —Several species.

(v)   Shrubs             —Tea, coffee, cotton, cassava, etc.

(ii) *Underground Competition*

Underground root competition for moisture, nutrients and space is relatively more important in agroforestry systems than above-ground crown competition. Under Indian conditions, light is relatively more abundantly available than moisture and nutrients. Before discussing the nature of underground competition, it is necessary to have information on the nature of root development in two types of crop plants.

Development of roots in plants depends upon several factors. These factors include genetic factors and site factors. Root systems vary considerably from one species to another and even from one variety to another in the same species. Several site factors, e.g., the physical and chemical properties of soil in the root zone, are very important. Important soil characteristics affecting the root growth include: soil texture, structure, water content, aeration, soil temperature, pH, pathogens and percentage of injurious salts. The roots of most agricultural crops are limited to the upper 50 cm of the surface soil. Only a few species, such as *Cajanus cajan*, sugar-cane, cotton, soyabean, sorghum, etc. might send their roots deeper than 50 cm. The root system of trees has not been extensively studied. There is a limited information on certain species under specific conditions. These data suggest that development of tree roots varies considerably from species to species and from site to site in the same species. The roots of several species go about 3 to 10 m vertically and extend 10–30 m horizontally.

Fine roots, i.e., those below 1 to 2 mm in diameter, are the main agents of nutrient uptake and constitute the important component of root length. In older trees the proportion of fine roots is about 90 per cent in the total root length. Most agricultural crops have higher root densities in surface soil in comparison to trees. Table 61 shows root abundance in different types of crop plants at different soil depths.

In a ten-year-old plantation of *Eucalyptus globulus* near Rome, the roots reached a depth of 4.2 m and extended a distance of up to 11 m from the trunk (Davidson, 1985). Root studies on *Eucalyptus tereticornis* indicated that the roots may reach up to a depth of 3 m with a lateral spread up to 3.5 m (George, 1977).

Another study has indicated that in 5- and 15-year-old *Eucalyptus camaldulensis* plantations, the roots may reach up to a depth of 3 m with a lateral spread of 9 m and 20 m respectively (Ram Prasad *et al.*, 1984). The lateral spread has been found to be affected by the spacing of the trees. Widely spaced trees possess roots with longer lateral spread.

On the basis of rooting depths and lateral spreads, different species could be grouped as under:

i) Medium—rice, millets, maize, groundnut, tobacco, cereals, banana, etc.
ii) Deep—cotton, gram, sorghum, sugar-cane, grain legumes, etc.
iii) Very deep—fruit trees, forest trees, etc.

It is argued that agricultural crops absorb nutrients from the soil usually from the top layer of a 30-cm soil depth and to some extent up to 50 cm soil

Table 61:  **Root abundance of some crop plants**

| Species | Age (yrs) | Soil depth (cm) | Root abundance cm/cm$^2$ | Source |
|---|---|---|---|---|
| **A.  Trees** | | | | |
| (i)  *Pinus radiata* | 3–4 | 0–10 | 0.13–0.18 | Nambiar (1983) |
|  |  | 10–20 | 0.28–0.34 |  |
|  |  | 40–50 | 0.03 |  |
|  | 14–26 | 0–10 | 2.0 | Bowen (1984) |
|  |  | 25–45 | 0.8 |  |
|  |  | 90–106 | 0.4 |  |
| (ii)  *Eucalyptus marginata* | 2 | 0–10 | 8 | Nambiar (1983) |
|  |  | 50–60 | 0.2 |  |
|  |  | > 60 | 0.005 |  |
| (iii)  *Eucalyptus tereticornis* | 2 | 0–24 | 4464 cm (root length) | Dabral *et al.* (1987) |
| (iv)  *Prosopis cineraria* | 3 | 0–30 | Large | Bhimaya and Kaul (1965) |
|  |  | 30–550 | Small |  |
| **B.  Grasses** | — | 0–15 | 0.50 |  |
| **C.  Cereals** | — | 0–15 | 5–25 | Several workers |
|  |  | 25–50 | 4 |  |

depth. A higher concentration of fine tree roots in the soil layer up to 50 cm suggests that trees also obtain most of the nutrient requirements from the soil layer up to 50 cm. The low concentration of fine roots below 50 cm soil depth suggests that the nutrient absorption from deeper soil layers may be small. The main function of the roots reaching greater depths appears to be water uptake, particularly during periods of water stress (Bowen, 1984). The proportional abundance of fine roots of agricultural crops, grasses and trees suggests that there is keen competition for nutrients between these crop plants when grown in mixture. In almost all species of trees, death and replacement of fine roots is a common phenomenon (Bowen, 1984). In some species, it has been found that the total fine root production in a season is several times more than that of fine roots present at any time. This suggests underestimation of the abundance of fine roots. This phenomenon further explains the likelihood of keen competition between different types of crop plants grown together.

When trees are grown on the bunds of agricultural fields, two important situations are encountered: (i) The adjoining fields are repeatedly ploughed for cultivation of crops which prevents the development of a tree root system in the surface soil. (ii) In the unploughed portion of the bund, it has been, found that the root development is more pronounced in the direction of the bunds. It would appear, therefore, that root development of trees is repeatedly disturbed due to ploughing, particularly in the surface soil.

The roots of *Eucalyptus* hybrid in combination with rain-fed maize and irrigated paddy and *Populus deltoides* in combination with irrigated paddy and sugar-cane were examined by Pin Board, Photographic and Density measurement methods by Singhal *et al.* (1987). It was found that the tree roots were generally far more extensive, more often than not reaching at least 80 cm in depth and 4 m into the field, though with a far less density (0.01 cm/cm³). As for the orientation and vigour of the tree roots, it was found that close to the tree, the roots radiated more in the tangential plane (parallel to the tree row) and only one third in the radial plane (right angle to the tree row). The majority of the roots descended steeply. This was because of the fact that cultivation had destroyed the superficial roots. However, fine roots again grew from the skeletal roots during the period of non-cultivation, but only below a depth of 20 cm. This was due to negative geotropism of fine roots, which is more pronounced in poplar than in eucalypt (Singhal *et al.*, 1987).

There is another aspect of root competition. It concerns the quantum of requirement and the process of absorption of nutrients and water. It is widely believed that the trees require a lesser quantity of nutrients and water for their optimum growth in comparison to annual crops. Table 62 gives a comparison of the nutrient requirements of fast-growing tree species such as *Eucalyptus* with those of agricultural crops (Ghosh *et al.*, 1978).

Table 62: Nutrient requirements in some crops (kg/ha/yr)

| Crops | $P_2O_5$ | $K_2O$ | CaO |
|---|---|---|---|
| *Eucalyptus* spp. | 1.4 | 58.6 | 34.8 |
| Rye (grain + straw) | 29.0 | 57.0 | 15.0 |
| Oats (grain + straw) | 34.0 | 69.0 | 15.0 |
| Wheat (grain + straw) | 30.0 | 53.0 | 15.0 |
| Potatoes (tubers + residue) | 45.0 | 210.0 | 79.0 |

Some trees are said to be great transpirers of moisture. Some trees, particularly the *Eucalyptus*, are accused of copious water consumption. The reliable measure of water consumption efficiency by plants is the weight of biomass produced per unit volume of water consumed. Some guide can be made from the ratio of rates of evapo-transpiration to pan evaporation for different species or vegetation types. *Eucalyptus camaldulensis* is reported to have a very low evapo-transpiration (Davidson, 1985). *Eucalyptus globulus* is reported to transpire about 3475 tonnes of water/ha which corresponds to about 347.5 mm equivalent; this is about 38 per cent of the total rainfall. Studies carried out regarding water consumption by different species indicate that the value of water consumption per unit of biomass produced (litre/gm) is lower in *Eucalyptus tereticornis* (0.48) and *Albizia lebbeck* (0.55) in comparison to *Acacia auriculiformis* (0.72), *Dalbergia sissoo* (0.77) and *Pongamia pinnata* (0.88) though per plant consumption of water is the highest in *Eucalyptus* spp. (Chaturvedi, 1983; Tiwari and Mathur, 1983).

When the trees are grown together with agricultural crops either as intercrop or on bunds, their adverse effects are visible on the agricultural crop in the vicinity of the trees. Such adverse effects are not recorded in shade-demanding crops such as tea, coffee, cocoa, etc. Other agricultural corps such as cereals, oils, pulses, etc. are adversely affected in the vicinity of the trees. The adverse effects on the growth and yield of agricultural crops are due to the competition for light, nutrients and moisture. The nature and quantum of these adverse effects depend upon (i) the age and size of the trees, (ii) nature of the tree species, (iii) nature of the agricultural crop, (iv) availability of water, nutrients, light, etc. The impact of the adverse effects is greatest in the close vicinity of the trees and diminishes as the distance increases (Dhillon *et al.*, 1979, 1984; Sharma, 1987; Dwivedi, 1987). Such effects were observed in different crops with a combination of different tree species. The effects of *Eucalyptus tereticornis*, *Populus deltoides*, *Dalbergia sissoo* and *Acacia nilotica* grown on field bunds were studied under different conditions on wheat, paddy, jowar, potato, etc., and it was found that *Populus deltoides* caused the least damage to the crops of the rabi season; damage by other species was higher (Sharma, 1987; Dwivedi, 1987). The yield in the vicinity of the trees was sometimes reduced to 50 per cent (Dhillon *et al.*, 1984; Sharma, 1987; Dwivedi, 1987). Some species, such as bajra, jowar, etc. have been found better competitors than groundnut, jowar, etc. (Dwivedi, 1987).

Khybri *et al.* (1983) reported that under rain-fed conditions in Dehra Dun, the effect of *Eucalyptus* trees was more on wheat than on paddy. In another study conducted in a 20-year-old plantation of *E. tereticornis* at Kota, it was found that the yields of rain-fed black gram, green gram and sorghum were adversely affected up to a distance of 2 m from the tree line (Shrivastav and Narain, 1980). In the humid climate of Dehra Dun, the effect of an 8-year-old *Eucalyptus* hybrid plantation on a kharif maize crop was observed and it was found that reduction in yield was not conspicuous because of moisture availability. But contrary results were observed on rain-fed maize and wheat (Sharma, 1988).

## 2. CHANGE IN MICROCLIMATE

Trees are reported to change the microclimatic conditions of the area. Trees in agroforestry systems are capable of increasing the precipitation and relative humidity of the area. They reduce air and soil temperature and evaporation from the surface soil. This results in decreased requirements of irrigation water. Some tree species whose crown is small are not effective in checking surface evaporation. In such cases competition for moisture is evident. The trees check wind velocity and provide shelter to the crops. The sheltering effect of trees in an agroforestry system has already been discussed. Trees reduce the impact of rainfall as about 25 to 30 per cent of the rainfall is intercepted by the tree crown, and 10–15 per cent of the rainfall reaches the ground as stem flow. Thus, only about 50 to 60 per cent of the rainfall comes to the ground directly. This reduces the total volume of water available for surface run-off. This leads to decreased loss of soil in form of soil erosion. Thus most microclimatic changes brought about by trees benefit agricultural crops. The extent of microclimatic changes due to trees has not been properly investigated under different conditions. Several variables, such as amount of radiation, temperature, relative humidity and soil water potential have received maximum attention in agroforestry investigations.

## 3. NUTRIENT CYCLING

Nutrient inputs from the atmosphere and rock-weathering are important to the long-term development of soils. Nutrient cycling, on an annual basis forms the major source of nutrients for plant use. This nutrient cycling involves an array of pools, transformation processes and various flows (Fig. 14). In an agricultural ecosystem, the major sources of input include decomposition of agricultural waste and application of manures and fertilisers. The output of the nutrients involves harvest of grass and grain, erosion and leaching. When trees are grown with agricultural crops, nutrient cycling is affected due to nutrient uptake and recycling within the trees and release in the form of litterfall. With their long taproot system, the trees are able to recover the nutrients which are otherwise lost due to leaching in the subsoil. Trees pump them to the surface through absorption and litterfall which decomposes and releases nutrients at the soil surface. The accumulation of decayed leaves and branches increases the organic matter content and improves the physical, chemical and biological properties of the soil. Thus the soil is enriched for better crop production. Intensive cultivation of shallow rooted annual food crops on hilly land has resulted in rapid soil erosion and reduction of farm productivity. This also leads to other effects, such as siltation of rivers and reservoirs, uneven stream flow and pollution of water. These adverse effects have been controlled to a great extent by planting trees for upland stabilisation (Vergara, 1982).

The conservation of soil and nutrients and release of nutrients through litter fall depends upon the agroforestry system, density of trees, species etc. Depending on the number of trees per ha and age, litterfall may vary from 0.5 to 6.5 tonnes/ha/year. Surprisingly very little difference has been observed in litterfall between evergreen and deciduous species (O' Neil and DeAngelis, 1981). On the basis of dry weight, leaf litter contains 0.5 to 1.5 per cent of $N_2$, 0.05 to 0.15 per cent of P, 0.25 to 0.75 per cent of K,

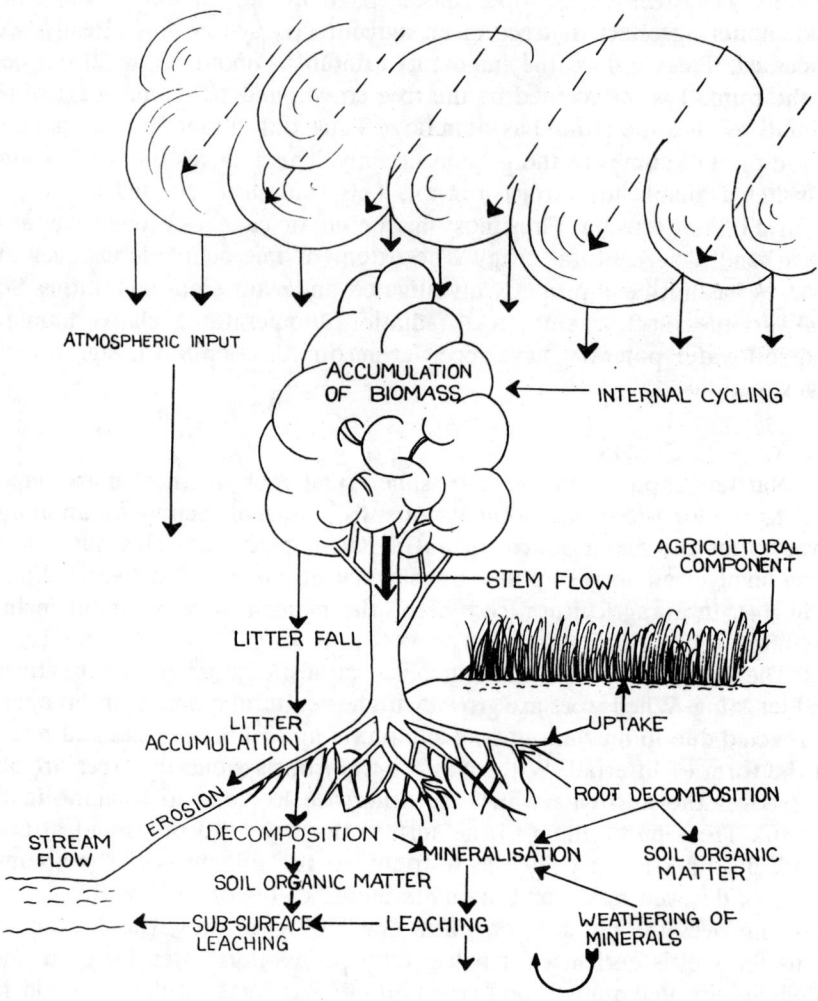

**Fig. 14: Nutrient cycling in agroforestry**

0.25 to 1.00 per cent of Ca and 0.10 to 0.20 per cent of Mg. After decomposition, a large proportion of these becomes available for plant growth. Sometimes the uptake of nutrients is considerably compensated for by the release of nutrients in the form of litterfall (Table 68). Agricultural crops are thereby benefited considerably. However, enough research evidence is still to be collected about the magnitude of nutrient cycling under different agroforestry systems and release of nutrients through litterfall, etc.

Some species of trees, especially nitrogen-fixing legumes, contribute substantially to sustaining an agroforestry system. Several leguminous trees, such as *Leucaena leucocephala*, *Acacia nilotica*, *Dalbergia sissoo*, *Gliricidia* spp., *Sesbania* spp., etc., and some non-legumes, e.g., *Casuarina equisetifolia*, *Alnus* spp., etc., are reported to fix about 50 to 500 kg of nitrogen per ha. Under some systems the leaves of trees could be used as mulch or green manure. This improves the site further.

### Effects of Agricultural Crops on Trees

The importance of trees in various agroforestry systems is basically due to their effect on the annual crops than *vice versa*. The agricultural crops and related cultivation practices do affect the trees in several ways. Some of these effects are as under:

 (i) Competition, especially during early stages
(ii) Effect of cultivation practices

1. COMPETITION
    Agricultural crops compete with the saplings for nutrients, water, light and space, particularly during the first few years. This competition, if intense, adversely affects the growth of tree species. Several agricultural crops, such as sugar-cane, arhar, jowar, maize, etc., overtop the tree species during the first one or two years. Trees interplanted with such crops usually suffer as light availability is considerably reduced. Several tree species interplanted with sugar-cane, maize and jowar in Haryana and western Uttar Pradesh showed considerable mortality; even saplings planted on the bunds suffered (Sharma, 1988). The rate of growth of saplings was also less in comparison to those growing in the open.
    Several agricultural crops, such as beans, cow-pea, etc., climb on the saplings and deform their stem, creating problems in growth. Trees interplanted with agricultural crops receive water, nutrients, etc., on the basis of agricultural crops. The requirements of trees are often neglected. Several tree species have shown a dislike for some agricultural crops having specific requirements. For example, *Populus deltoides* should not be grown with paddy as it dislikes excessive moisture and water inundation, which are liked by the paddy crop.
    When the trees are a few years old and their crown and root systems are well developed, the effect of agricultural crops on the trees is considerably reduced. The trees become more dominant and their effect on the agricultural crops becomes more relevant.

2. EFFECT OF CULTIVATION PRACTICES
    The cultivation practices of agricultural crops also affect the trees. Intensive working of surface soil often destroys the surface root systems of the trees. Since the surface soil receives a good amount of water and nutrients,

the trees having a lower root density in the surface zone, are not able to fully utilise them.

When machines are used for cultivation of agricultural crops, the spacement of trees is often dictated by the space required for the movement of these machines. Use of several machines causes severe disturbances in the growth of trees and sometimes branches require pruning.

### Effects of Animals on Vegetation

In several agroforestry systems, animals are an important constituent. Animals graze and affect the grasslands and fodder trees considerably. The effects of animals include direct effects, e.g., mechanical damage, browsing, reduction in combustible material and indirect effects, such as effect on regeneration and effect on soil properties, which affect tree growth. These effects are very much related to the number of animals allowed to be grazed within the carrying capacity of the land. The quantity of grass and fodder trees available in the area determine the number of animals which could be permitted to graze in that area. The amount of grazing that an area can support under controlled conditions without deterioration is usually called optimum grazing incidence. It is usually expressed in terms of number of animal units per unit area. Very little information is available on the carrying capacity of various categories of land. Most grazing lands in India are under heavy pressure from a large number of animals, resulting in very high grazing incidence. The effects of animals on such conditions are very serious and cause rapid degradation in vegetation and land. Even under normal grazing conditions certain effects of animals on vegetation are quite pronounced. Some of these effects are as under:

1. SEED DISPERSAL

Animals play an important role in seed dispersal. The seeds and fruits may be dispersed by animals, either by adhering to them or by being eaten by them. Sticky hairs on the fruits of *Plumbago* spp. cause them to adhere to the animals which take them to long distances. Hooked spines on the fruits as in *Achyranthas, Tribulus, Medicago, Chrysopogon* etc. become attached to animals and are transported to other places. Succulent fruits, drupes, pomes and berries are eaten by animals and seeds are dispersed through dung at other places. Several species, e.g., *Ficus, Fragaria, Acacia*, etc., are transported by this method.

2. REDUCTION OF COMBUSTIBLE MATERIAL

In several areas, the grasses become dry during summer. There are other inflammable materials such as leaf litter, etc. Fire in such areas causes extensive damage. If grazing is allowed, the grasses are grazed and leaf litter is mixed with mineral soil and this protects grasslands from fire.

## 3. MECHANICAL DAMAGE AND BROWSING

Grazing animals cause damage to young plants due to trampling and browsing. The young plants of several species are browsed. Only a few species, such as *Eucalyptus* spp., *Cassia siamea*, *Gliricidia* spp., *Prosopis chilensis*, etc., are not browsed. Other species are mostly browsed by cattle, sheep and goat. Browsing by sheep, goat and camel is more harmful as these animals not only eat the leafy portion but also the twigs and the tender branches. There are only a limited number of studies on the browsing effects of cattle, goat, sheep and camel. There is a need to work out browsing management practices for different categories of animals for different vegetation types. There are serious methodological problems in carrying out such investigations.

## 4. EFFECT ON SOIL

Several changes in soil properties are brought about due to grazing and browsing. The most important change in physical properties is the soil compaction which results in reduction in pore spaces and soil aeration, infiltration and permeability and disturbance in the soil structure. Hoofed animals disturb the soil which leads to soil erosion. If the land is sloping these problems become very serious. All these factors cause degradation in soil which results in poor vegetation and degradation of pasturelands.

**Effects of Trees on Animals**

Trees affect the animals directly by way of providing leaves and fruits for animal nutrition. Trees also bring about microclimatic changes and improve the carrying capacity of the pasturelands. Some of the effects of trees on animals are as under:

## 1. FOLIAGE CONSUMPTION

The leaves of several trees are good fodder. Some of them have a high nutritive value. The fodder yield from the trees is quite substantive and annual dry matter production may vary from about 10 kg/tree to 40 kg/tree (Singh, 1982).

Tree fodder is available in those periods when agricultural fodder is scarce. The utility of fodder trees has been amply demonstrated during drought years when a large number of cattle and other livestock were saved from starvation. In fact, in the hills, fodder tree species such as *Grewia optiva*, *Celtis australis*, *Quercus* spp., etc., have been a traditional source of fodder for livestock. Lopping of several trees for fodder purposes is still common.

## 2. Fruit Consumption

The fruits of a large number of species are eaten by animals. These fruits are usually richer in nutrients than leaf fodder. Some of these fruits/seeds are used as concentrates. For example, mahua flower and fruits and pods of several trees are very much liked by animals. These fruits sometimes work as fodder substitutes but the contribution from the fruits of trees has never been given more than marginal importance. No work has really been done regarding the contributions of fruits in production of fodder quality and quantity.

## 3. Improvement in Carrying Capacity

Agroforestry systems improve the carrying capacity of the grazing lands in a number of ways. Considerable experimental evidence exists to suggest that windbreaks and shelterbelts in temperate and tropical arid areas improve the pasture growth. Available information regarding silvipasture and pastoral silviculture with the introduction of certain nitrogen-fixing trees suggests a great future potential which can increase the productivity and sustainability of the grazing systems. It is widely recognised that the fine roots of trees which periodically die and decompose in the surface soil, help in reducing the effect of soil compaction as they aid the infiltration rate and soil aeration.

## 4. Microenvironment Modification

Growing of trees in the grazing lands brings about improvement in micro-climatic conditions. Further, scattered trees in a dry tropical pasture provide the animals a much needed place for rest. The utility of such trees in improving the quality of pasture has not been researched. In high altitudes, windbreaks and shelterbelts protect both pasture and animals from strong cold winds.

**Other Effects**

One of the important results of the agroforestry systems is the diversification of the produce. Diversification helps in reducing various kinds of risks. Multiple-use and multiple-product agroforestry systems help farmers to satisfy their various kinds of needs. The degree of mechanisation of various farm operations is sometimes controlled by the agroforestry systems. Physical and socioeconomic life is considerably improved by the agroforestry systems. Windbreaks and shelterbelts and home gardens have demonstrated the effects of trees on the comforts of the households.

Agroforestry systems affect various spheres of activities. The supply of raw material on a sustained basis creates a favourable climate for the industrial development of the region. Several other effects, which agroforestry systems have in the socioeconomic spheres are still to be properly evaluated.

## CHAPTER 13

# Choice of Species

One of the important decisions in agroforestry is the selection of species. Generally, species useful to the growers are preferred. People plant trees for fruit, fodder, fuel, shade and various other purposes. Since a large variety of trees are available which could grow in different edapho-climatic conditions, people can select trees of their choice very easily. The plantation of trees and subsequent maintenance also depend on the agroforestry system. For example, in shifting cultivation, which is the most primitive form of agroforestry, only fruit trees are not cut, while others are cut, dried and burnt to provide manure for agricultural crops. In other agroforestry systems, however, trees are given more importance. They are deliberately introduced into land use. More recently, due to the generation of market opportunities for tree products, greater attention has been paid to the selection of tree species in agroforestry in proper proportion.

In general, the following factors govern the choice of species in agroforestry:

   (i) Site factors
  (ii) Effect on annual crops
 (iii) Purpose of plantation
  (iv) Agroforestry system
   (v) Marketability of produce
  (vi) Effect on site
 (vii) Ease of establishment
(viii) Resistance to attack of insect pests and diseases

**Site Factors**

Site factors include: (i) climate, (ii) soil, (iii) physiography and (iv) biotic factors. In any plantation, the species selected should first be adapted to the site conditions (Anderson, 1950). The climate, the soil and the biotic factors affect the growth and performance of trees, shrubs and other forms of vegetation directly while the physiographic factors affect the climate and the soil and thus affect the vegetation.

From the point of view of selecting species for various agroforestry systems, several climatic parameters should be considered, such as total annual rainfall, humidity, number of rainy days, mean minimum and mean maximum with extreme ranges of temperature, incidence of natural calamities, e.g., frost, drought, floods, etc. Every locality has specific climatic parameters which dictate the choice of a limited number of species.

On the basis of temperature, the climate of locality may be classified as under (Table 63).

**Table 63: Temperature zones of the country (Champion and Seth, 1968)**

| Zone | Mean annual temperature °C | Mean January temperature °C | Remarks |
|------|----------------------------|------------------------------|---------|
| Tropical | Over 24 | Over 18 | Cold season short or none. Frost and snow absent. |
| Subtropical (montane) | 17 to 24 | 10 to 18 | Cold season definite but not severe. Frost during winter. |
| Temperate (montane) | 7 to 17 | −1 to 10 | Winter pronounced with frost and snow. |
| Alpine (montane) | Below 7 | Below −1 | Winter long and severe, snow common. |

Indian sub-continent mainly falls within the tropical zone. Only hill ranges constitute subtropical, temperate and alpine zones. Every zone has almost a specific flora and choice of species is limited within that flora.

On the basis of rainfall, the climate of a locality could be classified into: (i) arid when annual rainfall is below 500 mm, (ii) *semi-arid*: when the rainfall is between 500 mm to 750 mm, (iii) *dry*: when rainfall is between 750 mm to 1250 mm, (iv) *semi-humid*: when rainfall is between 1250 mm to 2000 mm and (v) *humid*: when rainfall is over 2000 mm. This classification is tentative and several variations may be found depending on other factors.

CLIMATE AND SELECTION OF SPECIES

On the basis of the above broad climatic classification, the selection of species becomes limited to those capable of growing in a particular climatic zone. Table 64 gives some important species which could be selected for different climatic zones.

The list in Table 64 is not exhaustive. However, it is necessary to prepare a detailed list of species which could be grown in a particular locality. Further selection of species will depend upon several other factors.

Soil is another site factor, which is quite important. Various soil conditions also affect the decision regarding choice of species. Several physical and chemical characteristics of soil, such as texture, stoniness, drainage, pan formation, soil depth, aeration, water-table, moisture regime, soil pH,

**Table 64: Climatic zones and suitable species for agroforestry**

| Climate | Examples of suitable species |
|---|---|
| 1. (a) Hot desert | *Prosopis cineraria, P. chilensis, Acacia tortitis, Capparis* spp., *Tecomella undulta*, etc. |
| (b) Cold desert | *Populus nigra, P. ciliata, P. alba, P. tremula, P. euphretica, Salix alba, S. angustifolia, S. fragilis, Juniperus* spp., etc. |
| 2. (a) Tropical semi-arid | *Prosopis* spp., *Acacia tortilis, A. nilotica, A. senegal, Albizia lebbeck, Eucalyptus camaldulensis, Azadirachta indica, Salvadora persica, Tamarix* spp., *Capparis* spp., etc. |
| (b) Subtropical semi-arid | *Pinus roxburghii, Acacia modesta, Albizia procera, Bauhinia variegata, Morus indica, Ficus* spp., etc. |
| (c) Temperate semi-arid | *Pinus gerardiana, Juniperus macropoda, Corylus colurna*, etc. |
| 3. (a) Dry tropical | *Eucalyptus camaldulensis, E. tereticornis, Albizia lebbeck, A. procera, Ailanthus excelsa, Cassia* spp., *Acacia nilotica, Hardwickia binata, Azadirachta indica, Mangifera indica, Aegle marmelos, Dendrocalamus strictus, Lagerstroemia parviflora, Dalbergia sissoo, Tectona grandis, Santalum album, Borassus flabellifer, Emblica officinalis, Madhuca indica* |
| (b) Dry subtropical | *Pinus roxburghii, Acacia modesta, Grewia optiva, Bauhinia variegata, Albizia chinensis, Grevillea robusta* |
| (c) Dry temperate | *Pinus gerardiana, Celtis australis, Cupressus* spp., *Juglans regia* |
| 4. (a) Tropical semi-humid | *Eucalyptus tereticornis, E. citriodora, Tectona grandis, Gmelina arborea, Casuarina equisetifolia, Dalbergia latifolia, Bambusa arundinacea, Bombax ceiba, Morus alba, Leucaena leucocephala, Dalbergia sissoo, Artocarpus heterophyllus, Anthocephalus chinensis, Adina cordifolia, Populus deltoides, Moringa oleifera, Anacardium occidentale*, etc. |
| (b) Subtropical semi-humid | *Albizia chinensis, Pinus roxburghii, P. kesiya, P. elliottii, Grewia optiva, Celtis australis, Morus indica, M. laevigata, Eucalyptus grandis, E. globulus, Toona ciliata* |
| (c) Temperate semi-humid | *Acacia dealbata, A. mearnsii, A. decurrens, Acer oblongum, Alnus nealensis, Cedrus deodara, Celtis australis, Quercus* spp; *cryptomeria japonica, Eucalyptus globulus, Fraxinus* spp; *Juglans regia, Robinia pseudacacia*, etc. |
| 5. (a) Humid tropical | *Terminalia myriocarpa, Duabanga grandiflora, Tectona grandis, Terminalia alata, Schima wallichii, Gmelina arborea, Dipterocarpus macrocarpus, Bombusa tulda, Dendrocalamus hamiltonii, Cocos nucifera, Areca catechu, Artocarpus heterophyllus, Mangifera indica, Pterocarpus santalinus, Chukrasia tabularis* |
| (b) Humid subtropical | *Eucalyptus globulus, Acer oblongum, Acrocarpus fraxinifolius, Aesculus indica, Terminalia crenulata, Pinus kesiya, Prunus* spp., *Quercus* spp. |
| (c) Humid temperate | *Acer campbelli, Abies spectabilis, Quercus* spp., *Robinia pseudacacia, Pinus patula, P. wallichiana, Alnus nitida, Populus ciliata, Cryptomeria japonica, Michelia doltsopa* |

soil nutrients, etc., are important for selection of species. The ability of a species to establish and grow well depends on its root system and its ability to absorb moisture and nutrients. Since trees are hardier, they are capable of growing in a variety of soil conditions. However, some species indicate a preference for specific soil conditions (Table 65).

Certain biotic factors, e.g., grazing, fire, insect pests, diseases etc. sometimes become important for selection of species. Where grazing is free and unrestricted, it is better to plant species which are not readily grazed, e.g., *Eucalyptus* hybrid, *Prosopis chilensis*, *Gliricidia muculata*, *Cassia siamea*, etc. Attack of insect pests is also important. For example, *Leucaena leucocephala* has been planted on a large scale in central and southern India. However, the appearance of Psyllid in some parts of the country on this tree now limits its selection in agroforestry plantations, without its control measures. Similarly, the stem borer in the case of *Populus deltoides* causes considerable damage. Where insect pests and diseases cause serious damage in some trees, it is better to avoid them as far as possible.

**Effect on Annual Crops**

In choosing tree species, one of the important considerations is their effect on the annual crops. Only such species are preferred which cause no or least damage to agricultural crops. Several factors are important from the point of view of their effect on annual crops. King (1979) suggested that the tree species chosen to be grown in conjunction with agricultural crops should have *inter alia* the following characteristics:

(a) they should be amenable to early wide spacement;
(b) they should possess self-pruning properties;
(c) if not self-pruning, they should be able to tolerate a relatively high incidence of pruning, that is, their photosynthetic efficiency should not significantly decrease with heavy pruning;
(d) they should have a low crown diameter to bole diameter ratio, that is, the width of their crown should be small, relative to the bole diameter;
(e) they should be tolerant of side shade, if indeed not of full overhead shade in the early stages of growth;
(f) their phyllotaxis should permit the penetration of light on the ground;
(g) their phenology, particularly with respect to leaf flushing and leaffall, should be advantageous to the growth of the annual crop in conjunction with which they are being raised;
(h) their rate of litterfall and litter decomposition should have a positive effect on the soil;
(i) their above-ground changes over time in structure and morphology should be such that they retain or improve those characteristics which reduce competition for solar energy, nutrients and water;

**Table 65: Soil types and suitable species**

| | Soil types | Important characteristics | Some suitable species |
|---|---|---|---|
| 1. | Desert soils | Sand, poor in humus and water holding capacity, high salt content | *Prosopis cineraria, P. chilensis, Acacia tortilis, A. senegal, A. nilotica, Salvadora* spp., etc. |
| 2. | Recent alluvium | Soils of recent origin, sandy in nature | *Acacia catechu, Dalbergia sissoo, Bombax ceiba*, etc. |
| 3. | Old alluvium | Soils are well developed | A large number of tropical species depending upon the climate. |
| 4. | Saline-alkali soils | High salt content, high pH | *Prosopis* spp., *Acacia nilotica, Azadirachta indica, Ailanthus* spp., *Eucalyptus* spp., *Tamarix* spp., *Pongamia pinnata*, etc. |
| 5. | Coastal and deltaic alluvium | Sandy in texture, high salinity, poor in nutrients | *Casuarina equisetifolia, Cocos nucifera, Areca catechu, Calophyllum inophyllum, Avicennia* spp., etc. |
| 6. | Red soils | Light in texture, red in colour, absence of lime and rich in iron | *Tectona grandis, Madhuca indica, mangifera indica, Dalbergia sissoo, Acacia nilotica, Leucaena leucocephala, Azadirachta indica, Eucalyptus hybrid, Pterocarpus marsupium, Adina cordifolia, Dendrocalamus strictus*, etc. |
| 7. | Black cotton soils | Clayey in texture, crack heavily during summer, presence of lime nodules and typical black colour | *Acacia nilotica, A. leucophloea, Tectona grandis, Hardwickia binata, Adina cordifolia, Tamarindus indica, Aegle marmelos, Leucaena leucocephala, Azadirachta indica, Bauhinia* spp., *Dalbergia latifolia, Phoenix sylvestris*, etc. |
| 8. | Laterite and lateritic soils | Brick in colour, gritty in nature, contain hard pan of vermicular structure | *Tectona grandis, Eucalyptus* spp., *Acacia auriculiformis, Azadirachta indica, Tamarindus indica, Anacardium occidentale, Dendrocalamus strictus, Emblica officinalis*, etc. |
| 9. | Peaty and organic soil | Found in humid regions, rich in organic matter with impeded drainage | *Syzygium cumini, Ficus glomerata, Bischofia javanica, Lagerstroemia speciosa, Gliricidia sepium*, etc. |
| 10. | Hill soils | High organic matter, low pH, brown to dark brown in colour | *Jugans regia, Alnus nitida, Toona serrata, Cedrus deodara, Quercus* spp., *Grewia optiva, Celtis australis*, etc. |

(j) their root system and root growth characteristics should ideally result in the exploration of soil layers that are different to those being tapped by the agricultural species and

(k) they should be efficient nutrient pumps.

These characteristics of an ideal tree for agroforestry are not complete and may change depending on the site and agroforestry system. In general, the ideal tree for an agroforestry system should be examined with respect to its: (i) root characteristics, (ii) crown characteristics, (iii) bole characteristics, (iv) phenological characteristics, (v) nutrition and water absorption characteristics, (vi) growth characteristics and (vii) shelter to insect pests and diseases.

## 1. ROOT CHARACTERISTICS

In a plant, roots perform two important functions: (a) they provide anchor and support to the plant and (b) they absorb water and nutrients from the soil. When woody perennials are grown with annual crops, it is necessary that root distribution of trees and annual plants should be such that they are distributed in different areas. Most annual crops have their roots distributed in the upper 40 cm soil layers (Singhal and Pant, 1989). Only some species, such as *Cajanus cajan*, gram, cotton, sugar-cane, etc. send their roots into deeper layers. The root system and distribution in the case of trees has not been studied in detail.

The growth and distribution of roots depend on species, soil characteristics and climatic conditions. The behaviour of the root system of some trees is indicated in Table 66.

Table 66:  Root system of some tree species

| Species | Age (yrs) | Length of taproot (m) | Maximum lateral spread(m) | References |
|---|---|---|---|---|
| *Albizia lebbeck* | 6 | 6.5 | 3.5 | Bhimaya and Kaul (1965) |
| *Prosopis cineraria* | 6 | 6.0 | 1.5 | |
| *Acacia senegal* | 6 | 6.4 | 3.0 | |
| *Eucalyptus tereticornis* | 2 | 2.3 | 3.5 | Dabral *et al.* (1987) |
| *Shorea robusta* | 35 | 1.2 | 8.3 | Dabral *et al.* (1984) |

In some species, such as *Eucalyptus* hybrid, *Populus deltoides*, etc., formation of the taproot takes place in the juvenile stage. It starts branching age advances. *Eucalyptus* spp. form strong lateral roots, perhaps due to their high water requirement as most roots run horizontal near the soil surface and extend up to 18 m within a soil depth of 30 to 60 cm (Dabral *et al.*, 1987). Perhaps that is the reason why *Eucalyptus* hybrid competes with annual crops for water and nutrients more strongly than other tree species and its adverse effect on agricultural crops is more visible in the case of moisture stress conditions (Dwivedi, 1987).

Since the roots of most agricultural crops lie in the surface layer of the soil, it is necessary that such tree species be preferred which have a deep root system. If tree roots are also distributed on the surface layer of the soil, they will not only compete with agricultural crops for nutrients and water, but also cause obstruction in soil working (Fig. 15a,b).

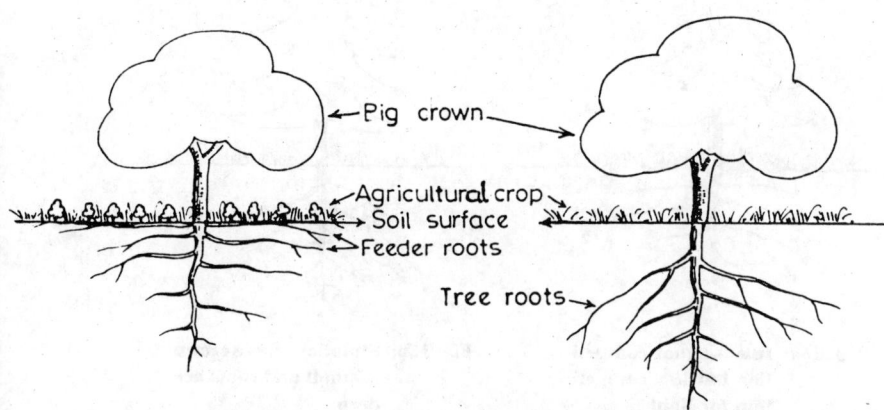

**Fig. 15a: Intense light and root competition**

**Fig. 15b: Intense light competition but less root competition**

Rooting depth alone is not necessarily the best indication of root activity. Probably, the degree of exploitation of the soil is a more correct guide. The degree of exploitation of soil is reflected in the *root length density* expressed as root length per $m^3$ of soil. The root length density in different soil depths has not been precisely studied in tree crops. Such studies would help in the proper selection of species.

## 2. CROWN CHARACTERISTICS

When trees are grown with agricultural crops either as an intercrop or on the boundary, the tree crown is of great significance. The tree crown consists of branches, twigs and leaves. The characteristics of a crown which are important from the agroforestry point of view include: (i) density, (ii) size, (iii) shape, (iv) height, etc. Trees having a dense crown do not permit enough light on the soil which adversely affects the process of photosynthesis in agricultural crops. The trees, therefore, should have a light crown which will permit enough light in the ground for photosynthetic activity of agricultural crops. Several trees, such as *Eucalyptus terericornis*, *Emblica officinalis*, *Acacia nilotica*, etc., have a lighter crown than *Mangifera indica*, *Tectona grandis*, *Dalbergia sissoo*, etc.

In agroforestry, the smaller the tree crown, the better for the mixture. The ratio of crown size and bole diameter should be as low as possible.

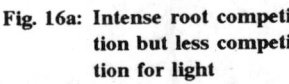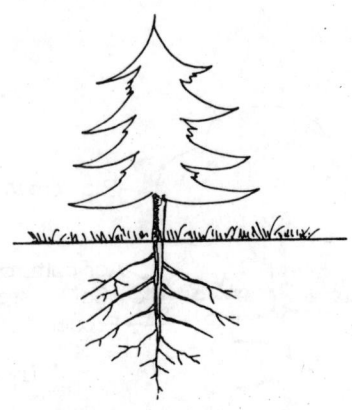

Fig. 16a: Intense root competition but less competi-
tion for light

Fig. 16b: Situation where crown
is small and roots are
deep

Larger crowns will have an effect on a larger area for shading. Smaller crowns also help in accommodating a larger number of trees per unit area in comparison to larger crowns. With size of the crown remaining the same, the trees with conical or cylindrical crowns should be preferred (Fig. 16a,b). Trees having an oval or umbrella-shaped crown will affect the annual crops over a larger area (Fig. 15a,b). The crown of the tree should be as high as possible. If the crown is placed quite high on the tree, the ground will be able to receive light from the sides. Low set and big crowns are not good for agroforestry (Fig. 17a,b). Several tree species, such as *Mangifera indica*, citrus, etc., have deep-set crowns which affect the crops adversely. An ideal tree for agroforestry should have a small conical crown with a long clear bole supported by deep root system (Fig. 18).

### 3. BOLE CHARACTERISTICS

The tree bole should be straight and long. The *clear bole* should be long enough and branches emerge at higher locations. Several trees, such as *Eucalyptus* hybrid, *Casuarina equisetifolia*, pines, etc., have a long clear bole. The bole should be upright and straight. In several species, however, branching starts early and the clear bole length is small. Most broad-leaved species have this tendency. Some species develop a long clear bole due to natural pruning if planted in close spacing. If trees do not have self-pruning characteristics, they should tolerate a high incidence of artificial pruning. Some species, such as *Populus deltoides*, *Dalbergia sissoo*, etc., tolerate pruning and develop a relatively good clear bole and produce quality timber. However,

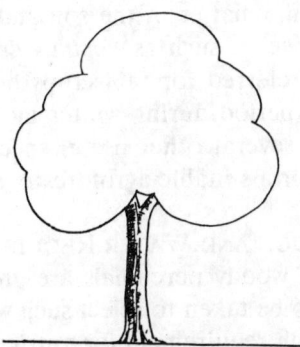

**Fig. 17a: Low crown is bad for agricultural crops**

**Fig. 17b: Crown of the same size located high allows sufficient light on the ground**

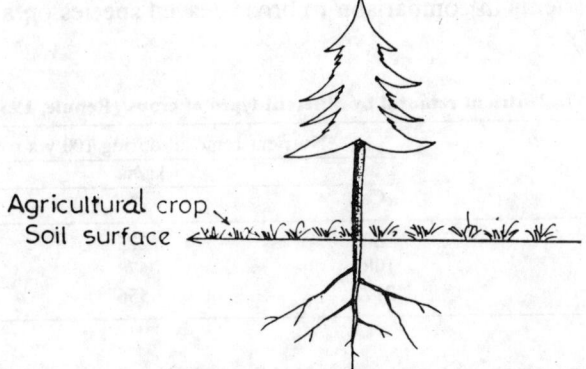

Agricultural crop
Soil surface

**Fig. 18: An ideal tree for agroforestry where crown is small and roots are deep**

several species are known for not tolerating pruning. These species include *Eucalyptus tereticornis*, *Pinus roxburghii*, etc.

Tree species which are planted for fodder have to be lopped. Therefore, these species should be able to tolerate lopping with no damage to the tree or disease infection.

## 4. PHENOLOGICAL CHARACTERS

Phenological characters, particularly leaffall, emergence of new leaf, phyllotaxis, etc., are also important. Deciduous species are preferred because they cause no competition for light, nutrients and moisture during the period of leaflessness. The longer the period of leaflessness, the better for the growth of agricultural crops. The leaf litter deposited due to leaffall adds organic matter in the soil and maintains the productivity of the soil. Several species, such as *Populus deltoides, Tectona grandis, Bombax cieba,* etc. are preferred for rabi crops because they are leafless for a considerably long period during winter months when water availability is limited. Similarly, several other useful species have this characteristic and can be selected for a suitable agroforestry system.

## 5. NUTRITION AND WATER REQUIREMENT

When woody perennials are grown with agricultural crops or grasses, care has to be taken to select such woody species which do not have unnecessarily high requirement for nutrients and water. This is necessary to avoid competition between agricultural crops and perennial crops. For example, *Eucalyptus tereticornis,* which has been grown on a large scale under agroforestry systems, has been accused of excessive consumption of water and nutrients adversely affecting the agricultural crops and soil productivity.

There have been a few studies regarding nutrient absorption by the tree crops. Some of the earlier studies indicated that pines removed a lesser quantity of nutrients in comparison to broad-leaved species on a long rotation (Table 67).

Table 67:  **Nutrient removal by different types of crops (Rennie, 1955)**

| Crop | Nutrient removal during 100 yrs rotation | | |
|---|---|---|---|
| | kg/ha | | |
| | Ca | K | P |
| Pines | 502 | 225 | 52 |
| Other conifers | 1082 | 578 | 101 |
| Deciduous trees | 2172 | 556 | 124 |

Some recent studies have indicated that trees removed quite a large quantity of nutrients, particularly when harvested on a short-rotation (Ovington, 1959; Faruqui, 1972; Singh, 1984; Malhotra *et al.,* 1987) as indicated in Table 68.

In general, uptake and release of nutrients are controlled by the species, density and age of the trees. When the crop is young, the uptake is more and the release is far less. As the age of the trees advances, the release becomes substantial. The release of nitrogen in *Pinus patula* at the age of six years is only 40 per cent of the total uptake while it becomes 60 per cent at the age of 14 years (Malhotra *et al.,* 1987). When trees are harvested at short rotation, they are likely to make heavy demands on the nutrients from

**Table 68: Annual uptake, retention and release of nutrients in different forests (kg/ha)**

| Forest type | Nutrient cycle | Nutrients kg/ha | | |
| --- | --- | --- | --- | --- |
| | | N | P | K |
| *Shorea robusta* | U | 339 | 36 | 380 |
| (40 yrs) Gorakhpur | R | 201 | 18 | 251 |
| | Re | 138 | 18 | 128 |
| *Tectona grandis* | U | 478 | 91 | 686 |
| (38 Yrs) | R | 243 | 58 | 321 |
| | Re | 234 | 33 | 351 |
| *Eucalyptus tereticornis* | U | 116 | 54 | 831 |
| (6 Yrs) | R | 102 | 46 | 732 |
| | Re | 14 | 8 | 89 |
| *Pinus sylvestris* | U | 139 | 11 | 56 |
| (24 Yrs) | R | 14 | 1 | 7 |
| | Re | 125 | 10 | 49 |
| *Pinus patula* | U | 40 | 1 | 24 |
| (6 Yrs) | R | 23 | 0.5 | 16 |
| | Re | 17 | 0.5 | 17 |

U = Uptake, R = Retained, Re = Released

the soil. Effort has to be made to select such tree species whose nutrient requirement is low so that it does not compete with agricultural crops.

Water is another important factor for plant growth. In the rabi season, when water is scarce, consumption of water by the trees is also important. Certain adaptations develop in plants to avoid water stress conditions. When moisture stress increases, photosynthetic activity is reduced as a result of closure of the stomata. However, the response varies with the species. $C_4$ plants such as *Zea mays*, *Sorghum bicolor*, *Panicum maximum*, etc., which are typically associated with semi-arid conditions, adapt to water stress more easily than $C_3$ plants. Several species, e.g., *Eucalyptus tereticornis*, *Populus deltoides*, etc., are believed to absorb a large quantity of water, depriving annual crops of water and creating water stress conditions. Research has indicated that *Eucalyptus tereticornis* does consume water in a large quantity when it is available in plenty; however, the plant tends to manage water scarcity situations through adjusting transpiration and stomata closure (Dabral and Raturi, 1985).

While selecting species for agroforestry, it is essential to know whether one species is more efficient in using water than another, particularly when water is scarce. The water-use efficiency is usually expressed in terms of *transpiration ratio*, which is the amount of water transpired per unit of dry matter produced. The water-use efficiency of some tree species has been studied. In the case of *Eucalyptus* hybrid, a higher water consumption in comparison to other species has been reported but all studies have conclusively proved that the species has the least water consumption per unit weight of dry matter produced (Chaturvedi, 1983; Chaturvedi *et al.*, 1984; Dabral and Raturi, 1985); study results are given in Table 69.

Table 69: Water consumption by some tree species

| Species | Total biomass produced per litre of water | Water consumed per gram of biomass (litres/gm) |
|---|---|---|
| *Acacia auriculiformis* | 1.39 | 0.72 |
| *Albizia lebbeck* | 1.83 | 0.55 |
| *Dalbergia sissoo* | 1.31 | 0.77 |
| *Eucalyptus* hybrid | 2.06 | 0.48 |
| *Pongamia pinnata* | 1.31 | 0.88 |
| *Syzygium cumini* | 2.00 | 0.55 |

The consumption per seedling, however, in another experiment was observed differently and details are indicated in Table 70 (Tiwari and Mathur, 1983).

Table 70: Potential water consumption per seedling

| Species | Water consumption (mm) per seedling |
|---|---|
| *Dalbergia latifolia* | 1143 |
| *Eucalyptus citriodora* | 5526 |
| *Pinus roxburghii* | 936 |
| *Populus euramericana* | 2704 |

In short, such species need preference whose water requirement is comparatively less and which can tolerate a drought scarcity situation. This is necessary because a large area of the country depends on rain-fed agriculture. Where land and water is scarce, simultaneous cropping of trees and food crops may offer considerable advantage.

Certain species of trees are able to fix nitrogen from the atmosphere with the help of nodules forming bacteria of the genus *Rhizobium* or actinomycetes of the genus *Frankia*. These are usually referred to as nitrogen-fixing trees (NFT). Some of these species are able to grow in a variety of edapho-climatic conditions. Since they are able to fix nitrogen from the atmosphere, their dependence for nitrogen in the soil is considerably reduced. Some details of these species are discussed later in this chapter.

6. GROWTH CHARACTERISTICS

The species selected should essentially be fast growing. It should be able to withstand competition during early stages, particularly when species are being selected for intercropping. The tree species if grown in wider spacing should not develop as a *wolf tree* and should be adapted to initial wider spacing. The tree must be able to withstand adverse conditions in the seedling and young plant stage. The tree should be capable of withstanding lopping, pruning and browsing. The tree should be adaptable to the difficult site conditions. Its roots should go deep and absorb nutrients from deeper layers.

### 7. SHELTER TO INSECT PESTS AND DISEASES

The tree should be such that it in no way provides shelter to insect pests and diseases of agricultural crops. The tree should not carry such pests and diseases which may damage agricultural crops. Fortunately, cases of significant damage have not been reported so far. However, it is necessary to be vigilant in this direction. Some common insect pests and diseases have been reported in the case of forest trees and fruit trees. The extent of damage, however, is not precisely known. These questions are becoming important gradually.

### Purpose of Plantation

The general criteria of site suitability and species characteristics will bring down the list of species and the choice of species will rest on the specific purpose for which the trees are being planted. Trees are planted for meeting one or several of the following needs.

### SOCIAL NEEDS

Social needs include the needs of firewood, fodder, small timber, poles, fruits, fibers, etc. The produce are needed to be utilised by the local people. Several species, such as *Azadirachta indica*, *Acacia nilotica*, *Mangifera indica*, etc., are planted in tropical areas on a large scale as these species meet several needs of the community. Every farmer would like to have a few trees of *Acacia nilotica* and *Azadirachta indica* as almost every part of the plant serves one use or the other. The leaves of these species are good fodder, the twigs are used as a toothbrush; barks are used as medicine; branches yield good firewood and the main stem is used as timber for agricultural implements and house construction and is considered durable with no treatment. The seeds of *Azadirachta indica* yield oil which is used as medicine and in soap making. The bark of *Acacia nilotica* yields tannin. The fruits are eaten by goats. The tree also yields gum. Therefore, multipurpose trees which can be used for meeting various social needs are preferred.

INDUSTRIAL NEEDS

In agroforestry, trees are also planted for meeting industrial needs. The Forest Policy of 1988, provides that industries should meet their raw material needs from agroforestry produce. Paragraph 4.9 specifically mentions that forest-based industries should establish a direct relationship with the individuals who can grow the raw material for them and industries should support individuals with inputs including credit, constant technical advice and harvesting and transport services. The policy also recommends that small and marginal farmers should be encouraged to grow on marginal/degraded lands the wood species required by the industries along with the species of fuel and fodder required for their own use.

Several forest-based industries have started helping farmers in some areas in planting of the species required by them. WIMCO started helping farmers in Uttar Pradesh and Haryana in planting *Populus deltoides* required for match industry. Some of the paper-mills have started buyback schemes for *Eucalyptus* hybrid. Similarly, other industries, e.g., plywood, furniture, packing cases, sports goods, etc., have also taken the initiative to meet their requirements. The farmers thus have a choice in selecting the species depending upon the merits of the alternatives available to them.

ENVIRONMENTAL PROTECTION

Trees are also useful for ensuring environmental protection. Trees are useful in checking water and wind erosion, stabilising landslides, landslips and other disturbed soil surfaces, controlling ravines and reclaiming several categories of wastelands. Afforestation of rail, road and canal sides could also help in protecting the environment. All tree species do not satisfy the above needs with an equal degree of satisfaction. Some species are more useful to check water erosion while others may be better where wind erosion is common. Some species are especially suited for providing protection to the soil and stabilising landslides and landslips. The species which can be raised by vegetative parts are better for afforestation of landslips and landslides. In such areas, therefore, *Populus* spp., *Salix* spp., grasses, vines, etc. may be grown. The characteristics of trees which are considered beneficial from an environmental point of view include continuity of plant cover, ability to utilise incoming solar radiation, the capacity to enrich soil by depositing litter and the capacity to modify micro-climate (Huxley, 1983). For environmental plantations a proper mix of trees and shrubs is usually recommended. In selecting species for environmental conservation, the following points need to be considered: (i) The species should be able to improve the site condition through conserving soil and water, improving the nutrient cycling and regenerating the site ecologically. (ii) The species should be able to favourably modify the micro-climate in the area. (iii) The species should be able to maintain the ecological balance in the system, should not disturb the vegetation complex and should not introduce undesirable insect pests and diseases. (iv) The species should be able to withstand adverse conditions.

## LAND USE NEEDS

According to land capability classification, certain categories of land, e.g., land classes V, VI, VII and VIII should necessarily be put under trees. In such areas selection of trees should be primarily decided by edaphoclimatic conditions. Sometimes, due to specific demands, tree planting is required for shade, shelter, soil productivity, fruit, wood, etc. Some trees are better in providing these needs than others.

### Agroforestry System

Choice of species will also depend upon the agroforestry system because the species requirement for different agroforestry systems is somewhat different. For example, under the agrisilviculture system such species are preferred which are multipurpose, fast growing and of commercial importance. For the silvipasture system, tree species which can yield fodder leaves are preferred. Similarly, for village wood lots, species having higher fuelwood and fodder values are generally recommended. There is no hard and fast principle about the species preference under various agroforestry systems. However, the present observations do suggest that some such species choice does exist.

For silviagriculture systems the choice of tree species is dominated by the silvicultural considerations and decision of the forest managers for plantation. Several species of trees, e.g., *Shorea robusta, Tectona grandis, Dalbergia sissoo, Acacia nilotica, Populus deltoides*, etc. have been raised in different parts of the country as taungya plantation. Here some restriction is applied on the cultivation of climbing agricultural crops.

For intercropping with agricultural crops several fast-growing multipurpose and commercial tree species are usually selected. The species which give sufficient economic returns are preferred. Several species, such as *Eucalyptus* spp., *Populus deltoides, Dalbergia sissoo, Mangifera indica, Acacia nilotica, Casuarina equisetifolia, Albizia* spp., *Prosopis* spp., *Leucaena leucocephala, Cocos nucifera, Areca catechu, Anacardium occidentale, Madhuca indica* etc. are preferred. For intercropping with plantation crops, such as tea, coffee, cocoa, etc., tree species such as *Albizia* spp., *Terminalia myriocarpa, Gliricidia* spp., etc. are usually selected. In home gardens, the most common species are *Mangifera indica, Acacia nilotica, Azadirachta indica, Dalbergia sissoo, Bambusa* spp., *Syzygium cumini, Cocos nucifera, Areca catechu, Artocarpus heterophyllus*, etc.

For pastural silviculture and silvipasture systems, fodder tree species are preferred, such as *Acacia nilotica, Dalbergia sissoo, Prosopis cineraria,*

*Hardwickia binata, Leucaena leucocephala, Dendrocalamus strictus, Ziziphus* spp., *Grewia optiva, Celtis australis*, etc.

It thus becomes evident that every agroforestry system has some specific requirement which can be satisfied by a limited number of species. The final choice of the species, however, has to be made taking into consideration other factors.

## Marketability of Produce

During earlier periods, generally such species were grown which were multipurpose. The produce was utilised in the family. If the produce was more than the requirement it was given to the needy persons in the same village or adjoining villages. The timber, firewood, fodder, minor forest produce, etc., produced on the farmer's field had no marketability, as prices were very low. During the past 20 to 25 years, the prices of firewood, timber, charcoal, minor forest produce, etc., have increased almost 20 times in comparison to the prices of agricultural commodities, which have increased only 4 to 5 times (Dwivedi, 1985). This abrupt increase in the prices of forest products, particularly timber and firewood, has created a sufficient market for these products, particularly near towns and cities.

In agroforestry, only such species have become popular which are fast growing and capable of giving a return within the shortest possible period. In some of the European countries, farmers wait for 60 to 100 years before harvesting their forest crop. However, under Indian conditions, most farmers are poor and cannot allocate their land for such use which is not able to give them a return within a short period, say 10 to 15 years. The rate of growth is important but marketability and return is more important. For example, in some parts of India, *Eucalyptus* hybrid, *Populus deltoides, Leucaena leucocephala*, and *Casuarina equisetifolia* became popular because these species could be harvested within a period of 8 to 10 years and the same provided a handsome revenue sometimes greater than obtainable under cash crops like sugar-cane, cotton etc. (Patel, 1988; Saxena, 1990). Other species, such as *Bombax ceiba, Grevillea robusta, Tectona grandis, Acacia nilotica, Dalbergia sissoo, Dalbergia latifolia, Santalum album*, etc., hold great promise.

It is necessary that market conditions are always favourable. If the market gets saturated for one reason or the other, the species is not preferred. For example, in Haryana, due to increased production of *Eucalyptus* wood, the market has become glutted and prices have gone down. This has adversely affected the plantation of this species (Malik, 1989). Even where these species have been planted, farmers have, in some cases, uprooted them because of the uncertainty of market conditions. It is, therefore, necessary to have a diversification in the species selection and species utilisation which might help in establishing a fairly stable market situation. It is also necessary to do demand studies so that the plantation of a particular species could be suggested depending upon the market requirements.

**Effect on Site**

The effect of a species on the site is important in the long run. The species selected should be such that it is able to protect and improve the site conditions. There is a growing awareness of a possible decline in soil fertility as a result of plantation of short-rotation, fast-growing species. When plantations are raised after clearing the natural vegetation, declining yields have been observed in several species. Reduced growth and yield in second and subsequent rotations in teak in India (Laurie and Griffith, 1942; Griffith and Gupta, 1948; Anon., 1974), *Pinus radiata* in South Swaziland (Evans, 1975), and *Eucalyptus* spp. in several countries (Champion and Brasnet 1958; NCA, 1976) are some of the examples. However, no decline in growth and yield have been observed in *Acacia* spp. (Lewis, 1967). However, there are no conclusive studies to indicate that these reduced yields were obtained due to change in the site conditions. In agrisilviculture systems where the lands are primarily under agriculture no such yield reductions are expected.

In agroforestry, such species need to be selected which produce leaves that are easily decomposed to form the right type of humus. If trees are nitrogen fixing through nodule bacteria, the situation may be better as some species are reported to fix a large quantity of atmospheric nitrogen. These species will enrich the soils. Species having allellopathic properties need to be avoided. Some species of *Eucalyptus, Juglans, Lantana*, etc., are known to have these properties (Del Moral and Muller, 1970; Rao and Reddy, 1984).

NITROGEN-FIXING TREES

Nitrogen-fixing trees (NFT), are those which show evidence of symbiotic nitrogen fixation usually through nodule-forming bacteria of the genus *Rhizobium* or *actinomyctes* of the genus *Frankia* (Fig. 19). More than 650 species of trees are reported to be nitrogen fixing (NFTA, 1989). The important families and genera of the nitrogen-fixing group are as under (Table 71).

There are several justifications for recommending nitrogen-fixing trees in agroforestry systems (Vergara, 1982; Brewbaker and MacDichen, 1985; Hughes and Styles, 1984). Some of them are as under:

(a) Nitrogen is one of the major essential elements for plant growth. Biological nitrogen fixation could be an important contribution to crop nutrition as production of chemical nitrogenous fertilisers requires large quantity of energy.

(b) Leguminous trees contribute to sustainable agroforestry production. Some of the nitrogen-fixing trees have several characteristics which help in

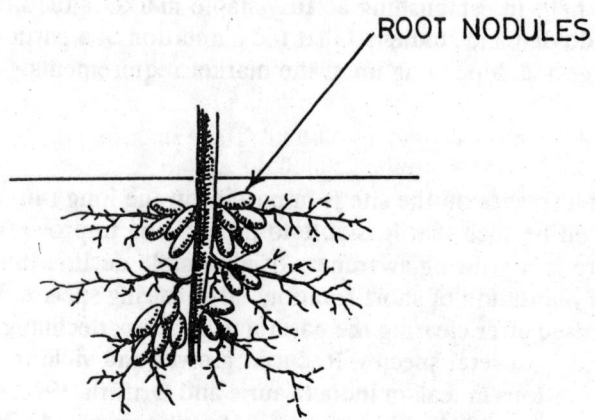

Fig. 19: Root nodules in leguminous plants

Table 71: Nitrogen-fixing families and genera

| S. No. | Family | Nitrogen-fixing genus |
|---|---|---|
| 1. | Betulacae | *Alnus* |
| 2. | Casuarinaceae | *Casuarina, Allocasuarina* |
| 3. | Coriariaceae | *Coriaria* |
| 4. | Myricaceae | *Myrica, Comptonia* |
| 5. | Rhamnaceae | *Ceanothus, Colletia, Discaria, Trevoa, Retanilla, Talguena* |
| 6. | Rosaceae | *Cercocarpus, Chamaebaria, Cowania, Dryas, Purshia* |
| 7. | Leguminosae | |
| | (a) Caesalpinioideae | *Chamaecrisia, Cordeauxia* |
| | (b) Mimosoideae | *Acacia, Albizia, Calliandra, Enterolobium, Leucaena, Mimosa, Paraserianthes, Pithecellobium* |
| | (c) Papilionoideae | *Cajanus, Dalbergia, Erythrina, Flemengia, Gliricidia, Pterocarpus, Robinia, Sesbania, Tephrosia* |

maintaining soil productivity. Some NFT, e.g., *Leucaena, Acacia*, etc. are known to have a long and deep taproot system. Therefore, they have a much better capacity to anchor and stabilise the soil and greater ability to recover and absorb moisture and nutrients from deeper subsoil. Some of the NFT have small leaflets, e.g., *Sesbania, Gliricidia, Leucaena Acacia*, etc. which decompose more rapidly and enable the nutrients to return more quickly to the surface to maintain the productivity of surface soil.

(c) Several NFT have multiple uses and provide a combination of benefits. Several species, e.g., *Acacia nilotica, Leucaena leucocephala, Calliandra calothyrsus, Dalbergia sissoo*, etc., provide fuelwood, livestock fodder, timber, green manure, living fence, etc. Some of them are good for alley farming. These trees could be used for green manure or mulch to replace inorganic nitrogenous fertilisers. Kang *et al.* (1981; 1985) showed that alley cropping of *Leucaena leucocephala* can support maize production levels at 2.0 tonnes/ha or approximately 80 per cent of the yield with nitrogenous fertilisers. *Flamingia macrophylla* has been used for alley cropping in the hills of the Philippines. *Inga* spp., *Gliricidia sepium, Erythrina indica*, etc. are used as shade trees over perennial crops such as coffee, cocoa, etc., with beneficial nitrogen inputs (Lindbald and Russo, 1986). Inclusion of NFT such as *Albizia falcataria* has been shown to improve the growth of *Eucalyptus* in Hawaii (Debell *et al.*, 1985).

Use of NFT in degraded sites and wastelands has a greater advantage. The evidence indicates that NFT can adapt themselves in wastelands and significantly improve the physical and chemical properties of the soil. Some of these studies indicate significant increased amount of nitrogen, higher cation exchange capacity, exchangeable Ca and K (MacDicken, 1981; FAO, 1988). A number of NFT are pioneer species which colonise disturbed sites.

(d) Some of the NFT are extremely fast growing although reliable yield estimates are not available. Available information suggests that *Acacia mangium, Leucaena leucocephala, Albizia falcataria, Caliandra calothyrsus*, etc., are quite fast growing and an annual yield of 30–90 m³/ha has been obtained in many experimental plantations (FAO, 1988). Several NFT show considerable tolerance to environmental conditions. Several species such as *Prosopis* spp., *Acacia tortilis*, etc., tolerate extremes of temperature and aridity. Some of these produce abundant seeds and are aggressive in nature and tend to occupy the blanks immediately. Therefore, these species are good for ecological restoration.

(e) Some NFT have the capacity to establish in areas despite heavy grazing pressures, because they exhibit a combination of unpalatable leaves, toxic chemicals, spines, etc. (Rehr *et al.*, 1973). Such species include *Prosopis chilensis, Acacia pennatula, Enterolobium* spp., *Cassia siamea, Acacia nilotica*, etc.

In spite of sufficient evidence of nitrogen fixation by NFT, some authors doubt the significance of nitrogen fixation, observing that these trees may make net nitrogen demand on the soil during their rapid juvenile growth (Wigston, 1985). This demand, however, is less compared to other fast-growing trees.

Several NFT are available for problem sites as under (FAO, 1988):

(a) NFT tolerant of saline soils: *Acacia nilotica, A. saligna, Albizia lebbeck, Casuarina equisetifolia, C. glauca, Parkinsonia aculeata, Pithecellobium dulce, Prosopis cineraria* and *P. tamrugo.*
(b) NFT tolerant of soil acidity: *Acacia auriculiformis, Acacia mangium, A. mearnsii, Albizia falcataria, Albizia lebbeck, A. procera, Calliandra calothyrsus, Casuarina glauca, Robinia pseudoacacia.*
(c) NFT tolerant of soil alkalinity: *Acacia auriculiformis, Acacia tortilis, Casuarina glauca, Casuarina equisetifolia, Prosopis cineraria.*

(d) NFT tolerant of poor soil drainage: *Acacia auriculiformis, Acacia nilotica, A. saligna, Alnus rubra, Casuarina equisetifolia, Dalbergia sissoo, Sesbania grandiflora, Sesbania sesban.*

(e) NFT tolerant of prolonged drought: *Acacia catechu, Acacia nilotica, A. saligna, A. senegal, Parkinsonia aculeata, Prosopis chilensis, P. cineraria.*

### Ease of Establishment

Agroforestry is done by millions of farmers, workers and others involved in agriculture, animal husbandry, agriculture and other related occupations. Several species of trees have been easily adopted in agroforestry because they are easy to raise. Good examples are *Eucalyptus* spp., *Populus* spp., *Dalbergia sissoo, Acacia nilotica, Salix* spp., etc. The species which can be established by cuttings or other vegetative parts are perhaps the best, e.g., *Populus* spp., *Salix* spp., etc. The trees which produce abundant seeds and can be raised by seed sowing are the next best. Such species include: *Acacia catechu, Acacia nilotica, Leucaena leucocephala, Dalbergia sissoo, Albizia lebbeck, Mangifera indica, Syzygium cumini*, etc. There are several other species whose nursery stock can be prepared more easily due to higher plant per cent, for example *Eucalyptus* spp., *Tectona grandis, Casuarina equisetifolia*, etc.

Some species pose a problem in plantation raising. Some species produce seeds which are poor in quality and quantity. For example, *Bambusa arundinacea* and *Bambusa vulgaris* are better bamboo species but these do not produce enough seeds. Therefore, these should not be planted on a large scale. Some species, such as *Shorea robusta, Dipterocarpus macrocarpus*, etc., have a small viability period which restricts the multiplication of these species on a large scale.

### Resistance to Attack of Insect Pests and Diseases

Susceptibility to attack by insect pests and diseases is one of the important considerations in selection of species in an agroforestry system. In agroforestry, usually fast-growing species are preferred. These species are more liable to attack by insect pests and fungi than slow-growing species. Fortunately no serious problems have been found so far in these regards. However, incidence of attack of sap-sucking psyllid (*Heteropsyla cubana*) on subabul (*Leucaena leucocephala*), stem borer (*Apriona cinaria*) in poplar and borer and defoliators in the case of *Gmelina arborea*, points towards serious limitation in future plantation programmes of these species. Similarly, pink disease in *Eucalyptus grandis*, spike disease in *Santalum album*, wilt diseases in case of *Casuarina equisetifolia*, etc. pose serious problems in plantation of these species. Species which are resistant to insect pests and diseases are preferred because farmers are sure of the rate of growth, survival, etc. They are also saved from adopting costly control measures. One of the reasons for the success of various species of *Eucalyptus* in India and elsewhere is the absence of widely prevalent leaf-eating insects, quite common in their native habitat (Pryor, 1978).

# CHAPTER 14

# Regeneration of Tree Crop

Regeneration is defined as the renewal of a forest crop by natural or artificial means. The words 'propagation' and 'reproduction' are also used as synonyms for regeneration. In forestry, however, the word regeneration is more widely used. In agroforestry, therefore, the word regeneration should be used for the renewal or introduction of a forest crop by natural or artificial means. There are two methods of regenerating a forest crop. These are:

 (i)  Natural regeneration
(ii)  Artificial regeneration

**Natural Regeneration**

Natural regeneration is the renewal of a forest crop by self-sown seed or by vegetative parts. When regeneration obtained from seed forms the crop, it is commonly called a *seedling crop*. When this seedling crop grows into a forest, it is called a *high forest*. When natural regeneration is obtained from a coppice or root suckers, the crop is usually called a *coppice crop*. When this crop grows into a forest, the forest is known as a *coppice forest*.

NATURAL REGENERATION BY SEED

Natural regeneration by seed is one of the common methods of regeneration of tree crops under agroforestry systems. The species which produce abundant viable seeds are likely to produce enough seedlings underneath. When the seeds are dispersed by the agency of wind, water, gravity, birds, animals, etc., the seedlings may appear in far off places provided suitable site conditions exist. When seeds fall on the ground just near the tree or away due to various dispersal agents, the seeds may germinate. The *germination* of seeds depends upon: (a) internal and (b) external factors. The *internal factors* include those factors which pertain to the seed itself, e.g., permeability to water and oxygen, development of embryo, size, after-ripening and viability of seed, etc. Water and oxygen are essential for seed germination. The embryo should be fully developed to be able to germinate. *After-ripening* refers to the biochemical or physical changes occurring in seeds which are necessary for germination. *Dormancy* is a condition of a mature viable seed in which germination is completely delayed even though the external conditions favour germination. All the seeds which fall on the ground do not germinate, because a large proportion of them is eaten by insects, birds and other animals. The *external factors* include moisture, air, temperature, light and seedbed. Proper conditions of these factors are required for germination of seeds.

The success of natural regeneration depends upon a chain of processes, viz., sufficient seed production, proper and efficient dissemination, good seed germination and establishment of seedlings in sufficient numbers. Establishment of seedlings in sufficient numbers is perhaps the weakest link in the whole process of natural regeneration. Large-scale mortality occurs during the succulent and juvenile stage of seedlings due to various factors. These factors may include climatic, edaphic, biotic and genetic. Every species has a certain degree of choice for different climatic and edaphic parameters. Several species have a wide range of choice for these climatic and edaphic variables. A large number of fungi and bacteria attack the seedlings causing death. Similarly, grazing, fire, etc. may damage the seedlings. Weed infestation also causes damage to the seedlings.

Several operations, such as manipulation of canopy, treatment of ground vegetation, soil working, control of grazing and fire and protection from insects and diseases may lead to better survival and establishment.

NATURAL REGENERATION BY VEGETATIVE PARTS

Natural regeneration by vegetative parts, e.g., coppice, root suckers, etc. is obtained in several forest crops. *Coppice* is defined as a shoot arising from an adventitious bud at the base of a woody plant that has been cut near the ground or burnt back (Anon., 1966). Natural regeneration by coppice can be obtained either by: (i) seedling coppice or (ii) stool coppice. *Seedling coppice* is defined as the coppice shoots arising from the base of the seedlings that have been cut or *burnt* back (Anon., 1966). This method is used for cutting back woody shoots and established shoots which are not making desired growth. After cutting back, these shoots produce vigorous growth. *Stool coppice* is the coppice arising from the stool or a living stump (Anon., 1966). In this method regeneration is obtained from the shoots arising from the adventitious buds of the stump of felled trees. The coppice shoot may arise either from the base of the stump or from its top (Fig. 20). The shoots arising from the top of the stump are likely to be damaged due to rotting as well as by wind. The shoots which arise from the base are capable of developing their own root system which provides support and protection against wind and pathogen attacks.

Coppice shoots are not produced by all species. The power of coppicing varies even in species that coppice. On the basis of their power

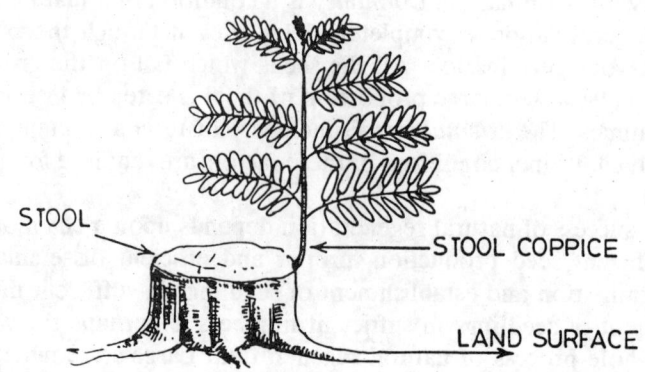

**Fig. 20: Stool coppice**

of coppicing, the species are classified into the following four classes (Dwivedi, 1990): (i) Strong coppicers: *Acacia catechu, Albizia lebbeck, Anogeissus latifolia, Anogeissus pendula, Azadirachta indica, Broussonetia papyrifera, Butea monosperma, Cassia fistula, Cleistanthus collinus, Dalbergia* spp., *Diospyros melanoxylon, Eucalyptus* spp., *Melia azedarach, Morus alba, Ougeinia oojeinensis, Prosopis chilensis, Robinia pseudacacia, Salix* spp., *Sapium sebiferum, Shorea robusta, Syzygium cumini, Tectona grandis,* etc. (ii) Good coppicers: *Aesculus indica, Chloroxylon swietinia, Hardwickia binata, Juglans regia, Pterocarpus marsupium, Quercus leucotrichophora, Q. semecarpifolia, Terminalia bellirica,* etc. (iii) Bad coppicers: *Adina cordifolia, Bombax ceiba, Casuarina equisetifolia, Madhuca indica, Populus ciliata,* etc. (iv) Non-coppicers: Almost all conifers.

It would thus be clear that for regeneration by coppice, one can depend only on species which are strong coppicers. In coppicing species, the power of coppicing depends upon the age of the tree. Younger saplings and poles coppice profusely. In older trees, the power of coppicing is reduced. The trees cannot keep on producing coppice shoots indefinitely. Therefore, the coppice forest crop needs to be replaced by the seedling crop after sometime. The older stumps suffer from insect and pathogen attacks. The vigour of the coppice crop is considerably reduced after a few rotations. For example, *Eucalyptus* hybrid or *E. tereticornis* is worked on the coppice system but the vigour of the crop after the second rotation starts showing a declining trend.

A shoot arising from the root of a woody plant is called a *root sucker*. Natural regeneration by root suckers is not practised on a large scale. The method is practised locally in some species, e.g., *Dalbergia sissoo, Boswellia serrata, Diospyros melonxylon,* etc. The roots when exposed or cut partially or wholly, produce shoots which are commonly called root suckers.

The trees are sometimes pollarded and continuous regeneration of useful branches is obtained through pollarding. *Pollard* is defined as a tree whose stem has been cut off in order to obtain a flush of shoots usually above the height to which the browsing animals can reach (Anon., 1966). Pollarding, therefore, is an operation in which the stem of a tree is cut off at a height usually above 1.5 or 2 m with the object of obtaining a bunch of new shoots. These new shoots are produced from the buds located just below the cut. *Salix* spp. in Kashmir valley is pollarded to produce shoots for wicker work. *Hardwickia binata* is pollarded in Andhra Pradesh to produce shoots for fibre extraction. *Grewia optiva* is pollarded in the Himalayan hills to produce shoots for fodder and fibre. In the plains, *Morus alba* is pollarded for production of shoots for fodder, fibre and basket making.

## Artificial Regeneration

Artificial regeneration is defined as the renewal of a forest crop by sowing, planting or other artificial means (Anon., 1966). The word plantation is used as a synonym for an artificially regenerated crop and is defined as a forest crop raised artificially either by sowing or planting. In raising a forest crop artificially either by seed or planting, the following major works are necessary:

(i) Seed collection
(ii) Nursery operations
(iii) Plantation
    (i) Soil working
    (ii) Plantation
(iv) Maintenance and protection

## Seed Collection

A seed is a miniature tree. A good seed means a good tree. The seed quality affects germination, growth of seedling, suitability to the site and the quantity and quality of products obtainable from the tree. The role of seed in agriculture was realised early and large scale selection and breeding work led to the development of high-yielding and disease- and pest-resistant strains of several crops. If good quality seeds are used in our plantations, the success is sure to be better. *Good quality* seed implies a seed which is highly viable and vigorous and is genetically well suited to the site and to the purpose for which it is planted. Agroforesters should collect the required seeds themselves. Important considerations in seed collection are: (i) the species and the provenance selected for planting, (ii) the quantity, (iii) the period of seed collection, (iv) method of seed collection and (v) storage of seed.

*Species and provenance*

The species to be planted will depend upon site factors, object of planting and several other factors which have been discussed in the chapter Choice of Species. It is also necessary to have information about different provenances in each selected species. *Provenance* means the place in which a stand of trees is growing. It refers to the area where the mother trees of the seeds are growing. Significant genetic variation is frequently found to be associated with geographic differences. Therefore, one must know which provenance is good for the locality. Seeds must be collected from known provenances as this is going to affect the quality of plantation.

*Seed quantity*

The quantity of seed required will depend upon: (i) area to be planted, (ii) spacement to be adopted, (iii) species—its seed weight, plant per cent, etc. Enough allowance must be given for replacement of casualty, nursery and transport losses, etc. The quantity of seed required for four different species has been worked out in Table 72. Such estimation should be made for other species selected for plantation.

Table 72: **Estimation of seed requirement**

| Item | *Tectona grandis* | *Dalbergia sissoo* | *Dendrocalamus strictus* | *Eucalyptus* Species |
|---|---|---|---|---|
| 1. Spacing (m) | 2 × 2 | 3 × 3 | 4 × 4 | 2 × 2 |
| 2. Plant/ha | 1670 | 1111 | 625 | 2500 |
| 3. Additional | | | | |
| (i) 30% for casualty replacement | 501 | 333 | 187 | 1850 |
| (ii) 70% Nursery, transport and other losses | 1169 | 778 | 438 | 1850 |
| 4. Total plant/ha | 3340 | 2222 | 1250 | 5000 |
| 5. Approximate number of seeds/kg | 2500 | 50,000 | 30,000 | 350,000 |
| 6. Number of germinated seedlings obtained from 1 kg of seeds | 500 | 20,000 | 10,000 | 100,000 |
| 7. Seed requirement (kg/ha) | 6.65 | 0.11 | 0.12 | 0.05 |

*Germination per cent*

The object of testing the germinative quality of seeds is to provide an indication of the percentage of seeds in a given lot that are expected to produce seedlings. It also provides comparison for quality index for one seed lot with another of the same species and one species with another. *Germinative capacity* means the total number of seeds that germinate in the test, plus the number of sound seeds remaining ungerminated at the end of the test expressed, as a percentage. For example, of a total 400 seeds, 220 germinated and after cutting open the remaining 180 seeds, it was found that 60 were sound, then the *germinative capacity* = $\dfrac{220 \times 60}{400} \times 100 = 70$ per cent.

*Germinative energy* is the percentage of seeds in a sample that have germinated in a test up to the time when the rate of germination (the number of seeds germinating per day) reaches its peak. The number of days required to reach this peak is called the *germinative energy period*. If for 400 seeds of a species daily rates of germination on the 7th, 8th, 9th, 10th and 11th day after sowing were 20, 60, 80, 70 and 55 respectively, then the *germinative energy* = $\dfrac{20 + 60 + 80}{400} \times 100 = 40$ per cent and the germination period is said to be 9 days.

*Field germination* is another word which is used to denote the percentage of seeds in a sample that germinate and appear above the seed bed surface. *Plant per cent* is the number of seedlings surviving at the end of the first growing season and expressed as a percentage of the number of seeds in the sample. For example, if 1000 seeds were sown in the nursery bed/pots and at the end of the growing season 500 plants could be raised, the plant per cent is 50 per cent.

Germinative energy is a measure of the speed of germination. The interest in germinative energy is based on the theory that only those seeds which germinate rapidly and vigorously under favourable conditions are likely to be capable of producing vigorous seedlings in field conditions. Very delayed germinants are automatically eliminated from the nursery.

## *Period of seed collection*

The colour of the fruits or cones is a good indicator for ripeness. Other methods of judging seed maturity, such as development of endosperm and embryo, specific gravity, moisture content of seed, etc. are available. But in the field, change in colour of the fruit and its readiness to fall are two important methods for deciding the period of seed collection. The period of ripening of fruits of some of the important species is indicated in Table 73.

Seed-bearing in trees is not regular every year. A *good seed year*, when tree species seed abundantly, is not common. Therefore, it is necessary to take advantage of a good seed year in seed collection. Enough seeds may not be available during poor seed years. Also, the seeds collected during poor seed years have a poor germination capacity (Turnbull, 1975). Damage from insect pests, birds, etc., is also more during poor seed years. A number of species, e.g., *Dalbergia sissoo, Tectona grandis, Gmelina arborea, Acacia cat-echu, Pinus kesiya, Acacia mearnsii, Delonix regia, Jacaranda mimosaefolia, Acacia nilotica, Eucalyptus* hybrid, etc., seed almost every year. However, species such as *Abies pindrow, Cedrus deodara* and *Pinus roxburghii* bear good seeds once in every 3 to 10 years. Bamboo species flower gregariously only once in their life which is 20 to 60 years (Dwivedi, 1980b).

**Table 73: Period of ripening of fruits in important species**

| Period of ripening of fruits/seeds | Species |
| --- | --- |
| 1. January–March | *Acacia catechu, Albizia lebbeck, Anthocephalus chinensis, Dalbergia sissoo, Eucalyptus citriodora, Emblica officinalis, Kydia calycina, Melia azedarach, Sterculia alata, Quercus leucotrichophora, Tamarindus indica, Terminalia bellirica, T. alata,* etc. |
| 2. March–June | *Acacia nilotica, Ailanthus excelsa, Albizia procera, Anacardium occidentale, Mangifera indica, Bombax ceiba, Cassia fistula, Casuarina equisetifolia, Dalbergia latifolia, Dipterocarpus alatus, Eucalyptus globulus, Gmelina arborea, Morus alba, Pinus roxburghii, Shorea robusta,* etc. |
| 3. July–September | *Juglans regia, Madhuca latifolia, Michelia champaca, Polyalthia longifolia, Robinia pseudacacia, Syzygium cumini,* etc. |
| 4. October–December | *Tectona grandis, Cedrus deodara, Lagerstroemia speciosa* |

*Selection of trees for seed collection*

In a stand or forest, when seeds are to be collected the selection of trees for seed collection is very important point. The following guidelines may be adopted for selecting trees for collection of seeds (Stenin *et al.*, 1974):

(i) Collect seeds only from healthy vigorous trees of reasonably good form that are making more than average growth.

(ii) Avoid collecting seeds from areas containing numerous poorly formed, excessively limby, abnormal, diseased and insect-infested trees.

(iii) Collect seeds from middle aged to mature trees. Young and over-mature trees should be avoided, as seeds from them may have low viability.

(iv) Avoid collecting seeds from isolated trees of naturally cross-pollinating species, since these are likely to be self-pollinated. The seeds are likely to be few and with low viability. The seedlings produced are frequently weak or malformed.

(v) Avoid collection of seed from vigorous wolf trees.

(vi) The seed should be collected from dominant or co-dominant trees.

*Methods of collection of seeds*

The method of seed collection will depend upon several factors, e.g., (i) characteristics of the tree, (ii) characteristics of the fruit/seed, (iii) relative size and numbers of natural dispersal units, (iv) characteristics of the stand, and (v) characteristics of the slope, etc. The following collection methods can be adopted for forest tree seeds: (i) collection of fallen fruits or seeds from the forest floor, (ii) collection from the crowns of felled trees and (iii) collection from standing trees.

Seeds and fruits can be picked from the ground after natural seedfall. This method of seed collection is usually followed in several species in which fruits are large. The method is cheap and does not require skilled labour. In species which have larger fruits, e.g., *Tectona grandis, Gmelina arborea, Emblica officinalis, Albizia lebbeck, Dalbergia sissoo, Acacia nilotica, Syzygium cumini, Terminalia arjuna, Terminalia alata, Shorea robusta*, and several others, this method is very useful. The main disadvantage of collection from natural seedfall are: (i) there is a risk of collecting immature, empty and unsound seeds, (ii) there is uncertainty about mother trees, (iii) of the seeds and fruits which fall naturally, a large number are defective and damaged by insects and rodents.

The seeds and fruits can be collected from the trees which are felled in normal fellings. This would require synchronising fellings with the seed-ripening season of the tree. Since seeds are to be collected from superior trees only, it is better to mark such trees before felling.

Seeds and fruits can be collected from standing trees by using several methods: (i) By hand; in shrubs and trees which have low branches. The collector while standing can collect seeds and fruits. (ii) The branches bearing seeds/fruits could be cut with the help of long handled tools and seeds/fruits could be collected. (iii) Fruits and seeds can be collected from standing trees with access by climbing.

SEED PROCESSING

In the majority of the tree species, the fruits are collected from the trees. The seeds are extracted from the fruits and dried and kept ready for sowing. Processing may involve precleaning, maceration and depulping, drying, separation, tumbling and threshing, dewinging and cleaning, etc.

SEED STORAGE

Storage means preservation of viable seeds from the time of collection until they are required for sowing. Storage of seeds is essential to meet the demands of seeds having the right physiological and genetic characteristics during all times. The object of proper storage is to maintain the viability of seeds. A *viable seed* is that seed which can germinate under favourable conditions, provided that any dormancy that may be present is removed. Storage helps to maintain the longevity of seeds. On the basis of natural longevity two classes of seeds are recognised (Robert, 1977): (a) *Orthodox seeds* which can be dried to low moisture content of about 5 per cent and successfully stored at low or subfreezing temperatures for long periods. (b) *Recalcitrant seeds* which cannot survive drying below a relatively high moisture content (20—50 per cent) and which cannot be stored successfully for long periods.

Several species of Leguminosae and some species of Malvaceae have high longevity. Seeds of several species of *Acacia, Albizia, Cassia, Leucaena, Caesalpinia, Prosopis, Ochroma* and *Hibiscus* can be stored successfully for more than 20 years (Dent, 1948). However, seeds of *Dipterocarpus, Swietenia, Hevea, Shorea, Hopea, Azadiracta*, etc. cannot be stored for more than three weeks without effect on viability.

*Factors affecting longevity of seeds*

Various factors influencing longevity of seeds in storage are type of seeds, stage of maturity, prestorage treatment, viability and moisture content, temperature, humidity and susceptibility to infection by fungi and bacteria.

The storage method of seeds may be such that there is control of moisture content of seed and/or temperature. If there is no control on moisture content of seeds and temperature of storage, seed longevity will be foreshortened. For simple storage conditions, the seed longevity is generally poor (Champion and Seth, 1968). The longevity period of some seeds under normal storage conditions is given in Table 74.

**Table 74: Seed viability under normal storage conditions**

| Period | Species |
| --- | --- |
| Two weeks | *Dipterocarpus, Shorea, Hopea, Michelia, Azadirachta, Tetrameles*, etc. |
| Three months | *Chloroxylon, Chukrasia, Mangifera, Podocarpus* |
| One year (sacks) | *Artocarpus, Cedrala, Mesua, Tectona, Terminalia, Bambusa*, etc. |
| One year (tins) | *Eucalyptus, Anacardium, Dalbergia, Pinus, Bombax, Dendrocalamus*, etc. |
| Two years or more | *Acacia, Albizia, Caesalpinia, Cassia, Leucaena, Ochroma, Prosopis* |

PRETREATMENT OF SEEDS

Pretreatment of seeds is done with a view to: (i) terminating dormancy in seeds and speeding up germination, (ii) giving protection against pests, diseases or adverse conditions and (iii) improving the uniformity of seeds or to render them more viable. Important types of pretreatments are: (i) treatment done to break the dormancy in seeds and to speed up germination, (ii) dressing, coating and pelleting of seeds with insecticides, fungicides and fertilisers for affording protection against pests and diseases and providing nutrients for growth in adverse conditions and (iii) other treatments. Pretreatment of seeds is not a prerequisite in the case of all tree species. Seeds of several species germinate without pretreatment. However, seeds of several species require pretreatment for breaking dormancy, speeding up germination and/or for other purposes. Common pretreatment methods include soaking in cold or hot water, scarification, etc.

Details of seed collection, seed longevity, presowing treatment, including method of propagation, and uses in respect of a large number of species are given in the Appendix (FRI, 1985).

**Nursery Operations**

Afforestation of wastelands and regeneration of agroforestry plantations may be carried out either by sowing seeds directly in the field or/and planting of nursery-raised seedlings, stumps, cuttings, etc. The easiest and the cheapest method of artificial regeneration is to sow the seeds of desired species directly in the field and allow them to grow after the seeds have germinated. But artificial regeneration by seed sowing has not been successful in the case of several species due to several problems. Planting of nursery-raised seedlings, stumps, cuttings and rhizomes offer several advantages over seed sowing. These advantages are: (i) Several species are initially slow growing. If seeds of these species are sown in the field, the seedlings are most likely to be swamped by weeds and killed by intense competition. The nursery-raised seedling is better equipped to compete with weeds and tolerate adverse site factors and therefore, better success is ensured in planting nursery-raised seedlings. (ii) Several species do not produce seeds every year. The nursery-raised seedlings provide planting stock, available all the years. The seeds collected during good seed years will produce seedlings in the nursery which may be used not only in the good seed year but also during years when seeding is poor or nil. (iii) Several species when grown by direct sowing are not successful compared to planting. In such species the nursery becomes an essential component of an artificial regeneration programme. (iv) Regeneration by nursery-raised seedlings is more dependable. Regeneration by direct sowing is very uncertain in most species. Nursery-raised seedlings tend to reduce the rotation as growth is faster. (v) Nursery-raised seedlings are capable of tolerating adverse conditions as their root and other systems are very well developed. Planting of wastelands and special sites can be done successfully only through nursery-raised seedlings. (vi) In areas where biotic pressure is more, e.g., road, canal and rail sides and areas adjoining villages can be afforested successfully only through nursery-raised seedlings.

Several tree species, e.g., *Acacia nilotica, A. catechu,* etc., have been successfully grown by direct sowing in several states. It is, therefore, unwise to raise them through nursery-raised seedlings.

SEED BED

A seed bed or a nursery bed is a prepared area in the nursery where seeds are sown or in which the transplants or cuttings are raised. These beds are a very important component of the nursery. In preparation of nursery beds, the points which need to be considered are: (i) shape and size of the beds, (ii) type of beds and (iii) surface of the beds. As a rule rectangular beds are preferred over other shapes. The width of the bed should be 1.2 to 1.5 m. The width of the bed should be kept such that it can be weeded by a labourer from both sides without entering the bed. The length of beds may be kept up to 12.5 m. Standard bed size is 12.5 m × 1.2 m (40' × 4'). The length of the bed may be kept smaller if sufficient space is not available.

Three types of beds, are usually made: (i) raised beds, (ii) sunken beds and (iii) level beds. *Raised nursery beds* are made in high rainfall areas. Raised beds 10–15 cm above ground level are made with the support of bricks, stones or bamboos or *ballies* which prevent edges of the beds from crumbling during rains or while giving irrigation to the beds. These beds tend to prevent water logging. Even during heavy rains the root zone is not flooded due to raised beds. Drainage from growing areas is also easy. *Sunken beds* are made in dry areas. The object of sunken beds is to avoid flow of water outside the bed. Sunken beds are made by excavating the soil in the bed area. These beds are usually 15 cm deeper than the normal ground level. It is better to connect such sunken beds to a common drainage line so that water does not stand during rains. *Level beds* are made in normal rainfall areas. These beds are easily irrigated by cans. The surface of the nursery bed should be perfectly flat or should have a slight camber. In order to enable good drainage in the beds, surface dressing should be given. If the soil is heavy, such dressings are more necessary.

SOIL WORKING

Soil working in beds is a very important operation. The object of soil working in beds is to provide the most optimum conditions for seeds to germinate and grow as healthy seedlings. The soil in the bed should be dug up to a depth of 40 to 50 cm. The soil should be taken out and left for 15 to 30 days for weathering. The boulders, large pebbles, etc. should be picked and thrown out. If soil is clayey, *surface dressing* with washed river sand should be done for improving the physical conditions of the soil. The soil in the pit should be properly rolled and packed. Before returning the soil to the pit it is advisable to sterilise the soil with formaline solution which has proved very effective in controlling a number of diseases. Addition of insecticides such as BHC for the control of termite and other insects is also useful. When seedlings are one week old they are transplanted in the polypot or transplanting bed. However, in seeds which are minute, e.g., *Eucalyptus*, *Adina* etc., where nutrient reserve in the seeds is little, the supply of some nutrients particularly nitrogen is found necessary (Paul, 1972; Jackson 1975). The question of soil mixture in pots or transplant beds is more complex. There is a great diversity in soil mixtures in transplant beds. It is generally agreed that the soil should be such that the proportion of silt and clay is between 10 and 20 per cent and it should be adjusted by adding sand. The addition of farmyard manure may be useful for *Eucalyptus* but not in the case of *Pinus caribaea* (Jackson, 1975).

INOCULATION WITH MYCORRHIZA

The mycorrhiza, a symbiotic association of root and fungus, plays an important role in plant nutrition. It is recognised that good seedling development is rarely obtained with several pines unless seedlings are inoculated with appropriate mycorrhiza. Suitable mycorrhiza can be easily inoculated by using natural inoculum, i.e., mixing of soil brought from the natural good quality forests of the species.

METHOD OF SOWING

Usually three methods of seed sowing in nursery beds are adopted: (i) broadcast sowing, (ii) dibbling and (iii) drill sowing. In broadcast sowing the desired quantity of seeds is sown by hand. In broadcast sowing, seeds should be larger in size. The minute seeds of *Adina cordifolia, Eucalyptus* spp. etc. should be mixed with earth or sand and then sowing should be done. Large-sized seeds can easily be dibbled by hand. Though *dibbling* is a good method because both spacing and depth of sowing are controlled, it becomes difficult to practise on a large scale. *Sowing by drills* is perhaps a very good method of sowing and can be practised for small and large seeds.

In seed sowing, another important point to decide is the *spacement* in sowing. In broadcast sowing usually no control on spacement is possible except sprinkling of seeds uniformly over the bed. Though spacing depends upon species the usual practice is to sow seeds in a spacing of 15 cm × 10 cm. Seeds should not be sown very densely because this results in poor germination, greater mortality, poor growth of seedlings, difficulty in weeding and more risk of damping off and other diseases. Closer spacing in seed sowing may be adopted when seedlings are to be pricked out soon after germination. Even in such cases spacing less than 5 cm is not advisable. For species which are allowed to remain in the germination bed for further growth without pricking out, spacing of 15 cm × 15 cm is necessary.

Depth of sowing is another point which needs consideration. Seeds should not be sown too deep as otherwise they may not germinate. On the other hand, if seeds are sown too shallow, there are chances of loss due to attack by ants, birds, etc. If seeds are not fully covered they may not even germinate. The usual practice is to have a soil thickness above the seeds equal to the diameter of the seeds in the case of medium and bigger seeds. In the case of minute seeds sowing at a depth of 2.5 to 5 mm is usually recommended. In light soils, seeds may be sown slightly deeper than heavy soils.

QUANTITY OF SEED

In order to determine the quantity of seeds to be sown in each bed, the knowledge of seed weight of the species, size of the bed, plant per cent of the species, etc., is required. The quantity of seed needed for a bed can be roughly worked out by the formula:

$$W = \frac{A \times D}{P \times N} \times 100,$$

where, W = weight of seed required in gms, A = area of the bed in square metre, D = number of plants required per square metre, P = plant per cent of seeds and N = number of seeds per gm. In practice, a larger quantity of seeds is sown than worked out by the formula given above. Usually the quantity of seeds sown is twice in the case of drill sowing and 3—4 times in broadcast sowing of the value obtained by the formula given above.

IRRIGATION

Irrigation in the nursery is an important aspect of nursery management. Irrigation in the nursery has to be planned most carefully as overirrigation is as harmful as underirrigation. Overirrigation also increases the costs of seedlings produced. Several species, such as teak, bamboo, etc., require little irrigation while several other species require more water. Overirrigation increases damping off, produces pampered and weak seedlings and increases cost of the plants. Therefore, even if cheaper water is available, the watering must be seriously planned. Irrigation in the nursery is usually provided by (i) automiser, (ii) watering can, (iii) percolation, (iv) flooding and (v) sprinklers.

Some species, such as *Eucalyptus, Populus,* etc., require more water than teak. Similarly, sandy soils require more irrigation than clayey soils. Soils rich in organic matter absorb more water and, therefore, less irrigation may be required in these soils. More irrigations are required in summer than in winter. In rains, usually no irrigation is required. Normally, 2 irrigations per month during winter and 3 to 5 irrigations per month during summer may be considered good for most species. Choice of irrigation method will depend upon source of irrigation, size of the nursery, availability of manpower, funds, etc. Every method has its own merits and demerits. Irrigation is usually recommended to be done in the afternoon but in places where frost and damping off are feared, it may be done in the morning.

PROTECTION OF SEEDS AND SEEDLINGS

Protection to seeds may be provided by: (i) covering the seeds with vegetative matter to protect them from birds, rodents, etc. (ii) seed dressing by fungicides and repellents, (iii) application of insecticide and (iv) soil sterilisation. *Shading* is done to provide protection to the plants from sun, frost, rain, hail etc. The requirement of shading depends on locality, species, time of sowing, etc. Shading is provided to the plants in hot and dry summer months for protection from intense solar radiation and high temperature. It also reduces evaporation and transpiration losses. Shading is necessary for several species of Lauraceae and Magnoliaceae which are sown during cold months and are unable to bear the hot sun from March—April onwards (Champion and Seth, 1968). Shading is also provided in nursery for protecting plants from frost. But shades in these cases are usually removed during the daytime. Shading also protects seeds and small seedlings from mechanical injury caused by rain and hail. Common types of shades are thatch, chics, grass and bamboo mats, palm leaf mattings, split bamboo, opaque polythene and sheet metal.

## WEEDING AND SOIL WORKING

Weeding is a very important operation and can be neglected only at the cost of growth of seedlings. Weeds must not be allowed to grow in the beds because they share food, water, light and space with seedlings. Every effort should be made to see that seeds are free from weed seeds and other impurities. Farmyard manure used in the nursery should be properly decomposed. Burning of leaf litter and other available debris over the seedbed results in lessening the weed growth.

First weeding should be done after the germination is over. It may be done earlier also if the germination period is long. In the case of broadcast sowing *hand-weeding* is done. The weeds have to be uprooted along with their roots. While weeding, the *soil working* consisting of loosening of soil around seedlings and *thinning out* of seedlings may also be done if seedlings are congested. If seeds are sown in lines, mechanical weeding and hoeing can be done.

## TRANSPLANTING

Transplanting is moving of plants from one nursery bed to another or into a pot for better root and shoot growth. The process is also called *pricking out* and *lining out*. Transplanting, though costly, has several advantages. (i) It increases root growth and produces better and efficient root system in several species. (ii) In transplanting, spacing is manipulated according to the requirement of the species. (iii) Transplanting makes the plant hardier. (iv) Transplanting beds provide a new environment for growth and results in better growth as better nutrient reserve is available.

Transplanting or pricking out is commonly done for coniferous species, e.g., *Pinus* spp., *Cedrus deodara*, *Abies pindrow*, *A. densa*, *Picea smithiana*, *Cryptomeria japonica*, etc. Transplanting has been found advantageous in several broad-leaved species, e.g., *Eucalyptus* spp., *Acer* spp., *Casuarina equisetifolia* and several others.

## USE OF CONTAINERS

Use of containers for raising seedlings has become very common with a large number of species. Either the seed is directly sown in the container or more commonly, the seedlings are transplanted into the container and allowed to grow for some period and thereafter these are planted in the field. The container planting gives better result than naked-root planting or planting with a ball of earth because: (i) this entails minimum disturbance to the root system, (ii) there is less damage to plants during transport and (iii) the plants in the containers can be kept in the field for a longer period with no loss in growth while naked-root plants require immediate planting as otherwise they start wilting. The plants with a ball of earth also need immediate planting for better results.

Several types of containers are used depending upon locality, climate and species, purpose of planting, etc. Some of these are: (i) dona, (ii) moss cylinder and fibre cubes, (iii) baskets, (iv) tubes made of earthern tiles, paper board, tin, etc., (v) earthern pots and (vi) polythene bags. *Polythene pots* of various sizes have gradually replaced most of the above containers because: (i) these are light in weight, (ii) have greater strength and durability, (iii) are easy to handle and (iv) comparatively cheaper. These polythene containers are made of 150–250 gauge polythene sheets. The containers are of various sizes for meeting diverse requirements.

## STUMP PREPARATION

In *Tectona grandis,* stump planting has been found to give better results than seedling or transplant planting. Stump is the cutting of root and shoot. For making stumps, plants are taken out of the bed with naked-roots. Plants which have a collar diameter equal to the thickness of the thumb (1.5–2.0 cm) and a taproot of not less than 30 cm are selected for preparation of stumps. Plants with a collar diameter of less than 1.5 cm or greater than 2.0 cm may be discarded. After selecting the plant, the shoot should be cut off with a sharp knife or pruning scissors leaving only 3 cm of the shoot portion. All the lateral roots should be pruned. The taproot should be cut at a distance of 20–22 cm from the collar. In moist areas, the root length may be reduced to 15 cm and in dry areas it may be increased to 30 cm. The cutting of shoot, root, etc., should be done with a very sharp knife or scissors. These stumps are ready for plantation in the field.

For some species such prepared stumps are grown in polythene pots for some months before they are planted out.

## ROOT PRUNING

The main aim of a nursery is to obtain healthy seedlings with a well-developed root system so that the plant is able to adapt to field conditions easily when planted out. Seedlings of several species, including *Eucalyptus, Tectona, Quercus* and almost all conifers, produce a single taproot with a few branches when grown in a seedbed. One of the objects of transplanting is to induce the development of a fibrous root system. If transplanting is not done, development of a fibrous root system can be induced by pruning the main root of the seedling.

VEGETATIVE PROPAGATION

Various vegetative parts of the plant can develop roots when they are planted, e.g., cuttings of branch and stem, root suckers, rhizomes in the case of bamboo, etc. Various horticultural techniques, such as layering, grafting, budding, etc., can also be used to a lesser extent in a forest nursery for propagation of some species. Vegetative propagation has several advantages: (i) A large number of plan.s can be obtained from a single stock. (ii) The plants will be genetically identical to the parent stock and consequently the growth and form will be uniform. (iii) A single rare tree can be propagated indefinitely. (iv) Vegetative propagation is more useful when trees are not producing viable seeds. (v) In some cases the method can be easier and cheaper. (vi) It permits multiplication of desirable hybrids without segregation of characters.

Vegetative propagation methods have several limitations also: (i) Very careful handling and better conditions are needed for the success. (ii) If cuttings are not skillfully selected, inferior growth may result. (iii) For some species elaborate nursery equipment and controlled conditions are required. (iv) A large number of species do not root even in controlled conditions or the rooting is not satisfactory.

*Small stem and branch cuttings*

*Populus* spp. and *Salix* spp. are raised by stem and branch cuttings. Other species which could be raised by stem and branch cuttings include *Ficus* spp., *Tamarix* spp., *Platanus* spp., *Bursera* spp., etc. Cuttings of stem and branch of *Lagerstroemia indica*, *Bougainvillea*, *Duranta*, etc. are also successful. Cuttings of 20–25 cm or longer are taken from terminal shoots of the main upper branches from specially raised stool beds. These cuttings are inserted in well-prepared nursery beds. The soil in these beds must be sandy loam with good fertility. Water supply should be sufficient. Spacing between the cuttings may be kept at 22 cm × 44 cm. These cuttings sprout and in *Populus* a height of about 1 to 1.5 m is obtained in the first year. In the second year they are cut back nearly to the ground and replanted in good soil at suitable spacing.

*Root suckers*

Root suckers are not being used on a large scale for raising plantations. Some species, such as *Populus* spp., *Millingtonia*, *Platanus*, *Boswellia serrata*, Palms, etc. produce root suckers which could be used for multiplication of the species on a small scale.

*Root and rhizome cuttings*

Root cuttings of *Bombax* spp., *Ailanthus* spp., *Robinia* spp., etc., may produce shoots. Similarly, bamboos can be propagated from the cuttings of *rhizomes*. In fact, in several species of bamboo, rhizome planting is the most common method of propagation.

### Layering

If a branch is bent down to the ground and/or a container and the soil is kept in contact with the branch, it will produce adventitious roots. The portion of the branch can then be severed from the tree and can develop into another tree. This method could be adopted on a small scale in the nursery.

### Grafting and budding

In grafting, a twig or a branch called the *scion* of a tree which is to be multiplied is grown on a rooted plant called the *stock* of another tree. In *budding*, a single bud with its surrounding cortex is inserted in the stock. In both cases, the top of the stock is cut back and the scion is allowed to grow. Grafting and budding are useful in raising seed orchards of genetically superior trees. Budding has been successfully used in teak trees for raising a teak seed orchard.

## Plantation

Plantation sites usually are of four types: (i) degraded lands where soil conditions are usually poor and soil erosion is rampant, (ii) wastelands where sites have one or several limiting factors, (iii) small areas consisting of homestead, around wells, ponds, courtyards, along paths etc., and (iv) agricultural lands. In most cases, the following points must be taken into consideration in selection of site: (i) The sites for plantation, as far as possible, should be easily approachable. If the sites are not approachable there may be problems in transport of planting stock, plantation work, weeding and other operations. There may be a problem in disposal of produce also. (ii) There must be enough area for undertaking plantation for several years. The area for each year should be continuous as far as possible. This facilitates supervision and protection. (iii) The site should be such that it is easy to identify and modify factors which lead to development of wastelands.

### SITE PREPARATION

The objectives of site preparation are: (i) to clear the site of existing vegetation by felling of trees and cutting of shrubs, etc. (ii) to facilitate planting and establishment, (iii) to reduce weed growth, (iv) to reduce soil erosion by creating a physical barrier to surface run-off, (v) to conserve soil moisture as much as possible in dry areas, (vi) to encourage root development, (vii) to work the soil and make it most receptive for promoting germination and growth of seedlings, (viii) to remove all surface and subsurface obstruction to plant growth and (ix) to improve drainage conditions in wet areas.

The degree of site preparation depends mainly on: (i) the species to be planted, (ii) existing vegetation, (iii) site conditions, (iv) finance and other factors. Most species grow well in well-worked soil where there is enough space for the seedlings to grow without competition and root development encounters no obstruction.

Site preparation consists of working the soil with planting hoe or digging pits or trench and pit. In waterlogged areas, soil working may involve preparation of *mounds*. More usually for planting trees, pits of 30 cm × 30 cm × 30 cm or 45 cm × 45 cm × 45 cm are dug. In low rainfall areas and poor site conditions, the trench-cum-pit method collects more moisture, which helps the growth of plants. For bund planting, deeper pits say, 60 cm × 60 cm × 60 cm are better.

## SOWING AND PLANTING

When the area to be planted has been properly prepared, the plantation can be raised either by *sowing of seeds* or by *planting* nursery-raised planting stock, e.g., seedling, stump, etc.

### Direct sowing

*Direct sowing is the practice where the seed is sown directly into the ground on the plantation site.* Direct sowing has been tried with several species but has been successful only with a few, particularly those which have large seeds. Direct sowing is successful in *Acacia nilotica, A. mearnsii, A. catechu, Prosopis* spp., *Pinus roxburghii* etc., in India. There are several advantages in direct sowing. These are: (i) Direct sowing is cheaper than planting as it avoids nursery operations and planting completely. (ii) Trees develop more naturally on the site since they are not disturbed after germination. This also helps in growth and stability of trees. (iii) It is difficult to raise planting stock of several species in the nursery and direct sowing gives comparatively better results. (iv) Several species, e.g., *Acacia nilotica*, and *Prosopis* spp., give equally good results in direct sowing; therefore, it may be unwise to raise them through nursery-raised seedlings.

There are, however, several disadvantages in direct sowing and they often outweigh its advantages. Important points against direct sowing are: (i) Relatively large quantity of seed is required to offset losses from seed-eating birds, rodents, insects, etc. Many seeds do not germinate due to adverse climate, soil and other factors. (ii) From direct sowing usually, we get irregular stocking. (iii) In a nursery, seedlings receive optimum conditions for growth and, therefore, the rate of growth is fast. They are able to compete with weeds and other adverse factors far better than new plants from direct sowing. (iv) Newly germinated seedlings are easily smothered by weeds or killed during adverse climatic factors. (v) Direct sowing requires more thorough soil working and weed control than planting.

Direct sowing has been successful only in cases where (i) seed size is large, (ii) cheap and abundant seed is available, (iii) weed growth is nil or little, (iv) soil is good and particularly fine textured, whereas root development is restricted in nursery-grown seedlings (Chapman and Allan, 1978) and (v) in species where planting stock cannot tolerate injury or shock during transport, planting, etc.

*Methods of Direct Sowing:* Direct sowing may be done by any of these methods: (i) broadcast sowing, (a) hand broadcasting, (ii) dibbling, (iii) line sowing, (iv) strip sowing and (v) patch sowing. *Hand broadcasting* of seed is done by hand after the land is prepared for sowing. The whole area needs preparation. If there is an experienced man, we can expect uniform distribution of seeds. *Line sowing* by hand or by seed drill is done in several species and the practice has been successful in *Acacia nilotica, A. catechu, Shorea robusta, Dalbergia sissoo* and a few other species in several areas of the country, particularly in taungya plantation. Soil working is done in lines by plough and seeds are sown and covered using a spade, etc. Very large seeds, e.g., *Juglans regia, Syzygium cumini, Shorea robusta, Mangifera indica,* etc., can be *dibbled* by hand. Sometimes soil working and sowing is done in *strips* usually 1 to 1.50 m wide and seeds are sown in these strips by hand broadcasting. Strip sowing is done in the case of *Shorea robusta* (Ghosh, 1977). *Patch sowing* or *spot sowing* is practised in which a small patch, usually 3 m × 3 m is cleared, the soil is worked and the seeds are sown. Such patches are located at regular intervals corresponding to the desired crop spacing. In all sowing methods, it is necessary that the seeds be in contact with mineral soil and that they are well covered. The depth of sowing of seed may vary with the species.

*Seed rate:* Seed rate is the requirement of seed for sowing a hectare of area. It depends upon the species and method of sowing. Seed rate is always higher in broadcasting than other methods (Table 75).

### Planting out

Establishment of plantation by nursery-raised seedlings, stumps, cuttings, etc. is the most common and successful method. The establishment of plantation depends upon several factors, e.g., adequate site preparation, correct choice of species, planting of adequate size of planting material at the proper time, control of weeds, protection against grazing, fire, etc. Planting has several advantages: (i) Planting is done at uniform and regular spacing which utilises the site properly, which facilitates weeding and other tending operations. (ii) Nursery-raised stocks adjust best to the site, grow fast and can establish at a shorter period. (iii) For difficult sites, plantation

Table 75: Seed rate for some species in kg/ha

| | Species | Broadcast sowing | Line sowing | Strip/patch sowing | Remarks |
|---|---|---|---|---|---|
| 1. | *Acacia catechu* | 100–125 | 35– 45 | – | Usually |
| 2. | *A. nilotica* | 400–500 | 150–175 | – | sown in |
| 3. | *Ailanthus excelsa* | 75–100 | 20– 30 | – | lines |
| 4. | *Cedrus deodara* | 25–35 | 15– 20 | 10–15 | 3–4 m apart. |
| 5. | *Dalbergia sissoo* | 100–120 | 35– 40 | 20–30 | |
| 6. | *Pinus roxburghii* | 30– 40 | 15– 20 | 10–15 | |
| 7. | *Terminalia myriocarpa* | – | 300–350 | 250–300 | Usually sown 3 m apart |
| 8. | *Shorea robusta* | – | 300–350 | 220–270 | -do- |

is perhaps the only way of afforesting them. (iv) It utilises the seed properly and, therefore, when seed production is costly, it ensures their proper utilisation. (v) When species produce little viable seed or the species is to be raised by cuttings, etc., plantation is the only method of regeneration.

*Size and age of planting stock*: One of the important factors for better survival and growth in plantations is the optimum size and age of planting stock. The optimum size and age of planting stock may depend primarily on: (i) method of planting, (ii) the species and (iii) the characteristics of the site. Several criteria are used to determine the optimum size of the planting stock. Some of these are: (i) height of planting stock, (ii) collar diameter and (iii) root-shoot ratio.

In India, for most species, the height of the planting stock is considered good at 50 to 100 cm. Experience has shown that for container-raised planting stock, the optimum height for most broad-leaved species is about 50 cm to 150 cm. For naked-root plantings, the planting stock may be 25 to 50 cm. Height is most often used as an index to grade the plants. Large size planting stock (90–120 cm) have been reported to give better survival and further growth in *Eucalyptus* hybrid (Mathur and Singh, 1973). Similarly, two-year rhizome shoots having a height of 75 cm to 100 cm are reported to be better in respect of survival and growth than one-year rhizome shoots having a height of 25 cm to 50 cm in the case of bamboo (Dwivedi, 1987). Root-shoot ratio of 1 : 2 is considered optimum for several species. Root-shoot weight is considered a better criterion (Chapman and Allan, 1978). Such criteria, however, are not available for different species under various conditions.

The optimum age of planting stock depends mainly on the species. In the case of deodar, fir, spruce, etc. the optimum age may be 2.5 years to 4.5 years, while in *Eucalyptus*, *Albizia* spp., *Azadirachta*, etc., good planting stock may be obtained within one year. In the case of teak and sissoo, seedlings of suitable size for preparation stumps, can be obtained in one to two years.

*Preparation of stock*: Preparation of planting stock includes those works which are done in the nursery for dispatch to the planting site. It includes taking out of plants from the nursery, cleaning, bundling, etc. Preparation of planting stock will differ according to the type of planting stock. Preparation for naked-root planting stock will involve taking out of plants from the nursery bed, shaking off or cleaning the plants, removing the leaves, bundling, etc. The plants are taken out from the bed in such a way that the roots are not disturbed. This can be done by digging a trench along the width of the bed up to a depth of root zone of the plants. The plants are taken out by pushing them into the trench gently. The leaves may be removed from the plants leaving only a few at the tip. They may be tied in bundles and should be kept in shade. Large plants of 1 to 2 m which are stripped of leaves to reduce transpiration losses are called *striplings*.

For taking out plants with a ball of earth, it is necessary that the nursery bed be irrigated one day before the plants are to be removed. The plants are taken out using *khurpas*, with sufficient soil left around the roots. The taproot may be cut with a sharp instrument. Such plants may be kept in grass or banana leaves wrapped round the soil and tied so that the plant is not damaged during transport.

Container plants require no special treatment. It is advisable to irrigate container plants before they are transported from the nursery. Stumps, cuttings, etc., need no special preparation. These stumps and cuttings should be tied in bundles and wrapped in wet gunny bags.

*Transport to planting site*: After planting stocks have been prepared they should be transported to the planting site as early as possible. They must be lifted and handled gently to avoid any injury to the plants. The naked-root plants and plants with a ball of earth should be kept in bamboo baskets and these baskets must be properly arranged in the trucks or tractor-trolly. At the planting site, the plants must be kept in shade near a water source so that they may be watered when need arises.

### Methods of planting

For all planting methods, the following general rules have to be followed (Dwivedi, 1990).

 (i) Place the plants in the pits, trenches or crowbar holes in such a way that (a) roots are kept in a natural position in the pit without coil and (b) the root should be in the soil up to the root collar (Fig. 21). After the soil is set, the collar of the seedling/root-shoot cutting should be in the same position or only slightly deeper with reference to the soil as it was in the nursery.

(ii) Avoid damaging roots and shoots by bending, breaking, crushing, etc.

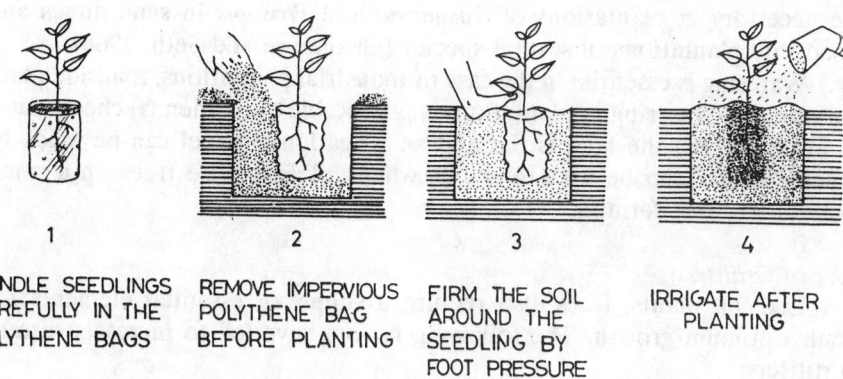

|   1   |   2   |   3   |   4   |
|-------|-------|-------|-------|
| HANDLE SEEDLINGS CAREFULLY IN THE POLYTHENE BAGS | REMOVE IMPERVIOUS POLYTHENE BAG BEFORE PLANTING | FIRM THE SOIL AROUND THE SEEDLING BY FOOT PRESSURE | IRRIGATE AFTER PLANTING |

**Fig. 21: Method of planting**

(iii) Only good soil should be placed in contact with the roots. The surface soil should be placed in the pit or hole. Leaf litter or underdecomposed organic matter should not be kept in contact with the roots.

(iv) The soil must be thoroughly firmed-up around the plant by heeling or foot pressure (Fig. 28).

(v) The impervious plant containers, e.g., polythene pots should be removed before planting.

(vi) The plant should be positioned according to the site. In dry sites, planting should be done deeper in the pits or furrow bottoms so that maximum water is available to the plant. The pit should not be filled up to the ground level so that a depression is left to collect rain-water. In wet sites planting should be done on raised mounds to avoid waterlogging.

(vii) Planting may be done in pits in well-drained soils. Making pits in heavy soil is not good, as they become nonaerated puddles.

Planting in pits may be adopted for naked-root plants or plants with a ball of earth, of container plants. Stumps can be planted in *crowbar holes*. In arid regions *deep planting* is better. The top layer of the soil is likely to be completely dry and, therefore, *deep planting* is helpful for tapping moisture in deeper layers of the soil.

*Replacement planting* or *in-filling* or *beating-up* are terms used for replanting in place of dead plants planted in the same year.

*Watering*

In plantations, watering ensures success. Since in most cases, plantations are done during rains, watering is not necessary. However, if watering is done it will ensure a greater survival percentage and early growth. If watering can be done during the dry season, it should necessarily be done, provided the cost of watering is not very high. Watering has been found to be necessary in plantations of *Casuarina* and *Prosopis* in sand dunes and road side plantations of several species (Champion and Seth, 1968).

Watering is essential in the case of industrial plantations, roadside plantations, etc. Watering in plantations is practicable only when (i) cheap water is available, (ii) the land is flat and an irrigation channel can be made or tractor-tanker can be used, and (iii) when villagers raise trees on a small scale under tree farming.

### Use of fertilisers

Like all plants, trees also require a supply of essential elements for their optimum growth. The following factors have led to increased use of fertilisers:

(i) The sites available for plantation, e.g., degraded lands, wastelands, etc., are poor in nutrients. Addition of fertilisers ensures better growth in plantations.

(ii) A recent trend is to adopt short-rotation crops. This makes fertilising more economical.

(iii) Nutrient drain by fast-growing short-rotation forest species, e.g., *Eucalyptus, Casuarina,* etc. is rapid and replenishment by nutrient cycling is not enough.

(iv) Plantation sites deficient in some nutrients may cause damage and reduced growth in plantations. Correction of such deficiencies is essential through addition of fertilising elements.

(v) Application of fertilisers in moderate to good soils which do not exhibit a deficiency in any nutrient, has been found to stimulate growth of plants. Application of NPK fertilisers even in good soils has been found to increase growth by about 50 per cent in Brazil (Mello, 1976). Significant increase in growth has been obtained in *Eucalyptus, Casuarina, Tectona grandis,* etc., in several areas in India (Champion and Seth, 1968).

Fertilisers are usually necessary during the time of establishment. Addition of fertilisers within three months of planting produces the maximum response (Chapman and Allan, 1978). Fertilisers should also be added when deficiency symptoms appear.

### Spacing

The distance between plant to plant in a plantation or in a crop is called *spacing*. Selection of proper spacing is perhaps one of the most important decisions in raising a plantation. Several factors affect the choice of spacing in plantations. Some of these factors are:

(i) *Rate of growth of species:* Slow-growing species may be planted in closer spacing than faster growing species.

(ii) *Growth form:* Some of the species are initially very branchy and need to be planted closely to promote self-pruning and formation of clear bole.

(iii) *Cost:* Closer spacing involves a higher number of plants, more soil working, etc. Therefore, costs are always higher in closer spacing.

(iv) *Availability of nutrients and moisture:* In sites where nutrients and moisture tend to be lower, spacing should be kept wider.

(v) *Weed growth:* Closer spacing tends to close the canopy earlier, which may reduce weed growth. But if weeding is to be done by tractor ploughing, a minimum spacing of about 3 m is necessary.

(vi) *Object of management:* If the object of plantation is to produce fuelwood, small timber and pulpwood, closer spacing may be adopted. Wider spacing is necessary, when the object is to produce structural timber. If intercropping with some other crops or cereals is proposed, wider spacing needs to be adopted.

Taking all these factors into consideration, optimum spacing for each plantation has to be decided. An optimum spacing may be, in theory, one which produces the greatest volume of desired size trees and the trees are dense enough to achieve full occupancy of the site. Too close spacing also exerts an adverse influence on the growth of individual trees and other crops. Wider spacings are also not desirable as they (i) reduce total production, particularly if rotation is short, (ii) tend to increase the taper which reduces the conversion and (iii) increase crown size. Since very close spacing is extremely expensive and very wide spacing grossly underuses the site, approximate spacings for different kinds of crops may be as in Table 76.

**Table 76: Spacing for different purposes (Dwivedi, 1990)**

| Sl. No. | Crops | Growth and management criteria | Spacing (m) |
|---------|-------|-------------------------------|-------------|
| 1. | *Firewood* | Maximum yield, short rotation, no limit for small size | 1 to 2 |
| 2. | *Pulpwood* | Maximum yield, short rotation (5 to 15 yrs), size limit 10–40 cm diameter class | 2 to 3 |
| 3. | *Saw-timber & veneer logs* | Long rotation, large log size, regular thinnings, loss in total volume is compensated by high value of wood | 2.5 to 4.5 |

The common spacing adopted for some Indian species is indicated in (Table 77).

Table 77: Common spacing for some species (Dwivedi, 1990)

| Spacing | Species |
|---|---|
| Up to 1.8 m | *Eucalyptus* spp., *Leucaena leucocephala*, etc. |
| 1.8–2.5 m | *Abies pindrow, Casuarina equisetifolia, Cedrus deodara, Cryptomeria japonica, Dalbergia sissoo, Eucalyptus* spp., *Mesua ferrea, Michelia champaca, Prosopis chilensis, Salix* spp., *Tectona grandis, Leucaena leucocephala* |
| 2.5–3.5 m | *Acer* spp., *Acacia nilotica, A. catechu, A. mearnsii, Adina cordifolia, Albizia lebbeck, Toona ciliata, Chukrasia tabularis, Gmelina arborea, Kydia calycina, Bombax ceiba, Santalam album, Terminalia arjuna* |
| 4–6 m | *Bombax ceiba, Hymenodictyon excelsum, Populus deltoides, Dendrocalamus strictus* |
| More than 6 m | *Anacardium occidentale, Bambusa arundinacea, Mangifera indica, Tamarindus indica,* etc. |

## Maintenance

After plantations are over, one finds several planted seedlings have died. Several factors affect initial death of seedlings. These factors include: (i) defective and careless planting, especially unfirm soil around plant, shallow planting, coiled roots, etc.; (ii) planting of defective and injured seedlings; (iii) poor soil conditions, especially waterlogging, or presence of injurious salts and hard pan below the soil surface; (iv) adverse post-planting weather conditions, particularly drought, high temperature, high intensity rain, floods, etc.; (v) attack of insect pests, e.g., termites; (vi) attack by diseases; (vii) competition by weeds; (viii) grazing, browsing, trampling by animals, domestic and wild, and (ix) fire, etc. If death of seedlings is proportionately high, they will have to be replaced. If weeds are more, weeding is necessary. If attack of insect pests, and diseases are noticed, plant protection measures have to be adopted. Thus, there are several silviculture operations which ensure the establishment of a plantation. These operations are included under maintenance operation of plantations. Some of the important maintenance operations are replacement of casualty, weeding, soil working, mulching, control on grazing, etc.

If seedling mortality is high, the plants are replaced by healthy seedlings. This operation is called replacement of casualty. This is also called *beating-up* or *'blanking'* or *in-filling* or *re-filling*. Seedlings used for beating-up should be healthy, vigorous, robust and taller than average so that they catch up with the initially planted seedlings. The time of this operation is very important. The operation should be done within a month of planting out so that seedlings receive sufficient period of rains for their establishment and growth.

The plantation should be weeded at least for 2 to 3 years. In a year,

2 to 3 weedings are necessary. Weeding around the plant is the common method usually adopted. When an agricultural crop is cultivated in the interspaces, weeding and soil working are not done separately. *Mulching* helps in checking moisture evaporation losses. Leaf litter or some organic matter is used as mulch which helps the growth of plants. The plantations have to be protected from grazing and browsing. For this, fencing has to be carried out.

CHAPTER 15

# Growth of Trees and Crops

## Silviculture

Silviculture is that branch of forestry which deals with the establishment, development, care and reproduction of stands of timber (Toumey and Korstian, 1947). The term silviculture refers only to certain aspects of the theory and practice of raising forest crops (Champion and Seth, 1968). Generally, silviculture is defined as the art and science of cultivating forest crops (Anon., 1966). Silviculture in forestry is somewhat analogous to agronomy in agriculture. It deals with mainly cultivation of forest crops. Silviculture has its origin in the word *Silvics*. Silvics is the study of the life history and general characteristics of forest trees and crops with particular reference to environmental factors, as the basis for the practice of silviculture (Anon., 1966). Silvics means the study of trees and forests as biological units, the nature of growth and development and the effect of environmental factors on them. Silviculture, however, does not exclude the economical aspects of growing trees. Silviculture includes propagation and regeneration of forest crops, selection of species, care and maintenance of forests, manipulating composition, improving site conditions, tending operations, etc.

## Forest Crop

There are about 199,000 species of angiosperms, of which 159,000 are dicotyledons and 40,000 are monocotyledons. The number of gymnosperm species is reported to be about 700. Of these, about 10 per cent of the species are reported to occur in India. A *species* is a group of individuals of plants having a close resemblance to one another structurally and functionally. The individuals of a species interbreed in nature freely and successfully among themselves and they produce the same type of progeny. They have normally the same number of chromosomes in their cells. A *genus* is a collection of species which bear a close resemblance to one another in the morphological characters of the floral and reproductive parts. Sometimes under the influence of external conditions certain individuals of a species may show a marked degree of variation in form, size, shape, colour and other minor characteristics. Such plants are said to form varieties. A species may have one or more varieties or none at all.

A forest crop consists of trees, shrubs, herbs and grasses. The grasses and annual herbs are small plants and constitute ground cover. *Herbs* are those plants whose height is usually not more than a metre and may be annual or perennial. *Shrubs* are woody perennial plants larger than herbs and smaller than trees usually below 4 meters in height. *Trees* form the upper strata in forests. A tree is a large woody perennial plant, usually with a single stem or trunk from which limbs or branches sprout at some distance above the ground to carry a spreading crown of leaves. According to another definition, trees are woody plants having one erect perennial stem or trunk at least 7.5 cm diameter at breast height, a more or less definitely formed crown of foliage and height at least 4 m (Anon., 1944). Tree as a legal term in forest laws includes not only trees as defined above but also such plants as shrubs, bamboos, canes and even stumps and brushwood. While framing these definitions, foresters have in mind the utility of trees as a source of commercial timber. For silviculturists, trees are important because they yield products and benefits in larger quantities than other forms of vegetation. For ecologists, trees are important because of their competitive ability and as a unit, capable of casting shade and sharing nutrients, water and space with other plant communities. From an anatomical point of view, trees are more rigid than other plants because of their capacity to produce legnin.

Trees are the most important elements in forests as they provide maximum quantities of products and benefits. Therefore, most forests of the world are managed primarily with a view to producing trees.

FORMS OF CROWN

A tree consists of a crown and stem above the ground and roots below the ground. The *crown* of a tree is the branchy part of a tree above the bole. It is formed by branches which originate from the bole. A crown has a height, length and width. *Crown height* is the height of crown as measured vertically from the ground level to the point half-way between the lowest green branches forming the green crown all around and the lowest green branches of the bole. *Crown length* is the vertical measurement of the crown of a tree from the tip to the point half-way between the lowest green branches forming the green crown all around and the lowest green branch on the bole. *Crown width* is the maximum spread of the crown expressed as its widest diameter. The crown length, height and diameter at various points determine the *shape and size of the crown*. Shape and size of the crown of trees vary with species and environment in which they grow. The crown is the result of the branching behaviour in the bole. In some trees, e.g., *Phoenix, Cocus, Borassas*, etc., there is no branching in the stem and the crown is formed by large leaves which come out from the top of the unbranched stems. In other trees, the crown may be: (a) *conical* as in the case of pines and deodar, (b) *cylindrical* as in fir, spruce, eucalypts, etc., (c) *spherical* as in mango, neem, mahua, imli, etc., (d) *broad and flat topped* as in *Acacia planifrons*, (e) *broom-shaped* as in babul and (f) *frondose crowns* as in *Prosopis*, etc. (Fig. 22).

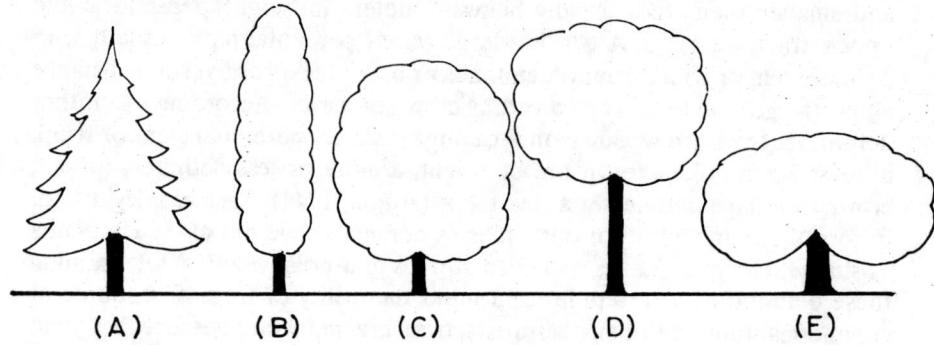

(A)        (B)        (C)        (D)        (E)

Fig. 22: Crown shapes

There may be several other shapes of crowns and in the same species crown shapes and size vary considerably with canopy density. Trees growing in a closed canopy tend to eliminate the lower branches due to natural pruning. This results in short and smaller crown forms (Fig. 23).

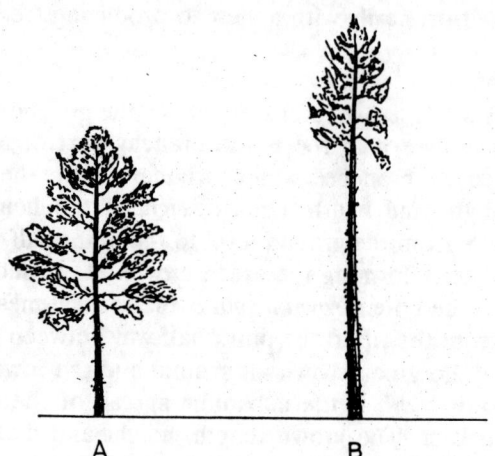

A                B

A – GROWING SOLITARY IN THE OPEN
B – GROWING IN A DENSE CROP

Fig. 23: Tree growth in dense and open crop

BRANCHING

The mode of branching of the stem is a species characteristic. Branching is probably always lateral, i.e., branching arises from lateral buds in the axils of leaves. The branching may be *racemose* or *cymose*. In racemose branching, at each node, there may be either a single branch or a series of two or more branches (*whorl branching*). In cymose branching, it may be uniparous, if only one daughter axis is given off at each branching, *biparous* if two daughter axes are given off and multiparous, if more than two daughter axes are given off at each branching.

In most trees, branches leave the axis at angles of 60°–70° but in some species, e.g., *Populus, Cupressus, Acacia* (Ramkanta variety of *A. nilotica*) branches arise at more acute angles, usually 25° to 40°. In a few species, branches may arise at a right angle, e.g., *Homalium, Casearia, Bombax*, etc. High level conifers usually have obtuse angle branching.

FOLIAGE

Several characteristics of the foliage are important which agroforesters should know. These include leaf shedding, colour, size, texture, shape, drip, twig shedding, etc. From the silviculture point of view, trees may be deciduous or evergreen. *A tree is called deciduous if it normally remains leafless for some time during the year.* The period of leaflessness varies from species to species and area to area in the same species. For example, sal remains leafless for a very short period, e.g., a week, while several species remain leafless for longer periods. *Hymenodictyon excelsum, Populus deltoides, Tectona grandis, Boswellia serrata, Lannea coromandelica, Garuga pinnata*, etc., remain leafless for longer periods, usually 2 to 5 months.

In the same species, the period of leaflessness depends on the environment. Abundant well-distributed rainfall, lower temperature and higher humidity, etc., tend to reduce the period of leaflessness. For example, in drier areas teak remains leafless for almost 4 to 5 months while in moist areas with high rainfall and higher humidity, it tends to become evergreen and remains leafless only for a very short period. Important deciduous species are khair, sissoo, teak, babul, sisham, semal, sain, gamari, bija, haldu, dhawra, etc.

*An evergreen species is a perennial plant which is never entirely without green foliage, the old leaves persisting until a new set has appeared.* The evergreen character depends upon the species and the environment in which it grows. *Pinus roxburghii* is an evergreen species but it tends to be deciduous in lower elevations. The old leaves are not necessarily shed annually in evergreen species. They are usually retained for longer periods. In *Pinus roxburghii*, they are retained for one-and-a-half year while they persist for 5 to 6 years in the case of *Cedrus deodara, Abies pindrow*, etc. In evergreen species, new foliage is fairly well developed before there is appreciable fall of the old leaves.

Important evergreen species include: *Dipterocarpus* spp., *Hopea parviflora*, *Michelia champaca*, *Pterocarpus acerifolium*, *Masua ferrea*, *Calophyllum* spp., etc. Coniferous species, e.g., *Abies pindrow*, *Cedrus deodara*, *Cupressus torulosa*, *Pinus wallichiana*, etc., are also evergreen species. *Santalum album*, a root parasite tree, is found to become evergreen or deciduous according to the nature of the host on which it is parasitic (Champion and Seth, 1968). In several species, twigs along with the leaves are also shed regularly, such as *Cryptomeria*, *Taxodium*, *Quercus* spp., etc. Mature leaves of most of the trees are green. The upper portion of the leaves is usually darker than the lower. In several species, the lower parts of the leaves are covered with rusty or white tomentum, e.g., *Oaks*. The new foliage in several evergreen species is bright red in colour. As leaves grow old, they turn green in colour. In several species, leaves turn yellow when they are about to fall.

Leaf size depends upon the species and site conditions. In a large number of species, the leaves are usually small and do not exceed 20 cm but there are several species with very large leaves. Species with large leaves include *Tectona*, *Dellenia*, *Sterculia*, *Steriospermum*, etc. *Oroxymum indicum* has very large compound leaves of 1 to 1.5 m in length. The size of the leaves in the same species varies with site factors. Areas with fertile soils and well distributed good rainfall support tree growth with comparatively larger leaves than areas with poor soils and low rainfall.

## STEM

A *stem* of a plant is the principal axis of it on which buds and shoots are developed; in trees, stem, bole and trunk are used synonymously (Anon., 1966). The bole is the main stem of tree. Sometimes it refers only to the lower part of the stem up to that point where the main branches are given off. A *clear bole* refers to that part of the bole which is free from branches. A *commercial bole* refers to the length of the bole that is ordinarily fit for utilisation as timber. A *standard timber bole* is the length of the bole from the ground level up to the point where the average diameter overbark is 20 cm.

Several tree species have an inherent tendency to form a tall, erect and straight bole with relatively small branches. This is a good characteristic and is usually found in most conifers, *Dipterocarpus*, *Eucalyptus*, *Bombax*, *Michelia*, etc. However, there are several tree species which have a small trunk, e.g., *Butea monosperma*, *Lagerstroemia speciosa*, *Acacia nilotica*, etc. In stems the characteristics which are important include: (i) taper, (ii) forking, (iii) clear bole, (iv) buttressing and (v) fluting.

### Taper

The decrease in diameter of the stem of a tree from the base upwards is called taper. Tree stems are thicker at the base and thinner at the top.

This is necessary to transmit the wind pressure from crown and stems to the root system and to ensure a sufficiently stable form. The shape of the stem is close to a truncated cone or paraboloid. The taper of the stem is usually expressed in terms of *form factor*, which is the ratio of the volume of a tree or its parts to the volume of the cylinder of the same height and the same base (usually taken at breast height).

It has been observed that in trees which have a smaller crown volume or the crown is conical, the taper is less, as in *Eucalyptus* and most conifers, particularly in fir, spruce and deodar. In broad-leaved species, the taper is usually greater. The trees of the same species growing in a dense crop taper less than those growing in the open.

*Forking*

Forking is a situation wherein a stem has more than one leader. It is an undesirable characteristic from the point of view of timber production. Forking is common in broad-leaved species and rare in conifers. It is caused due to injury to the growing apical meristem. Forking in conifers has been observed in areas where snow damage is more.

*Clean bole*

That part of a bole or stem which is free from branches is called a clean bole or clear bole. The length of a clean bole is important for quality timber production. The formation of a clean bole depends upon the shedding of the side branches. A larger clean bole is formed when side branches are shed early. If branches become large, they are likely to produce knots even if they are shed after sometime. In several species, side branches persist for longer periods even if they are growing in a dense crop, e.g., *Agathis*, *Bombax*, etc. In order to obtain a clean bole, early pruning of branches is necessary. If such pruning is delayed in nature, it may be done artificially. Pruning in some species, e.g., poplars, helps to obtain longer clear bole and larger timber volume.

Some times trees with a clean bole, in the later part of their life, due to some adverse factors, develop small branches which are called *epicormic branches*. Usually these branches develop in clusters from dormant or adventitious buds on stems due to the influence of adverse factors, e.g., excessive light, fire or suppression.

*Buttressing*

The buttress is an outgrowth from the base of the tree connecting it with the roots, especially common in tropical rain-forest species. Buttresses are common in several wet tropical species, e.g., *Lophopetalum*, *Sterculia*, *Pterocarpus*, *Acrocarpus*, *Bombax*, *Terminalia myriocarpa*, etc., and may extend up to 5 metres in the stem.

Buttressing is common in wet and moist tropical forests and is practically absent in trees of drier areas. Buttressing is found to be associated with the absence of a pronounced taproot system (Richards, 1952). The position of the buttress is usually vertically above the large lateral roots to provide additional support for the stability of the tree.

*Fluting*

When a stem shows irregular involutions and swellings, it is called a *fluted stem*. It is considered a serious defect for timber production. Fluting is usually associated with buttress formation. It also develops below the base of large branches. Several species, e.g., *Tectona grandis*, *Mitragyna parvifolia*, etc., have a marked fluting tendency.

ROOT

Roots are an important portion of plants. The roots support the plants firmly on the ground and absorb soil moisture containing nutrients and assist in translocation to stems, branches and leaves. The root system consists of a *taproot system* which is formed by the primary root and its branches. The taproot grows vertically downward to longer depths while branched roots grow obliquely downward or horizontally outwards.

A root consists of a (i) root cap, (ii) region of cell division, (iii) region of elongation and (iv) region of maturation. The *root cap* occurs at the apex of the root system. The *region of cell division* is the growing apex of the root near the root cap. *The region of elongation* lies above the region of cell division. *The region of maturation* lies at the top of the root system. This region has clusters of root hairs which absorb water and nutrients.

When seeds germinate, roots develop from the radicle. When roots grow from any part of the plant body other than the radicle, they are called *adventitious* roots. Taproots also become modified due to site conditions. Several plants, e.g., *Rhizophora*, *Heritiera*, etc., growing in marshy places and occasionally places inundated by tides as in Sundarbans, develop special kinds of roots, called *respiratory roots* or *pneumatophores* for *respiration*. These roots grow from the underground roots but rise vertically. They grow in large numbers near the trees. The upper end of such roots have numerous pores for respiration. Some trees with large branches and crowns, e.g., banyan, India-rubber plant, screw pines, etc., produce a number of roots from the main branches to grow vertically downward to penetrate the soil. Such roots are called *prop* or *stilt* roots. In large trees, some of the stout roots around the base of the main trunk in large trees from an abnormal growth which emerges above the soil surface and is called a *buttress root*.

In several tree species, minor roots are found to be associated with fungi, called mycorrhiza (fungus + root) which help trees in their nutrition. Roots of several leguminous species have swellings called *root nodules*. These root nodules provide shelter to nitrogen-fixing bacteria.

## GROWTH REGIONS

Growth in plants is a complex phenomenon associated with several physiological processes. Growth results in formation of cell substances and new cells and tissues. *Growth means a permanent and irreversible increase in size and form attended by increase in weight*. Growth does not take place throughout the whole length of the plant body, but is localised in *meristematic* regions. Meristematic regions are those which are composed of cells that are in a state of division or retain the power of division. On the basis of position, meristems may be: (i) apical, (ii) intercalary and (iii) lateral. (i) The *Apical meristem* lies at the apex of the stem and the root, representing their growing regions. Growth in the length of plants is due to the gradual enlargement and elongation of the cells of the apical meristems of roots and stems. (ii) *The intercalary meristem* lies between masses of permanent tissues, either at the base of the leaf, as in pine, or at the base of a node or internode, as in grasses, and gives rise to permanent tissues. (iii) *The lateral meristem* is the cambium of the stem which lies laterally in the strips of elongated cells, extending from the apical meristem as in stems of dicotyledons and gymnosperms. It divides in a tangential direction giving rise to the secondary permanent tissues and is responsible for growth in thickness.

## PHENOLOGY AND GROWTH

*Phenology refers to the seasonal changes in the development of foliage, flowering, fruiting*, etc. Phenological changes are controlled by species, locality, climatic conditions, etc. The time of development of *new foliage* differs from species to species and within the same species; it depends upon locality and climatic factors. Evergreen species are usually never leafless. They continue shedding leaves for long periods. Some of them start shedding their leaves during winter and re-equip themselves with a new flush of leaves during February–March, as in *Hopea, Mesua, Michelia*, etc. On the other hand, there are a large number of deciduous species which are leafless during winter for a considerably long period, e.g., *Bombax ceiba* (Dec.–March), *Populus deltoides* (Nov.–March), *Gmelina arborea* (Dec.–March), *Tectona grandis* (Jan.–March), etc. When the species is leafless, it is physiologically not active. The growth periods start soon after the new leafing.

The periods of bearing new leaves, flowering and fruiting depend upon the area and the climate. For example, in mango, flowering takes place during January–February in Southern India and during March–April in northern India. It has been observed that low temperature, high relative humidity and high rainfall tend to decrease, the period of leaflessness. For example, in dry areas, teak remains leafless for 4 to 5 months but in high rainfall areas it is leafless for a period of less than a month.

Growth in most trees starts after the end of the winter season and at the beginning of new leafing. It usually corresponds to Feb.–March and continues up to September–October. The peak growth period may depend upon species and locality but it is usually during June to September when sufficient moisture is available and other growth conditions are favourable.

## Growth rings

The main stem and branches of a tree consist of the central wood (xylem), the cambium and the bark (phellem). The xylem tissues in gymnosperms consist of tracheids, wood ray cells, and wood parenchyma. The tracheids are densely packed interlocking elongated cells. Wood ray cells serve as a route for transverse translocation of water from the xylem to the cambium and phloem. The wood parenchyma consists of parenchymatous cells which are of frequent occurrence in the xylem. They help in translocation and food storage. In the angiosperms, the xylem tissues consist of vessels, tracheids, fibres, parenchyma strands, wood rays, gum ducts, etc. The vessels are the most conspicuous.

In gymnosperms, the xylem consists mostly of tracheids which occupy about 90 per cent of the volume. The tracheids are vertical, thick-walled, dead cells of 0.5 mm to 15 mm length. There is a marked difference in size, shape and cell wall thickness of the early and late wood tracheids which produce growth rings or growth marks, as in chir, kail, deodar, fir, spruce, etc. In several broad-leaved species, several anatomical features are responsible for the production of growth rings as detailed in Table 78 (Chaudhari, 1939, 1940).

**Table 78: Anatomical features and growth rings in species**

| | Characters | Species |
|---|---|---|
| 1. | Ring porous | *Tectona grandis, Morus alba, Melia azedarach* |
| 2. | Semi-ring porous | *Juglans regia, Toona ciliata, Pterocarpus marsupium, P. dalbergioides* |
| 3. | Initial parenchyma | *Terminalia alata, Chloroxylon swietenia* |
| 4. | Terminal parenchyma | *Michelia champaca, Magnolia campbelli* |
| 5. | Differences in early and late wood fibres | *Schleichera oleosa, Anogeissus* spp., *Albizia* spp., *Grewia* spp. |
| 6. | Differences in size and frequency of vessels in early and late wood | *Boswellia serrata* |

These growth rings are distinct in some species and not so distinct in other species. Several species, e.g., *Pinus* spp., *Cedrus deodara, Abies pindrow, Acacia catechu, Anthocephalus chinensis, Artocarpus chaplasha, Be-*

*tula* spp., *Toona ciliata, Gmelina arborea, Lagerstroemia speciosa, Michelia champaca, Tectona grandis*, etc., have growth rings which are distinct. In several species growth rings do exist but they are not so distinct. Such species include *Adina cordifolia, Albizia lebbeck, A. procera, Dalbergia sissoo, Duabanga sonneratioides, Bombax ceiba, Schima wallichii, Sterculia villosa, Trewia nudiflora*, etc.

The formation and width of growth rings depend upon several factors. Several species of Pinaceae and Magnoliaceae show the presence of growth rings. Sometimes these growth rings are correlated with seasonal rates of growth. Therefore, one would expect that annual growth rings must be present where there is marked seasonal variation with a period of unfavourable growth due to drought or cold. But this is not the case. Some evergreen species where growth conditions are almost uniform also possess growth rings.

The width of annual growth rings primarily depends upon growing conditions. Width of growth rings is usually greater in the case of large crowned trees in favourable growing conditions. The average width of the ring correlates with climatic fluctuations. Wider growth rings indicate a faster rate of growth and are associated with poor timber properties.

*Heartwood*

In a large number of species, there is often differentiation into an inner core of *heartwood* and an outer ring of *sapwood*. The heartwood is often denser, darker and more durable than the sapwood. The heartwood differs from sapwood as the pores in heartwood are more or less blocked with cellular growths called *tyloses*.

The heartwood is darker in colour and is easily identified. However, there are several species, e.g., *Mangifera indica, Abies pindrow, Anthocephalus chinensis, Sterculia villosa, Tetramelis nudiflora*, etc., where heartwood does not develop a dark colour and, therefore, it is not distinguishable from sapwood. Heartwood in several species, e.g., *Tectona grandis, Dalbergia sissoo, D. latifolia, Albizia lebbeck, A. procera, Acacia nilotica, A. catechu, Shorea robusta*, etc. has a commercial value several times more than its sapwood. The sapwood is related to the demands of the crown for water. The area of sapwood is smaller in slow-growing and short-crowned trees and bigger in fast-growing and large-crowned trees. The period required for the formation of heartwood differs from species to species. Formation of heartwood takes 20–30 years in deodar, 18–25 years in teak, 20–30 years in sal, 15–20 years in sissoo, etc. In fact, it is not the formation of heartwood but its total quantity which is important because heartwood is several times more durable and commercially more important than sapwood. It should be noted, however, that strength and other physical properties of sapwoods are only slightly, if at all, inferior to those of heartwood (Champion and Seth, 1968).

### Growth of Trees and Crops

Growth means increase in size and the formation of new tissues. Growth an important phenomenon, common to all living beings, follows certain basic principles. Within the species, plants have a certain inherent control on the pattern of growth. Three stages of growth in plants are easily recognised: (i) formative period, (ii) grand period of growth, i.e., period of rapid increase in size and (iii) a period of slow growth, i.e., period of maturity (Fig 24). We are mainly concerned with the growth of trees and the crop. The process of growth in plants starts with the process of germination of the seeds. It is governed by various factors, e.g., site conditions, inherent capacity of the tree for photosynthesis, availability of light, water, temperature and nutrients. The formation of new tissues in trees depends upon the total assimilate. A small part of the assimilate is consumed in respiration and the balance is used for the growth of shoots, roots, branches and replacement of worn-out tissues (Table 79).

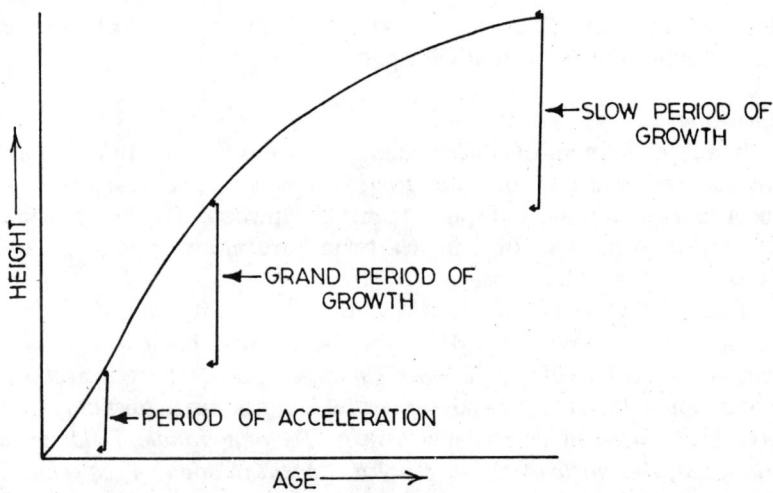

**Fig. 24: Pattern of height growth**

The different stages of growth of a tree are generally recruit, seedling, sapling, pole and tree. The use of these terms has been standardised for India as follows (Anon., 1966):

*Seedling:* From germination up to a height of 1 m.

*Sapling:* From the time the young tree reaches 1 m height until the lower branches begin to fall. A sapling is characterised by the absence of dead bark and its vigorous height growth.

*Pole:* From the fall of the lower branches to the time when the rate of

**Table 79: Approximate distribution of assimilate in a middle-aged tree**

| Sl. No. | Item of consumption | Percentage |
|---------|---------------------|------------|
| 1. | Respiration | 10 |
| 2. | Growth | |
| | (a) Roots | 10 |
| | (b) Bole and branches | 40 |
| | (c) Flower and fruit | 1 |
| 3. | Replacement of worn-out tissues, mostly leaves | 39 |
| 4. | Total assimilate | 100 |

increase in height begins to fall and the crown expansion becomes marked.

*Tree:* After passing the pole stage.

These terms are widely used in forestry to describe the stage of development of a tree. It is obvious that clear-out distinction is not possible between the sapling and pole and the pole and tree. From the point of view of practical forestry, the following types of growth in a tree are important:

(i) Height growth

(ii) Diameter growth

HEIGHT GROWTH

The meristematic cells at the tip of a shoot divide and elongate which causes height growth in the shoot. The height growth of trees generally has a brief period of juvenile acceleration followed by a very rapid growth in the sapling and pole stage and a long period of maturity in which height growth is usually very slow (see Fig. 24).

The period of height growth in a year usually differs from species to species and with climatic conditions. Most trees grow in height during March to September in northern India. Growth in winter months may also be significantly high in areas receiving winter rain. At Dehra Dun the trend of height growth indicates that in the case of teak and haldu, the maximum height growth corresponds to the period of maximum temperature while in some species the maximum growth period corresponds to the period of maximum rainfall and temperature. The rate of growth in a season mainly depends upon the geographical character of the species and availability of water and nutrients.

GROWTH IN DIAMETER

The thickness of stems and branches is increased by the growth in lateral meristematic tissues which produce xylem and phloem. The cambium, which is a self-perpetuating layer, consists of two types of cells, viz., fusiform initials and the ray initials. Fusiform initials elongate and divide in a tangential plane into longitudinally arranged elements which usually make up the fibres, vessels, etc., of the wood. The ray initials give rise to ray tissues in xylem and phloem. Due to division in the cambium cells, the diameter of the cambium increases, which results in an equivalent amount of increase in the diameter of the tree. The rate of diameter growth is not uniform throughout the year.

Diameter growth is very important from the utilisation point of view. A tree cannot produce logs for sawing and plywood until it is of sufficiently large diameter. The rate of diameter growth is similar to that of height, i.e., the rate of growth is faster during early and middle age and decreases at maturity. The rate of growth in diameter differs from species to species. In the same species, it differs with site quality and crop density. Fast-growing species will usually have a thicker diameter than the slowgrowing.

*Site and diameter growth*

The poorer the site, poorer is all growths, viz., diameter, height, etc. The poorness in site decreases the growth in diameter considerably. There seems to be no general rule as to whether poorness of site depresses diameter growth more than the height growth or vice-versa. But observations indicate that height growth is more easily affected by site quality than diameter growth. This does not, however, mean that the effect of site on diameter is less important. The diameter growth is reduced considerably with decrease in the site quality.

*Crop density and diameter growth*

Diameter growth in individual trees is greatly affected by the density of the crop. The effect of crop density on height growth is not very prominent unless the tree is suppressed or the crop is too dense or very open. Diameter growth in an open crop is greater than in a dense crop. This may be due to the fact that more light falls upon the crown and more water and nutrients become available to the roots, resulting in greater production of photosynthate and dry matter.

A greater rate of diameter growth is usually obtained in trees after the crop has been thinned. But it is wrong to presume that thinning should cause dominant crop trees to show a marked increase in annual growth. Acceleration in growth is likely to occur only when the growth of a tree has been seriously restricted by competition due to higher density. In fact, greater response is likely to occur in trees of co-dominant and dominant crown classes, provided that they have not been suppressed beyond the point of recuperation. In a dense crop, tapering is also reduced but in an open crop, higher tapering is often encountered. The degree of response to the density may differ from species to species. Less hardy species are usually affected more intensively than hardy species due to competition. The shade intolerant species under extreme competition suffer so severely that their whole pattern of life becomes disorganised; when the suppression is removed, some may show recovery to their normal growth while others may not. Due to higher density, root competition for nutrients and water as well as competition for light is increased and only hardy species may tolerate the competition. But how they behave after the density is opened may depend upon a number of factors.

**Age of the Forest crop**

On the basis of age, forest crops are usually classified into: (i) even-aged and (ii) uneven-aged. An *even-aged* crop is that in which trees are almost of the same age. The character and form of an even-aged crop vary largely on the fact of whether the crop comprises a single species or a number of species. With the same number of stems per unit area, the competition is keener when a single species is present in comparison to a mixture of a number of species. This is due to the fact that the tree crowns and roots remain almost on the same level. The even-aged crops are often classified into young, middle-aged, mature and overmature.

Uneven-aged crops consist of trees of different ages. The uneven-aged crops, on the basis of distribution of trees, are classified into: (i) balanced uneven-aged and (ii) irregular uneven-aged. A balanced uneven-aged crop consists of trees of all age classes and each of them occupies an almost equal area. An irregular uneven-aged stand does not contain trees of all age classes.

ADVANTAGES OF EVEN-AGED CROP

Several advantages and disadvantages are attributed to even-aged crops. The main advantages of even-aged stands are: (i) There is greater uniformity in size. (ii) There are more trees of the same size. (iii) Bole formation is better, trees grow to larger heights and length of clear bole is also more. (iv) Natural pruning is better. (v) Crowns are small. (vi) Wood is usually of better quality; taper is less. (vii) Silvicultural operations are easier and cheaper.

An even-aged crop has certain disadvantages. These disadvantages are: (i) The soil is better and more uniformly protected in an uneven-aged crop because the canopy is deeper, multilayered and denser. (ii) In an even-aged crop, there is little natural regeneration while in an uneven-aged crop natural regeneration is plentiful. (iii) The even-aged crop is more adversely affected by wind, pest, diseases and other natural factors.

**Composition of Crop**

The tree crop may consist of a single species or a mixture of several species. The crop is called *pure* when it consists predominantly of a single species.

The crop is called *mixed* when more than one species form the crop and the contribution of a single species is not sufficient enough to classify it as pure. For practical purposes, if a species constitutes more than 75 per cent in the overwood it may be called a pure crop.

Although pure crops are ecologically less desirable, there are several advantages with pure crops. These advantages are: (i) Management of the crop is simple. Various silvicultural operations require less skill. (ii) Natural pruning in a pure crop is more uniform than in mixed stands. Even-aged pure crops result in timber production of high quality with less taper. (iii) The crop can be harvested more economically. (iv) Only the species which is in demand would be grown. (v) The crop in unit area would be more valuable. (vi) Raising plantations of a single species is much easier and less expensive. (vii) The resultant crop is more uniform.

The advantages of a mixed crop when correctly maintained are: (i) When the mixture is suitably arranged, the site is more efficiently utilised. (ii) A mixed crop is considered better for its effect on site than pure crops. There is greater risk of soil deterioration under pure crops. (iii) Mixed crops are better protected from the epidemics of insect pests and diseases (Beason, 1941; Bakshi, 1976). Sporadic natural mixtures of a large number of species are better than artificial line or block mixture. (iv) A mixture of shallow-rooted with deep-rooted species protects the crop from wind damage. (v) Mixed crops are better adapted to poorer sites than most pure crops. Mixed crops maintain the soil properties much better than pure crops. (vi) Mistakes made in selection of species in plantations are more easily corrected in mixed crops than in pure crops. (vii) A mixed crop is better from an aesthetic point of view. (viii) A mixed crop is capable of meeting the diverse needs of the family/community. (ix) Natural regeneration of mixed crops is much easier than pure crops.

# CHAPTER 16

# Management of Trees in Agroforestry Systems

## Management

*Management* of any undertaking includes the organisation and conduct of all operations that are needed to fulfil the purpose of the owner. This definition of management however, excludes any mention of the policy which governs the purpose of any undertaking. The management of any undertaking embraces three main interlinked functions, namely: (i) deciding the purposes and main policy to be pursued, (ii) the planning and organisation of activities and (iii) the conduct of operations (Osmastan, 1968).

The main purpose of agroforestry systems is to produce enough food-grains, fodder, fuel, timber and other products so that the country is able to meet the needs of a large human and livestock population. At the same time, the production system and the crop-mix should be such which are able to maintain the productivity of the land on a sustained basis. Within this general framework, the purpose of an owner may be to maximise his profits from his piece of land. The policy should determine which part of his land is to be allocated for agriculture, agroforestry, forestry, buildings, roads, and similar other activities. If there is a question of preference from one type of crop to another, the same should be decided taking a broader objective into consideration. For example, the National Forest Policy, 1988 while emphasising the need for increasing the forest area in the country, recommends that good agricultural lands should not be diverted for forestry purposes. Therefore, decision about the policy will require assessment of purposes of the landowners. The purposes may be decided taking into consideration the availability of land, labour, and capital, the trend of demands for various kinds of commodities and technology available along with personal commitments.

The planning and organisation component decides how the resources of the undertaking should best be utilised to attain the purpose and policy. Planning and organisation are required for buildings, machines, capital, labour, etc., as are arrangements for providing maximum output. The products, both material goods and services, need proper planning for storage, consumption, market and disposal. Similarly, organisation of personnel is required to be built up for conducting various duties and responsibilities.

Conduct and control of various operations are also an integral part of management. This aspect of management concerns the daily running of the undertaking. It is also called administration of the undertaking. All aspects of management are closely interlinked and each one is essential for having proper management. Without a well-defined policy and purpose, proper planning is almost impossible. There can be no purposeful structure and organisation if planning is not proper. Therefore, the policy objectives, planning, organisation, implementation, control, etc., are important components of management.

## Management of Agroforestry

Indian agriculture is a small-scale undertaking. The holdings are small, the inputs are few and the profits are low. The family is the owner and also the worker. Except in some agroforestry systems, there is no clear distinction between the labourer and the manager. The products are primarily directed towards sustaining the family. The farmers are poor and do not have purchasing capacity. There is an acute land hunger as the land/man ratio in India is one of the lowest in the world. Most of the farm operations are performed manually. Only about 30 per cent land has irrigation facility and the remaining suffers the vagaries of nature.

Several agroforestry systems have been practised in India through the ages. Some of these systems were considerably modified. However, trees have always found a place in the agricultural systems. Some trees are worshipped and grown traditionally. However, more recently trees have found a place because they yield a variety of products and give a handsome revenue.

Though the policy and general purpose of agroforestry systems continue to be maximisation of production of agricultural commodities, some minor deviations are often observed wherein emphasis is given to maximising total production and net returns. The recent trend in forest farming towards conversion by absentee landlords of good and productive agricultural land into forest plantations in some parts of the country alarmed the policy makers and therefore the Forest Policy of 1988 recommends that no good agricultural land be diverted for forest plantations. The planning, organisation and control of different management practices depend upon the agroforestry system. The technology also varies greatly with the system.

### 1. SHIFTING CULTIVATION

Shifting cultivation is a very primitive agroforestry system. The system is of the nature of rotational agrisilviculture. The technical details of the system have already been discussed. The main purpose of the system is to provide sustenance to the people practising it. Some people believe that shifting cultivation is the most efficient method for converting forest resources into food, clothing, etc. This was true during the period of primitive culture. The system involves no inputs. Under the present situation, the system is not sustainable due to its short cycle. This system has been the main cause of destruction of forests in several areas.

The important management consideration is selection of area for shifting cultivation. The practice of selection of area varies from state to state. In several states, the land is owned by the village community and the area for cultivation is made available to the individual farmer on an annual basis. If the shifting cultivation cycle were made longer, there would be greater chance of recuperation of soil productivity.

The selection of agricultural crop depends upon the food habits of the local community. The tendency is to produce everything, i.e., cereals, oilseeds, pulses, cotton, vegetables, spices, etc., which a family needs to sustain itself. Paddy is the most widely planted cereal. Individual cultivators have no specific right over the land which they cultivate. Therefore, they tend to keep moving from place to place. Several schemes for control of shifting cultivation have been initiated which aim at permanent cultivation by giving the cultivator right over the land and adopting proper land-use techniques.

## 2. TAUNGYA

The taungya system is a method of raising forest plantations in combination with agricultural crops. The system has been tried in several states for raising plantations of forest species. Successful plantations of teak in Kerala, cashew-nut in Andhra Pradesh, poplar, sal, khair and sissoo in Uttar Pradesh and a few other species in West Bengal, Karnataka, and Madhya Pradesh have been raised through taungya during different periods. However, in some areas there were serious problems in raising plantations of forest species and taungya areas were permanently diverted to agriculture due to socioeconomic and political reasons. This is perhaps the reason that the taungya system has been given up in several states, although the system is economically most desirable.

Successful implementation of the taungya system depends upon several management considerations. The purpose of the system is to provide employment and livelihood to the taungya cultivators and to raise forest plantations during initial years when the trees are not able to utilise the site effectively. Successful implementation of taungya requires: (i) proper planning, (ii) selection of suitable crops, (iii) adjustment in species and plantation, (iv) proper organisation of operations, etc. It is essential to plan the operation in such a way that there are contiguous areas available for plantation and cultivation. The cultivators should raise the plantation but be willing to shift to another area when the plantation has reached the stage where further cultivation would prove injurious to the trees. If the cultivators were assured of getting equally good, if not better areas, they would move and leave the first area. It has to be borne in mind that the cultivators are basically land hungry and their objective is to cultivate foodgrains as long as possible. In the new area, they have to first clear and make the area fit for cultivation. Whenever possible, the forest department should help in clearing the area from bushes to make it fit for cultivation. If the area is dry and the cultivators are used to eating rice, it is necessary to give them some wet area for cultivation of rice, etc.

The agricultural crop selected for taungya cultivation should be such that it has the least adverse effect on the trees. It is better to regulate to some extent the different crops grown there. Such crops which would unduly shade the seedlings or would climb over or smother them should be discouraged. It might be necessary to understand the interactions between the crops and seedlings under different edapho-climatic conditions. It is better to have a detailed agreement with the cultivators outlining details of the crops grown, operations permitted and species and other specifications of plantations.

In managing the taungya, adjustments of species and plantation techniques are necessary. The species which are easy to raise should usually be preferred. Cultivation of agricultural crops for at least a few years is possible only when the spacings are comparatively wider. This facilitates various farm operations. Experience with teak, khair, sissoo, etc., in different parts of the country raised in taungya plantations has indicated that wider spacings resulted in poor bole formation and more branchiness.

Some of the experiments recently conducted indicate that good yields of turmeric, ginger, black pepper and other shade demanding crops can be obtained under established plantations of teak, subabul and poplars (FAO, 1981; Lahiri, 1983). This opens up a new field of understanding between the taungya cultivators and forest departments. Wherever cultivation of these crops is possible under established plantations, the cultivators may be allowed to continue the cultivation in old plantations. This will infuse a new confidence among the poor cultivators. In tarai area of Uttar Pradesh, taungya plots are sold in the auction, generally well-to-do farmers purchase these plots who engage labourers for doing the agricultural and forestry works. This practice needs to be curtailed. Only poor and marginal farmers should be allowed to work as taungya cultivators so that these plantations could work as an instrument for improving the economic conditions of rural poor.

For successful taungya management, it is necessary to have some organisational network. There should be specific criteria for being a taungya cultivator. As far as possible, the forest workers employed in forestry operations, e.g., nursery, plantation, harvesting, etc., should be given preference for taungya cultivation. This would provide round-the-year employment to the workers and improve their socioeconomic conditions.

Taungya cultivation overcomes the labour problem. It optimises land use and the plantation is raised without investment. However, the system involves almost clear-felling and sometimes leads to deterioration of site conditions. In several areas, sociopolitical conditions have led to permanent diversion of forest land. Therefore, the system once widely practised is now localised in a few areas.

In taungya, cultivation of agricultural crops is allowed only initially for a few years. When the tree canopy closes, cultivation of agricultural crops is not possible. The plantation afterwards is managed as other plantations of the forest department. Regular tending operations, e.g., weeding, improvement fellings, climber cutting, thinnings, pruning, etc., are usually done as per practice and plan of the forest department.

### 3. AGRISILVICULTURE SYSTEMS

The management considerations in agrisilviculture systems aim to provide maximum attention to the agricultural crops. At the same time efforts are usually directed to grow trees in such a way so that the production is maximised and farm output is increased. Management of the agricultural component is dictated by several considerations, such as site conditions, availability of inputs, demand of foodgrains and other agricultural commodities, market prices, etc. The forestry component finds second place in these systems. Nevertheless, several decisions are taken keeping in view the production from trees.

In agrisilviculture, the selection of agricultural crops, application of doses of manures and fertilisers, irrigation schedule, sowing and planting, post-harvesting technology, etc., are mainly governed by the agricultural component in the system. The management practices for agriculture are usually adopted without concern for the trees. This is more apparent when the trees do not constitute such an important part in the system. On several occasions, the selection of tree species and management practices have been dictated by the cropping system prevalent in the region. For example, in areas where paddy is the main crop and the area remains submerged for quite sometime, only such trees which could tolerate these conditions were grown. Management of the agricultural component in the system is practised as per its requirement.

## 4. SILVIPASTURE AND PASTURAL SILVICULTURE SYSTEM

In the silvipasture system, the trees are the main component. In pastural silviculture, the pasture is more important. The trees in pastural silviculture are usually scattered and do not receive sufficient attention in management. In these systems, however, the main concern is the grazing management. Grazing management in India has several complicated issues. The large population of low-value livestock, socioeconomic conditions of the people, absence of stall-feeding, heavy pressure on grazing lands and their low productivity are some of the important issues. It is recognised that management of grazing is incompatible with the management of trees.

The introduction of fodder trees in pasture lands with a view to maximising fodder production has been taken up in some areas. This practice requires sufficient research work for deciding the appropriate mix of different plant components. The practice of pruning and lopping would also necessitate proper orientation to maximise fodder production.

### Silvicultural and Management Operations

After trees are planted, they require continuous care and maintenance. There are several operations which need to be done to boost the growth of the trees. Several operations are required to be done during different periods of the life of trees; some of these operations are as under:

  (i) Ground operations
 (ii) Tending operations
(iii) Silvicultural systems
(iv) Harvesting, utilisation and disposal

### 1. GROUND OPERATIONS

Ground operations include soil working, mulching, management of soil water and nutrition, controlled burning, mixing of humus with mineral soil, and other similar operations done to improve the growth conditions of the planted tree species. During the initial years of plantation, soil working has been found to be very useful for several species. It increases the rate of growth, particularly of species such as *Populus deltoides*, *Eucalyptus* spp., etc. Soil working improves aeration and offers several other advantages. *Mulching* is the application of plant residue and other materials used for covering the soil surface to conserve moisture, reduce evaporation, run-off and erosion, check weedgrowth, protect from winter climate and improve the soil. Mulching materials include grasses, plant residue, straw, compost, wood chips, sawdust, papers, sand, stones, etc. The main purpose of mulching is to conserve soil moisture and to check its evaporation from the soil. The importance of mulching is very well recognised in horticultural trees. Some of the experiments conducted indicate that mulching is a useful operation to boost growth of trees, particularly in arid and semi-arid areas.

Management of soil water is another important aspect. In areas of water deficiency, irrigation helps in boosting the growth of trees to a great extent. It is the reason that trees grown on farm lands show a much faster rate of growth in comparison to trees grown in the forest. In areas of waterlogging, removal of soil water with proper drainage helps to increase the rate of growth of trees.

Management of nutrition of trees is also very important to ensure a fast rate of growth, particularly in areas which are poor in nutrients. Application of manures and fertilisers in the initial years of plantation helps in maintaining a proper rate of growth. However, more research is required to suggest proper doses of fertilisers and irrigation during different ages for different species.

Controlled burning is done to reduce the fire hazard in the plantation. In plantations, inflammable material accumulates which should be collected and burnt. It also checks weed growth. Several methods of controlled burning are prevalent. (i) In *broadcast burning* the debris is burnt under controlled conditions wherever it occurs. (ii) In *spot burning* the inflammable material is burnt in patches and spots where there is a concentration of debris. (iii) Burning of *piled debris* may be carried out where debris can be piled at a few places. In several areas, a considerable quantity of leaf litter and other debris accumulates due to the slow rate of decomposition. In such conditions the situation can be improved either by controlled burning or by resorting to *scarification*. Scarification consists of removing organic matter from the soil or mixing it with the mineral soil.

## 2. TENDING OPERATIONS

Tending is defined as an operation carried out for the benefit of the forest crop at any state of its life between seedling and maturation; it essentially covers operations both on the crop itself and the competing vegetation, e.g., weeding, cleaning, thinning, improvement fellings, pruning, climber cutting, girdling of unwanted trees, etc. However, it does not include ground operations such as soil working, drainage, irrigation and control burning, etc. Tending operations are carried out for the benefit of the trees by creating the best possible conditions for growth. These operations are required to be carried out every year during initial years and at suitable intervals thereafter. Tending operations help in producing good quality wood and increasing the rate of growth. They also help in maximising returns per unit area. Though tending operations are very important they are often neglected.

(a) *Weeding:* Weeding is an important operation which consists of removing weeds. Any unwanted plant that interferes or tends to interfere with the growth of individuals of the favoured species is called a weed. Weeding, therefore, is an operation done in the seedling stage in the nursery or in the forest crop, that involves the removal or cutting back of all weeds (Anon., 1966). The objects of weeding are to reduce the root competition and improve the light conditions. The plants interfering and likely to interfere with the planted trees are necessarily to be removed in weeding. The plants can be uprooted or cut back or killed by insect pests, chemicals, etc. The common methods of weeding are: (i) physical or mechanical, (ii) chemical and (iii) biological. The removal of weeds by uprooting them physically or with the help of some tools is called the *physical method* of weeding. The common methods of physical weeding are: (a) spot weeding, (b) line weeding, and (c) complete weeding. In *spot weeding*, the weeds are uprooted or cut back around the planted seedlings. In *line weeding* the weeds are removed in line or rows along the planted seedlings. It is also called strip weeding. When the whole area is weeded, it is called *complete weeding*. Uprooting of weeds is better than cutting back. The uprooting involves soil working also. The weeds when cut back sprout easily and again cause competition. The *biological methods* of weed control may involve use of cover crops, pest and pathogen, controlled grazing, controlled burning, etc. The *chemical method* of weed control involves the use of weedicides, such as 2-4 D, 2, 4, 5-T, etc.

(b) *Cleaning:* Cleaning is defined as a tending operation done in a sapling crop, involving the removal or topping of inferior growth including the individuals of favoured species, shrubs, climbers, etc., when they are interfering with the better grown individuals of the favoured species. It merges with thinning as saplings grow into poles (Anon., 1966). The object of cleaning is to improve the environment for the growth of better grown individuals of the desired species. Cleanings are done more to reduce light competition than root competition.

Cleaning may be done selectively around the desired stems or over the whole crop. Cleaning usually involves the following operations: (i) cutting back of shrubs, herbs and other vegetation interfering with the growth of saplings of the desired species, (ii) cutting back the individuals of inferior species when interfering with the growth of desired species, (iii) cutting back of malformed or less vigorous individuals of the desired species and (iv) cutting of climbers.

In order to provide maximum benefit to the desired individuals, it is recommended that cleanings be done during the season which is the growth period of the favoured species. Often when tree plantations are done, a large number of useless shrubs such as *Lantana camara*, *Dodonea* spp. and several other trees also grow naturally in the crop. These cause intense root and light competition with the individuals of the desired species and their removal becomes necessary. In several species, such as *Eucalyptus* spp. *Lagerstroemia parviflora*, *Shorea robusta*, etc., a number of coppice shoots come up after felling the main stem. Retention of one or two best coppice shoots and cutting back the remainder is necessary. This operation is called *singling*.

(c) *Climber cutting:* Climbers are herbaceous or woody plants that climb up trees or other support by twining round them or holding onto them by tendrils, hooks, aerial roots or other attachments. Large woody climbers in tropical areas are called *liana*. Climbers depend on other plants for support. They usually twine round the stems of the trees to get access to the light. They are a menace to the tree crop. They sometimes completely cover the young plants resulting in complete shading. The young plant may die or deform badly. In bigger trees climbers make grooves on the stem and reduce the quality of the wood.

The climbers should be cut regularly. Thick climbers are generally cut at two places. One near the ground and one metre above it and this piece is removed. This is done in order to ensure that the climber has been really cut. Thin climbers should be uprooted.

(d) *Thinning:* Thinning is defined as a felling made in an immature forest crop for the purpose of improving the growth and form of the trees that remain without permanently breaking the canopy (Anon., 1966). Thinning is based on the imitation of nature. Forest crops during an early age have a large number of trees per unit area; this number continuously decreases and at maturity only a few hundred trees are left. This gradual reduction in number of trees per unit area is due to the fact that the area has limited moisture, nutrients, light, etc. In the process of development, keen competition between the individuals of the same species takes place and only the *fittest* survive and the less vigorous and weak individuals die due to suppression. In a thinning operation all such trees which otherwise would have perished are removed and utilised.

Thinning helps to improve the growth of trees which are retained. The specific objects of thinning are: (a) to improve the hygiene of the crop by removing dead, dying and diseased trees, (b) to salvage anticipated losses of the merchantable volume of wood, (c) to ensure the best physical conditions of growth, (d) to obtain a desired type of a crop, (e) to improve the stand composition and afford protection from the spread of insects and diseases, (f) to improve the quality of wood and (g) to increase the net yield and financial return from the crop. Thinning in forest crops is an essential and important tool which helps in improving the growth of the trees which are retained.

For thinning purposes, trees are usually classified into various classes depending upon their crown class. The tallest trees which determine the top level are called *predominant trees*. The rest of the dominant trees, which are generally 5/6 of the average height of the predominants, are called *co-dominants*. Trees which do not form part of the upper canopy but in which the leading shoots are not overtopped are called dominated trees. Trees which are about 1/2 of the tallest trees and in which the leading shoots are overtopped are called *suppressed trees*. There are dead and diseased classes also. In the forest crop, various types and grades of thinnings are prescribed. *Mechanical thinnings* are prescribed in young plantations. *Ordinary thinnings* in which inferior and suppressed trees are removed are practised mostly in light demander species. In *crown thinnings*, the dominant trees are removed and the dominated trees are retained. The practice is recommended for moderately shade-tolerant species. Some other methods such as free thinning, elite thinning, etc., are also practised.

Agroforesters should know when tree crops require thinnings. They should be able to decide which method of thinning would be suitable for the crop. For forest plantations of teak, sal, chir, etc., thinning methods, intensity and regime are available from forestry experience. However, under agroforestry conditions, it is necessary to work out suitable thinning methods, intensity, cycle, etc., for different species in various agroforestry systems.

(e) *Improvement felling:* Improvement felling is identified as the removal or destruction of less valuable trees in a crop in the interest of better growth of more valuable tree individuals. It is usually applied to mixed uneven-aged forest. It includes thinning in well-stocked groups and cleaning in young crops. It also includes removal of inferior trees. The following operations are done in improvement fellings: (i) removal of dead, dying and diseased trees; (ii) felling of overmature trees; (iii) felling of unsound or badly-shaped mature and immature trees for the benefit of better trees; (iv) thinning in the congested groups of poles and trees, (v) cutting back and removal of undesirable vegetation, damaged trees, etc. It also includes climber cutting.

(f) *Pruning:* The removal of live or dead branches or multiple leaders from standing trees for the improvement of the tree or its timber is called *pruning*. Pruning may be natural or artificial depending on the mode of pruning. *Natural pruning* is the natural death and fall of branches of standing trees grown closely due to deficiency of light or decay, etc. It is also called self-pruning. When the trees are grown closely, their branches rub each other and die. Fungi and insects further damage these branches and they fall in due course. The process continues and the boles of trees become cylindrical and clean. *Artificial pruning* is the removal of branches artificially with the help of sharp tools without waiting for nature to do it in a dense natural crop or where natural pruning is not feasible due to wider spacing in artificially raised plantations. Pruning should be started early in the life of the crop so that the branches to be removed are thin. The operation is done with a sharp tool so that knot-free timber is formed. In agroforestry plantations, pruning is more important, particularly in agrisilviculture systems where trees are grown widely spaced. *Populus deltoides*, which has been grown in northern India, tends to be branchy and artificial pruning is quite important and widely practised.

(g) *Girdling:* Girdling is cutting through the bark and outer living layers of wood in a continuous incision all round the bole of a tree. This operation is done to kill trees of inferior or undesired species where their removal by felling is either uneconomical or not desirable for fear of causing damage to house and other property. Girdling can be done by axe and other sharp tools. Girdling is usually done on trees which are more than 20 cm in diameter as trees below this can easily be felled.

(h) *Cultural operations:* Cultural operations are undertaken to assist or complete existing regeneration, to promote proper development of the crop or to minimise the aftereffects of felling damage. Cultural operations include subsidiary fellings, weeding, cleaning, improvement fellings, thinning in groups of advance growth, girdling, climber cutting, control burning, etc. Cultural operations do not include ground operations and pruning. Cultural operations are associated with the management system involving natural regeneration. Cultural operations are done in the year following the main fellings or any other fellings. They consist of removal of left-over trees, damaged trees, cutting back malformed growth, and inferior species, thinning the even-aged groups of poles and climber cutting.

## 3. SILVICULTURE SYSTEMS

Silviculture systems refer to the procedure worked out in accordance with the accepted sets of silvicultural principles, by which crops constituting forests are tended, harvested and replaced by new crops of distinctive forms (Anon., 1966). In forestry, silviculture systems are classified into two main classes: (i) *High forest systems* where the regeneration is normally of seedling origin either natural or artificial and rotation is generally long. (ii) *Coppice systems* where the crop originates mainly from coppice and where the rotation of coppice is short. In high forest systems, the important systems followed in forestry are: (i) *clear felling* system, when the old crop is removed in one operation; (ii) *shelter wood* system, when the old crop is removed in a series of felling operations and (iii) *selection* system, where fellings are distributed in the entire area and only selected trees are removed. In forestry, several well-recognised modifications also exist to suit the local environmental situation.

Under agroforestry systems, the situation is not forest-like and these management systems are not practised in their totality. Nevertheless, these systems do find application in agroforestry systems in one form or another. In village wood lots and social forestry plantations, some of these systems are being practised in their proper form.

*Selection system*

The selection system by far is the most common system adopted under agroforestry conditions. Other common systems are clear-felling and coppicing. The selection system is common in home gardens, trees on agricultural crops, trees with plantation crops, silvipasture, shelterbelt and other systems. In all these systems, generally dead, dying, diseased, over mature and mature trees are removed. If natural regeneration is available, the same is encouraged and tended. Some examples where the selection system is common are as under:

(a) In several states, *Acacia nilotica, Eucalyptus* spp., bamboos, etc., have been planted in village wood lots. Some of these village wood lots are being worked under the selection system. The trees which reach the merchantable category are harvested. Natural regeneration is encouraged and tended. Natural regeneration, if inadequate, is supplemented by artificial regeneration.

(b) In home gardens, the selection system is the most common in different states. The utilisable trees are harvested and new trees are tended and encouraged. Usually, there is no fixed exploitable size of the tree. The harvesting is done depending upon the need of the family.

(c) Several fruit trees, such as *Mangifera indica, Tamarindus indica, Emblica officinalis, Ziziphus* spp., etc., grown on farm lands, when, they become overmature, have a reduced fruit yield. These trees are then harvested for their wood.

(d) Trees raised on farm lands, on bunds or scattered are also harvested on the basis of selection fellings. Several trees, such as *Dalbergia sissoo, Tectona grandis, Dalbergia latifolia, Santalum album*, etc., have value due to their heartwood. These trees are felled when the heartwood is fully developed. Though there is no prescribed size for harvesting, local knowledge is a great help in deciding the harvestible size for different species.

(e) In Gujarat, *Eucalyptus* hybrid is planted in agricultural land by the farmers in closed spacings, usually 1 m × 1 m. After 5–7 years, the plantation is inspected and well-formed stems which may be about one-fourth to one-fifth of the total number are harvested (Patel, 1988). The coppice which is obtained is tended properly so that it forms the future crop. In the next year another one-fourth of the trees are harvested. After the original planted trees are fully harvested, the coppice crop becomes harvestible.

(f) In several states, bamboos are commonly grown in home gardens. *Bambusa arundinacea, B. vulgaris* and *Dendrocalamus strictus* are grown in central India, *Dendrocalamus hamiltonii* is common in the south and Assam. These bamboos are worked on a culm selection system.

*Coppice system*

The *coppice system* has been extremely popular with some species. Several valuable species, e.g., *Tectona grandis*, *Eucalyptus* hybrid, *E. globulus*, *E. tereticornis*, *Prosopis cineraria*, *Lagerstroemia parviflora*, *Leucaena leucocephala*, etc., are good coppicers. These species are first harvested at the age of 5 to 10 years and subsequently 2 to 3 coppice crops are taken. The following are some of the important examples of agroforestry systems managed under the coppice system:

(a) *Eucalyptus* spp. have been planted under various agroforestry systems. Several species of *Eucalyptus* are grown in home gardens, farm bunds, village wood lots, and other systems. *Eucalyptus* spp., being very good coppicers, are worked under the coppice system. In Gujarat, *E. tereticornis* is raised in dense spacing usually 1 m × 1 m. About one-fourth of the trees which develop into good pole size are harvested after 6 years, another one-fourth trees harvested the following year, etc., until the whole seedling crop has been harvested by the 10th year. The first coppice crop becomes harvestible during the 11th year and the process is continued for at least two coppice rotations. When it is observed that the vigour of the coppice crops has reduced, the same is replaced by another crop. *Eucalyptus* spp. have been widely grown on farm bunds in rows and in farm lands as block plantations, particularly in the states of Haryana, Punjab, Uttar Pradesh, Gujarat, Karnataka, etc. The tendency of the farmers is to grow them in close spacing. On bunds, the trees are grown at a spacing of 1 m to 1.5 m and in block plantations, the common spacing is 1 m × 1 m or 1.5 m × 1.5 m. The trees are harvested from the 6th year onwards, but more commonly around the 10th year. Two to three coppice crops are taken from these areas before any other species or *Eucalyptus* itself is planted again.

(b) In Gujarat, teak is planted on farm lands sometimes at close spacing having about 5000 to 10,000 trees/ha. About half the trees develop into good poles after 5–6 years. The crop is thinned and these poles are harvested. The remaining trees are harvested at the 10th year when these trees are able to yield wood for furniture, door and window frames (Patel, 1988).

(c) Subabul (*Leucaena leucocephala*) has been grown in the central and southern part of the country on a large scale under agroforestry conditions. The species is a good coppicer. Poles can be produced within 5 to 6 years. The species yields good fodder and is sometimes repeatedly lopped and maintained in the form of bushes. The species is also good for hedgerow cropping. The species is good for fuelwood, fodder, small timber and green manure. It is often grown in high density and harvested after 2 to 4 years for fuelwood and fodder purposes. However, if poles are required, longer rotation is necessary.

(d) *Calliandra callothyrsus, Prosopis chilensis, Robinia pseudacacia, Albizia* spp., *Azadirachta indica, Cleistanthus collinus,* etc., are also good coppicers and are worked under the coppice system with varying rotations. *Morus alba, Salix* spp., *Hardwickia binata, Grewia optiva,* etc., are worked and managed under the pollard system. Pollarding is done usually above 1.5 m height. Numerous shoots arise from these which are utilised for wicker work, basket making, fibre extraction, etc. The leaves have good fodder value.

### Clear-felling system

*Clear-felling and plantation* is another system quite common under agroforestry conditions. Several important species are not good coppicers and, therefore, when these trees become harvestible, they are clear-felled. The area available is either used for planting trees again or for some other purpose. Some examples of the clear-cutting system are given below.

(a) In many areas, *Mangifera indica* has been planted by seed. These trees or orchards when they become old do not yield enough fruits. Also the owners are aware of the advantages of grafted varieties of mango. Therefore, there is a tendency among the farmers to clear-cut the old overmature mango orchard and either convert it into a mango orchard of improved grafted varieties or to use it for agriculture or some other purpose. Mango trees yield valuable timber. The timber is used for plywood and door and window shutters in rural areas. Mango trees from seed become mature after about 100 years.

(b) *Populus deltoides* has been widely planted in the northern region. The wood is used for match splints and for manufacture of plywood. WIMCO is running a buy-back guarantee scheme in which the company gives technical and material help to the farmers and guarantees some specific price for the trees which have a girth of above 90 cm at bh. Farmers have raised block plantations of *Populus deltoides* in the Punjab, Haryana, and Uttar Pradesh which attain the desired size at about 10 years. The trees are then harvested and the area is clear-felled. The area is sometimes again planted with poplar or it is used for agriculture.

(c) *Casuarina equisetifolia* has been widely planted in the southern part of the country, particularly along the coast. The species is a fast-growing one and is harvested at the age of 8 to 10 years by clear-cutting. The species produces poles and firewood which are in great demand in the area. After clear-felling the area is again planted with *Casuarina* or the area is brought under agriculture. The village wood lots and social forestry plantations are worked comparatively on a longer rotation.

These systems have been evolved due to experience and local needs. No research data are available for comparison among the systems regarding growth, outturn and economics. The present practices need proper study for specifically suggesting suitable management systems under agroforestry conditions.

## 4. HARVESTING AND UTILISATION

Harvesting of trees in an agroforestry system is done to meet the requirement of the family for timber, fuel, etc. The excess available is sold usually in the market or to neighbours. When the produce is just enough to meet the domestic requirement of the family, there is no problem of disposal. It is, however, not correct to assume that trees are always harvested first to meet the needs of the family. On several occasions trees are harvested for cash benefits. Several agroforestry systems have been widely adopted by resource-rich farmers. In the case of poor farmers agriculture is the first priority for sustenance. Trees are planted to meet their own requirements. There is hardly any surplus with poor farmers. Rich farmers, however, have taken to tree farming in different forms as cash crops, particularly in northern India (Saxena, 1989).

Agroforestry can play an important role in the wood supply of the country. The example of Punjab, Haryana and western Uttar Pradesh needs to be multiplied in all the states. However, enthusiasm for planting trees, particularly *Eucalyptus* spp., has been diminishing due to the crash in the market price. *Eucalyptus* wood, once selling at the rate of Rs. 55 to 60 per quintal during 1986–87 is now (1989) available at a much cheaper rate. The papermills paid only Rs. 38/quintal during 1989. This has slowed the progress of tree planting in these areas as farmers did not get the kind of returns they expected (Saxena, 1989; Malik, 1989; Kapur, 1989). Some farmers have even uprooted their *Eucalyptus* stumps and gone back to annual crops (Saxena, 1989; Malik, 1989; Raizada, 1989).

Some attempts are needed to improve the situation. The harvesting of trees from forest department plantations needs to be stopped or slowed down. Since farmers do not have the capacity to store their produce, government should purchase the produce from the farmers and sell it when there is demand. Some people have advocated having a support price fixed for wood, such as is available for agriculture crops (Sharma, 1989). It is reported that the government of India imported timber worth about Rs. 2000 million during 1989. The trend is likely to continue. Therefore it is expected that there would be no problem for disposal of wood and other forest produce in the near future. However, local price variation might exist depending upon the supply position in the market.

# CHAPTER 17

# Economics of Agroforestry Systems

Economic issues in agroforestry systems are very important and have received considerable attention lately (Arnold, 1983; Hoekstra, 1985). There are several complicated factors such as demand, supply, land-use policies, market forces, etc., which are important in deciding the economics of agroforestry systems. Studies regarding investment and outturn would greatly assist the decision makers in assessing the economic worthiness of the system. During the early period of the twentieth century, tree growing was not very popular. Agriculture received the highest priority. Wherever trees adversely affected the growth of crops, such trees were often removed. Only fruit trees, e.g., mango, ber, mahua, imli, etc., were retained. However, during the past few decades, several developments took place and the demand for timber and firewood rose while the supply reduced. Trees, therefore, again formed an important component in farm lands. The economic considerations, though quite important, of agroforestry systems cannot be fully realised as some systems have several social, cultural and ecological ramifications. Some factors which affect the economics of a system are discussed below.

**Affective Factors**

1. LAND-MAN RATIO

Land resource is very scarce in India. India has only 2 per cent land resource to support about 15 per cent of the human and 16.5 per cent of the livestock populations of the world. Therefore, pressure on the land is very high. Per capita land area is one of the lowest in the world (Table 80). In India, about 76 per cent of the population is rural and 148 million people are actively engaged in agriculture. The number of persons actively engaged in agriculture is one of the largest in the world. Therefore, land area in relation to the active person in agriculture is also very low. Though land area under agriculture is 155 million ha, since a large number of persons are actively engaged in agriculture, the ratio is very small (Table 80).

Table 80: Comparative land/man ratio in India (ha)

| Item | World | Asia | India |
|---|---|---|---|
| 1. Per capita/total land area | 2.85 | 1.00 | 0.40 |
| 2. Total land area/active persons in agriculture | 15.74 | 4.34 | 2.20 |
| 3. Per capita/agricultural land | 1.01 | 0.41 | 0.23 |
| 4. Agricultural land/active person in agriculture | 5.58 | 1.79 | 1.00 |

Most holdings in India are small. Holdings less than 1.0 ha constitute about 56.6 per cent of the total number of holdings. Small holdings (1–2 ha) are 18.1 per cent, medium (4 to 10 ha) to large (more than 10 ha) holdings are only 25.5 per cent. About 40 million ha, i.e., 25 per cent, of agricultural land is irrigated and the remaining faces the vagaries of nature. Most farmers and agricultural workers are poor; a large number of them live below the poverty line. Agriculture to them is subsistence. A large number of them are not able to use costly inputs such as fertilisers, insecticides, fungicides, irrigation, etc. Since cow dung is used for fuel, the productivity in agricultural lands is continuously decreasing. The main aim of small and marginal farmers is to produce enough foodgrains for their sustenance. They do not think of diverting any piece of agricultural land for any other purpose.

## 2. SHORTAGE OF WOOD AND STEEP PRICE RISE

Some decades ago, there were good forests. Both the human and the livestock populations were manageable. People could meet their wood requirement easily. However, during the past few decades, the human and the livestock populations have risen considerably, with increased pressure on the land for more food, fodder, wood and other products. Agricultural land, only 118 million ha during 1951, increased to 142 million ha during 1981, indicating that 24 million ha more land was brought under cultivation. During the past 25 years, 4.3 million ha forest land have been diverted to agriculture and another 1.0 million ha is under encroachment. Due to unrestricted grazing and fire coupled with other pressures, the good forest cover remains only over 38 million ha, as indicated in a recent report (FSI 1989). The yields from the forest have reduced and the demands have increased considerably. As against our requirement of 157 million tonnes of firewood and 27 million tonnes of timber, the present recorded production is only about 40 million tonnes and 12.5 million tonnes respectively (Anon., 1988b; Anon., 1989). Because of the increasing gap between demand and supply, the price of wood has been rising steeply.

The prices of agricultural commodities, e.g., rice, wheat, millet, pulses, oilseeds, etc., have increased by 3 to 5 times over the past 25 years, while the prices of firewood, timber and charcoal have increased by 15 to 20 times

(Dwivedi, 1985). Growing of trees, earlier not so economical, became a quite remunerative land use in comparison to agriculture, particularly in areas where demands of poles and firewood were high. This began to have an impact on land use. Growing of trees in farm lands was taken up in almost all areas but became very popular in areas near urban centres with easy accessibility. Since people want a quick return, they have preferred fast-growing species, such as *Eucalyptus* spp., *Populus deltoides*, *Casuarina equisetifolia*, *Leucaena leucocephala*, etc.

## 3. ENVIRONMENTAL AWARENESS

There is increasing awareness for environmental conservation. Some of the forests which were being harvested to meet the market demands are now not being harvested either because the forest has degraded and nothing is available for harvesting or because of restriction on felling of green trees. The Forest Policy of 1988 says that natural forests are to be managed primarily for conservation of the environment. More and more forest areas are being declared as National Parks and Sanctuaries. The total area at present under National Parks and Sanctuaries is about 14.1 million ha, which was only 2.6 million ha during 1970 (Anon., 1989). Some other areas are being declared biosphere reserves or natural reserves. There is a blanket restriction on felling of green trees above 1000 m elevation. Even in normal forests, there is a greater emphasis on environmental conservation. Clear-cutting systems are being discouraged as they do not provide enough environmental safeguards. All these factors have led to a reduced timber availability in the market. This awareness has led to afforestation of wastelands and planting of trees on farm lands and on sides of rails, roads and canals. People have paid greater attention to developing home gardens. More and more trees are being planted in farm lands and other areas.

## 4. WOOD CONSUMPTION PATTERN

There has been a gradual increase in wood consumption in the country. Wood consumption studies carried out in different parts of the world indicate that with an increase in income, there is almost always an increase in wood consumption. With increase in general standard of living, literacy and per capita income, there is proportionate increase in the consumption of commercial and industrial wood as consumption of paper, plywood, furniture, etc., increases considerably. The use of wood as fuel is dictated by the availability of alternative cheap fuel. In some areas, there is a large demand for wood for construction and repair of houses and agricultural implements. Also, there has been an increase in demand for wood in rural areas for non-traditional items such as furniture and industrial wood product. Fire wood scarcity is being acutely felt in different parts of the country. These trends have increased the demands of wood in the market. There is a wide gap between the present demand and production of various categories of wood in the country, as indicated in Table 81 (Anon., 1988b).

**Table 81: Demand and supply of wood in India**

(In million m³)

| Item | Demand | Production | Gap |
|---|---|---|---|
| Timber | 27.58 | 12.00 | 15.58 |
| Firewood | 235.00 | 40.00 | 195.00 |
| Bamboo | 35.00 | 15.00 | 20.00 |

Estimates by NCA (1976) under high income growth considerations indicate that the requirement for industrial wood by 2000 AD will be 65 million m³. Firewood needs are expected to go above 300 million m³ (Anon., 1981; Anon., 1982). The demand position indicates that India is likely to face an acute shortage of timber and firewood in the coming years. Timber import in a considerably large quantity has already begun and during 1989–90 timber worth Rs. 2000 million was imported. Plantation and agroforestry efforts might be able to ease the demand for wood locally as in Haryana, Punjab and Gujarat, but the general trend of deficit will continue for a considerably long period.

## 5. POLICIES

The draft land-use policy recognises the importance of achieving self-reliance particularly in food production. Improved efficiency, productivity, equity and social justice have given high priority to promoting optimum land use leading to higher productivity in terms of food, fodder, fuelwood, industrial raw material, economic independence and better social environment (Anon., 1988). It is suggested to develop proper land use according to land suitability, sustainability for different types of utilisation and incorporate principles of national land-use safeguards, conservation and management of soil and water resources.

The Forest Policy of 1988, aims to ensure the environmental stability and maintenance of ecological balance, including atmospheric equilibrium, which are vital for the sustenance of all life forms—human, animal and plant. The policy recommends to increase the productivity of the existing forests. Recognising the importance of food production, the policy recommends that diversion of good and productive agricultural lands to forestry should be discouraged. One of the important recommendations of this policy relating to forest-based industries and farm forestry says: as far as possible, a forest-based industry should raise the raw material needed for meeting its own requirement, preferably by establishment of a direct relationship between the factory and the individuals who can grow the raw material by supporting the individuals with inputs including credit, constant technical advice and finally harvesting and transport services. The farmers, particularly small and marginal farmers, should be encouraged to grow, on marginal degraded lands available with them, wood species required for the industries.

The present policies therefore support growing of trees by individual farmers not only to meet their own needs, but also the needs of the local industries. However, there is no mention about the pricing policy of the produce. The farmers have been left at the mercy of the industries. Farmers are forced to sell their produce to industry at a much cheaper rate in the event of high production and market saturation, as has been the case with *Eucalyptus* in Haryana.

In some states, the trade in respect of certain tree species is under the control of government. For example, in Madhya Pradesh trade of teak and bamboo is under the control of government. Similarly in Tamil Nadu and Karnataka, trade of sandal wood is controlled by the state governments. Such restrictions are not good and adversely affect the price of the produce. In some states, transit of almost all wood, including other forest produce, is restricted. Such restrictions work as disincentives to better price of the produce.

## 6. MARKETING

In the case of agricultural produce, the government has four main mechanisms for controlling the prices in the market: (i) the government has a support price for each produce and in the event of bumper production government resorts to purchase of these commodities; (ii) the government maintains a huge store of foodgrains; (iii) it has a large network of public distribution systems and (iv) some of the inputs for agriculture are subsidised.

In the case of fodder, wood and other produce, such price-control mechanism does not exist. Therefore, the farmers are likely to be exploited by powerful industrial sectors. The experience of Haryana, Punjab and western Uttar Pradesh suggests that farmers did not get an appropriate price for eucalyptus wood from the paper-mills in the event of sufficient production from agroforestry. There is no facility for the storage of wood in India. Easy marketability of the produce is one of the key factors which governs the plantation of species. The species which are established in the market are preferred. Poor accessibility and high transport cost work as disincentives to agroforestry. The wood marketing in India has several distinctive features (Saxena, 1989). (i) *Inadequate information*: Farmer awareness about buyers, the prevailing market price and the government rules is inadequate and weak. (ii) *Market access*: The farmers' contact is limited to village buyers only. There is always uncertainty in the minds of the farmers about the price they will get for their produce. (iii) *Margins*: Middlemen's margins are large, leading to high price differentials between producers and consumers. (iv) *Interlocking of credit and output markets*: Credits are not available on the basis of wood resources. If credit is advanced by the merchant who ultimately buys the product, farmers are not likely to get a good price for their produce. (v) *Rigid laws*: Farmers' rights to trees on their own lands are vague, ill-defined and curtailed by rigid laws, which bring uncertainty in marketing.

The underdevelopment of wood markets is also the cause of low price for the producers, even when consumers continue to pay substantially higher price (Chambers *et al.*, 1989). There are a large number of middlemen. Prices paid by various categories of buyers for eucalyptus poles in West Bengal in 1989 are indicated in Table 82.

**Table 82: Price of eucalyptus poles to intermediaries (IMRB, 1989)**

| Category | Price paid in Rs./pole for diameter of poles in cm | | |
|---|---|---|---|
| | 10 | 12.5 | 15 |
| Agent to producers | 18 | 40 | 70 |
| Merchants to agents | 22 | 45 | 85 |
| Businessmen to merchants | 30 | 65 | 100 |
| Retailers to businessmen | 42 | 75 | 120 |

The consumers paid much higher prices than paid by the retailers. It would appear that the producers got much less while the consumers paid much more.

## 7. TIME SPAN

Another important consideration in working out the economics of various agroforestry systems is the long time span involved with tree species. In several tree-based agroforestry systems, the outputs are usually obtained after several years of giving inputs. This is a serious disadvantage for several reasons. (i) The available inputs could have been used for producing more immediate outputs. (ii) The proposed production may not be available in desired quantum due to several risks during the gestation period. (iii) The produce finally obtained may no longer be in demand due to the availability of various kinds of substitutes. Therefore, the procedure of discounting normally used in evaluating delayed returns does not really reflect the benefits in true measure.

## 8. RISKS

Risk means the uncertainty which exists with regard to yield or return due to physical and biological factors. The risk in agricultural production systems is sometimes greater than tree-based production systems as trees are less susceptible to drought and floods. At the same time, trees are long-duration crops and the return and prices, etc., are subject to fluctuation.

The factors affecting the prices of agricultural products are not the same as those which affect the prices of products from trees. It has also to be borne in mind that the market infrastructure for agricultural commodities is well developed in comparison to timber and fuel market. Fodder markets are almost non-existent in several areas. All these factors are important in the economics of agroforestry systems.

## 9. VALUATION OF OUTPUTS

Agroforestry systems yield two types of main produce, i.e., material goods and services. One of the important services of the agroforestry system is said to be *sustainability* besides some other environmental values due to trees. The *sustainability* of the system is attributed to biological nitrogen fixation, nutrient cycling, sheltering effect of windbreaks, contour planting, alley cropping, green manuring, etc. There is no precise method for estimating the value of these services in spite of the fast-developing area of environmental economics. The valuation of these services is based on a number of assumptions which vary considerably with the site and its physical and biological factors. The valuation of material outputs such as fuelwood, fodder, timber, fruits, etc., will differ with public and private economies. It will also differ with subsistence and commercial objectives. When the produce is marketed, the actual value can be realised. However, in areas where significant trade does not take place and marketing is not developed, the value of the produce can not be realised by the grower. For example, the plantations for production of fuelwood and pole have been very profitable near cities like Delhi, Ahmedabad, Baroda, etc., where there is a great demand. However similar plantations in rural areas could not be sold due to undeveloped marketing conditions.

**Economics of Individual Systems**

Agroforestry systems vary considerably in their crop composition, ecological interaction, labour requirements, investment and nature of returns. Different systems have different components. In some systems, where trees are dominant, it is ecologically safer. The investment is made during the initial year and it takes comparatively a longer period for giving return. In systems where annual crops are dominant, the investment and returns are of an annual nature. The trees in these systems are capable of affecting the returns however. It appears necessary to discuss the nature of different agroforestry systems, therefore, with regard to economic considerations.

## 1. SHIFTING CULTIVATION

In shifting cultivation areas, the land is plentiful and the labour is scarce, i.e., the land/man ratio is wide. Transport and market facilities are least developed. The system pertains to subsistence level. There is usually no demand for the tree products. The products obtained are meant for the household and not for sale or marketing. Economic evaluation of the system for different areas is completely lacking when land is available in plenty, the system works efficiently both economically and ecologically.

The adverse effects of the system include: declining soil productivity, increased weed growth and ecological degradation due to destruction of forests. Under the increasing population pressure, the system will collapse due to decreasing soil fertility and ecological degradation.

## 2. TAUNGYA SYSTEM

The taungya system is advantageous in one sense because it saves completely the plantation costs. The system involves intensive use of land. The system can be popular in areas where land is scarce and the people are poor but not very poor. In a worldwide study, it was found that the system was not generally practised if the per capita income was below 150 pounds sterling per annum (King, 1984). In India, however, the system has been popular in areas where the forest land is fertile, cleaning is easy, and acute land hunger exists. The system once practised in several states has been abandoned in many areas. The system is still prevalent in the terai area of Uttar Pradesh where the cultivators not only raise forest plantations free of cost, but also pay some lease rent to the forest department.

The economic evaluation of the system should take into consideration not only the investment from the government source and returns to it, but also the investment from the cultivator and returns to him. In the final analysis it may come out that the system is exploitative. There is need therefore that government, which is the main beneficiary in this case, should realise the social pressure and should take care of health, sanitation, education, etc., of the taungya cultivators. The farmers may be given a liberal share in the additional benefits.

## 3. AGRISILVICULTURE SYSTEMS

Among agrisilviculture systems the common systems for which some data on economic issues may be available include: trees on agricultural land, trees with plantation crops, home gardens, village wood lots, alley cropping, etc. Precise information on the inputs and outputs of these systems is not available. Therefore, there is always a problem in working out the economics of the system. Much information is required to be collected for economical analysis.

### (i) *Trees on farm land*

The most common agrisilviculture system is growing of trees on farm land. The trees are grown on the farm lands in various geometry. They are grown scattered all over the farm without much consideration for their effect on crops. They are also grown in blocks, rows or strips in or on the boundaries of the field. The economics vary considerably with the species mix, duration of the association and other physical and biological factors. When trees are grown or tended in scattered form, they are meant for produce usually for the consumption of the household. A few trees here or there do not produce enough produce for marketing. However, if their concentration is greater, enough produce might be available. One common geometry of tree plantation on farm land is growing on field bunds. Several trees, such as *Eucalyptus* hybrid, *Dalbergia sissoo, Populus deltoides, bamboos, Acacia nilotica, Casuarina equisetifolia, Leucaena leucocephala* etc., have been widely grown in different parts of the country. These trees are generally planted with no specific additional inputs. Water, nutrients, etc., are available to these trees along with the crops. Usually, 100 to 150 trees/ha are easily grown on bunds. The production from fast-growing species such as *Eucalyptus* spp., *Populus deltoides*, etc., may be 25 to 50 tonnes/ha (Dwivedi and Sharma, 1990). These trees have also been raised in the form of block plantation particularly by absentee landlords.

The rate of spread of agrisilviculture in the form of farm forestry has been phenomenal in several states (NCAER, 1988; USAID, 1988). Most of the farmers in several parts of the country, particularly in Punjab, Haryana, Uttar Pradesh, Gujarat and Karnataka, have transformed their farms and cropping systems to include a large number of trees in their farm lands and farm forestry has emerged as the most significant land use in these states.

In several areas, high cash returns were obtained from trees during the 1980s (Jain, 1988; Singh, 1988; Patel, 1988). Some farmers in Gujarat obtained an income of about Rs. 317,500 per ha in five years, against a total investment of Rs. 72,500/ha, giving a cost-benefit ratio of 1 : 5. This ratio for other cash crops, such as groundnut and cotton in that region, was only 1 : 2 (Jain, 1988; Saxena, 1989). In Haryana, the cost benefit ratio for *Eucalyptus* spp. was calculated as 1 : 3.3 in 1986 as against only 1 : 1.2 for good crops in the same area (Singh, 1988). A financial analysis carried out in Gujarat with eucalyptus cultivation indicated that farm foresters who intercropped eucalyptus and cotton during the first year, made an investment around 1700 US dollars per ha and total returns after 5 years were 5900 US dollars/ha; the internal rate of return was calculated at 129 per cent for the first rotation and estimated to increase to 213 per cent for each successive coppice crop (FAO, 1985).

However, these situations have been changing very rapidly. A study conducted in Gujarat indicated that of 45 farmers who sold trees only 9 made relative profits; 36 would have been better off sticking to pure agriculture (Saxena, 1990). In Punjab and Haryana, a 7 to 8-year tree which gave about

Rs. 100 a few years ago, sold for Rs. 35 in 1990. In Gujarat, the price of trees with 12.5 cm diameter fell from Rs. 60 in 1986 to Rs. 23 in 1988 (Bhattacharjee, 1988). The market collapse has led to indifference on the part of the farmers towards eucalyptus in particular and trees in general. Lifting of eucalyptus seedlings by the farmers has reduced considerably. Table 83 indicates the number of eucalyptus seedlings distributed during different years in some states (Saxena, 1990; Kapur, 1989; Gaonkar, 1989).

Table 83: Number of seedlings distributed (million seedlings)

| State | 1984 | 1986 | 1988 |
| --- | --- | --- | --- |
| Gujarat | 134 | 35.0 | 22.0 |
| Haryana | 20 | 15.3 | 4.0 |
| Punjab | 8.7 | 3.5 | 2.6 |
| Karnataka | 94.2 | 85.0 | 49.0 |

The obvious explanation for the farmer's reluctance to continue with tree farming is the fall in the market price due to overproduction of eucalyptus poles and market saturation in some states. Oversaturation of the market has apparently occurred for a specific type of produce, i.e., pulpwood and poles. Several more recent developments could offset this problem: (i) Use of the eucalyptus is being diversified. The species is being used for furniture, doors and windows and other constructional purposes. (ii) The demands of other species is gradually picking up and several fast-growing species are being planted.

The introduction of trees on farm land and replacement of agriculture with tree cultivation are two different issues. In some farms, agriculture was being replaced. There are several socioeconomic factors responsible for these developments. (i) The wages in these areas were quite high and unemployment was comparatively low. Large farmers were having labour problems. Tree farming, which is less labour intensive, was preferred. (ii) Intensive agriculture demands intensive supervision and care. This tied the farmers to the land and did not permit them to take other occupations. The landowners in government service or business, found tree farming a better compromise. (iii) The prices of agricultural produce have not been favourable during the past few years. The prices of land have increased but there has been no proportionate increase in the prices of agricultural commodities. In areas where cheaper labour is available, people are poor and large inputs in agriculture are not available, trees with a long gestation period are not a viable option for small farmers (Saxena, 1990).

Besides eucalyptus, other trees which have been grown are poplar, teak, subabul, casuarina, bamboos, etc. However, no cost-benefit analysis is available for the varying conditions. Plantation of *Populus deltoides* in the terai area of Uttar Pradesh has indicated higher profitability. A cost-benefit analysis of some plantations with agricultural crops such as wheat, vegetables, pulses, berseem, soybean, has indicated higher profitability (Chaturvedi, 1981; Mathur and Sharma, 1983). Patel (1988) has reported that teak plantations raised with cotton and other agricultural crops showed much greater profitability than pure agriculture, usually 1.5 to 2 times. Similarly, casuarina and bamboo plantation with agricultural crops reportedly give a higher net present value under certain conditions.

Another important aspect of agroforestry is employment. Tree cultivation requires low labour input in comparison to agriculture. Saxena (1989a) has reported that groundnut cultivation requires about 112.5 man-days per year while in the case of cashew plantation (30-year felling cycle) and eucalyptus (10-year felling cycle) the labour requirement is 62.5 man-days and 40 man-days per year respectively. Thus it is clear that if the tree component dominates over the agricultural component, some labour is likely to be surplus. Therefore, it is necessary to adjust the components in such a way that labour opportunities are not reduced. In agriculture, the requirement for labour is annual or seasonal while in trees plantations it is periodic, usually at long intervals during the time of formation and harvesting. However, if trees are grown with agricultural crops in proper proportion, employment could be guaranteed all year round and also there would be no danger of diminishing the employment opportunity.

## (ii) *Trees with plantation crops and commercial crops*

In India, trees are grown with several crops of commercial value. In plantations of tea, coffee, cocoa, rubber, coconut, etc., several trees are usually planted depending upon site and local preferences. In plantations of tea and coffee, such trees not only provide necessary shade for their optimum development, but also yield firewood and timber and improve the profitability of these plantations. Many tea, coffee and cocoa gardens earn handsome profits due to trees. In plantations of rubber, additional income can be obtained by rearing animals and birds. About 3–4 sheep/ha could be managed in rubber plantations. Several cash crops and species, e.g., betel, pepper, ginger, turmeric, etc., could be grown in these plantations which would increase the profitability of the system.

## (iii) *Home gardens*

Home gardens are fast developing in almost every part of the country. Every village house has almost a small area where are grown some trees, vegetables, fruit trees and other crops such as spices and even common crops. The villager stores hay, makes his livestock pens, rears poultry, and does several miscellaneous works. The compound area is well protected. Since the area remains close to habitation, there are distinct features of this system: (i) The area is usually small, usually less than 0.1 ha. (ii) Availability of inputs, manures of livestock kept in pens, household residue and residual water is plentiful. For giving the same inputs in some other land the farmer would have to walk a considerable distance and transport would be necessary. (iii) Supervision and protection are maximum. (iv) Production mix is highly diversified and meant for household consumption.

Among the various agroforestry systems, per unit area production is perhaps maximum in home gardens. Small home gardens, however, have a higher production per unit area than large home gardens. Home gardens involve small land area, large labour and material inputs and usually higher returns per unit area. In Indonesia, it is reported that cassava production per unit area increased when home garden size decreased; the labour input was 3 times higher in small home gardens of 0.1 ha as compared to large home gardens of 0.3 ha (Stoler, 1978). Though home gardens are found in different regions of the country, details of inputs and outputs are completely lacking, mainly because a large part of the produce is consumed in the family.

### (iv) *Alley cropping*

Alley cropping is a new system which is yet to be practised on a large scale. The system holds promise for higher economic returns in the form of improved soil fertility and higher fuelwood and fodder yields. The system has been introduced in the semi-arid areas of Maharashtra and Madhya Pradesh where subabul (*Leucaena leucocephala*) has been grown as a hedge in some farms. If leaves are lopped for fodder, they cannot be available as mulch or green manure. In such cases, improvement in soil fertility is not achieved. Since the practice is still in an experimental stage, data on inputs and outputs are still to be collected under different agroclimatic conditions. At this stage nothing can be said about the economic viability of the system.

### (v) *Live fence*

Live fencing is useful and beneficial because: (i) it protects the crops from damage by stray animals, wild animals, etc., and (ii) it provides poles, firewood and fodder. Some work on an economical analysis of live fencing has been done in the case of *Agave* spp. and the results indicate an encouraging trend. The protection provided by the fence is of great importance and tilts the balance in favour of a live fence in comparison to other fences. Other types of fencing, e.g., stone wall, barbed wire, cattle-proof trench or electric fence have several limitations. Also, they are costly. A live fence, once established, is very effective. The thorny species, e.g., *Acacia nilotica, Carissa spinarum*, etc. can be used for fencing. The firewood, poles, fodder, fruits, etc., obtained from a live fence provide additional advantages.

### 4. SILVIPASTURE AND PASTURAL SILVICULTURE

In silvipasture and pastural silviculture systems, the trees are retained and nursed to provide fodder, fuelwood and timber. The fodder production per unit area with fodder grasses and fodder trees is almost always higher than fodder production from fodder grasses alone (Patil *et al.*, 1979; Pathak, 1989; Deb Roy, 1990).

Sufficient data on various densities of different species along with different combinations of fodder grasses are still lacking. However, preliminary estimates indicate that pastural silviculture or silvipasture systems are economical in comparison to simple pasture management. In arid and semi-arid areas, the presence of *Prosopis cineraria* up to the extent of 250 trees per ha gives an average yield of 35 to 40 kg leaf fodder per tree at about 15 years of age and 2.3 kg of pods. The pods of this tree are used as vegetables. The total benefits are much larger in silvipasture system in comparison to pure grass production. In several studies carried out in arid areas in Rajasthan indicated that silvipasture system is 2 to 3.5 times more beneficial than simple grass production system (Mathana and Shankarnarayan, 1978; Shankarnarayan *et al.*, 1987). After seven years of establishment of *Acacia tortilis* with *Cenchrus ciliaris*, the following yields and revenue are reported (Table 84).

**Table 84: Economics of the silvipastural system (*Acacia tortilis* and *Cenchrus ciliaris*)**

| Treatment | Fuel Yield (Q/ha) | Grass Yield (Q/ha) | Revenue (Rs.) Fuel wood | Grass | Total |
|-----------|-------------------|--------------------|-----------------------|-------|-------|
| 1. Trees only (10 m × 5 m) | 60 | — | 3000 | — | 3000 |
| 2. Trees only (10 m × 10 m) | 32 | — | 1600 | — | 1600 |
| 3. Trees with grass (10 m × 5 m) | 50 | 55.8 | 2500 | 1395 | 3895 |
| 4. Trees with grass (10 m × 10 m) | 28 | 52.9 | 1400 | 1323 | 2793 |
| 5. Grass only | — | 46.0 | — | 1150 | 1150 |

Similar higher returns have been reported in other studies carried out in different edapho-climatic conditions (Mann and Daulay, 1981). Along with fodder, much needed timber and fuelwood are also available. In Rajasthan, the leaf fodder of *Ailanthus excelsa* is sold and fetches sometimes as high as Re. 1.00/kg. A medium size tree is capable of producing about 50–75 kg of leaf fodder. Marketing of fodder leaves is a real problem. Until now the practice has been limited near cities. Fodder production in rural areas is consumed locally and, therefore, evaluation in terms of money is not always possible. Economic evaluation of tree growing in marginal lands in Rajasthan indicates much higher returns from tree growing than traditional agriculture (Gupta and Mohan, 1982).

# References

Albrecht, R.W. (1976). Drought prone area programme—development of deccan plateau range lands. Seminar on Management of Forests and Pastures in DPAP. IGFRI, Jhansi.

Albrecht, R.W. (1979). Regulation of livestock numbers. *Proc. Seminar on Peoples Participation in Social Forestry, Fodder and Pasture Development*, Indore (M.P.).

Anderson, M.L. (1950). *The Selection of Tree Species: An Ecological Basis of Site Classification for Conditions Found in Great Britain and Ireland.* Oliver and Boyd, Edinburgh.

Anon. (1944). Forestry Terminology. *Soc. of Amer. For.*, Washington, D.C.

Anon. (1947). Effects of different Kumri crops on teak. *For. Res. in India and Burma*, Part III, 53 pp.

Anon. (1949). Effects of different Kumri crops on teak raised with them. *For. Res. in India and Burma*, Part II, 53 pp.

Anon. (1952). The National Forest Policy of India. Manager of Publications, Delhi, 53 pp.

Anon. (1955). Annual Report for Silviculture Research for 1953–55, Madras.

Anon. (1962). Wealth of India. Publication and Information Directorate, Govt. of India, Vol. VI: pp 208–215

Anon. (1965). Progress Report, 1965. Central Arid Zone Research Institute, Jodhpur.

Anon. (1966). *Abridged Glossary of Technical Terms* (for use in forest colleges). F.R.I., Dehra Dun, 266 pp.

Anon. (1972). Report of the National Commission on Irrigation. Ministry of Agriculture and Irrigation. Govt. of India, New Delhi.

Anon. (1973). Report of the Task Force on Integrated Rural Development. Planning Commission, New Delhi.

Anon. (1974). Progress Report, 1966–72. Tenth Commonwealth Forestry Conference, U.K.

Anon. (1975). Report, Soil and Water Conservation Research 1956–71. ICAR, New Delhi.

Anon. (1978). Report of the Working Group on the Integrated Action Plan for Food Control (in the Indo-Gangetic Basin). Ministry of Irrigation, New Delhi.

Anon. (1979). Report of the Working Group on Energy Policy. Planning Commission, New Delhi.

Anon. (1980). Report of Rashtriya Barh Ayog, Vol. I, Ministry of Irrigation, New Delhi.

Anon. (1980a). *Firewood Crops: Shrub and Tree Species for Energy Production.* National Academy Press, Washington, D.C. 263 pp.

Anon. (1981). Report of the Fuelwood Study Committee. Planning Commission, New Delhi.

Anon. (1982). Soil Conservation Problems, Approach and Progress in India. Ministry of Agriculture & Co-operation, New Delhi.

Anon. (1982a). Annual Report (1981–82). Govt. of India, Ministry of Agriculture. New Delhi.

Anon. (1982b). Report of the Advisory Board on Energy (ABE). Planning Commission, New Delhi.

Anon. (1983). West Bengal Social Forestry Project Evaluation: an Interim Report. Forest Department, West Bengal.

Anon. (1984). Report of the Sub-group on Forests and Soil Conservation for Formulation of VII Five Year Plan. Ministry of Agriculture and Co-operation, New Delhi.

Anon. (1984a). Report of the Society for Promotion of Wastelands, New Delhi.

Anon. (1984b). *Leucaena : Promising Forage and Tree Crop for the Tropics*. National Academy Press, Washington, D.C. 100 pp., 2nd ed.

Anon. (1984c). Report of the Committee for Review of Rights and Concessions in the Forest Areas. Ministry of Agriculture, New Delhi.

Anon. (1985). Tree Plantation in Alkali Soils. Central Soil Salinity Research Institute, Karnal.

Anon. (1986). Indian Agriculture in Brief. Ministry of Agriculture, New Delhi, 394, 21st ed.

Anon. (1986a). Report of the Committee on Fodder and Grasses. Ministry of Environment and Forests, Govt. of India, New Delhi.

Anon. (1987). Evaluation of Orissa Social Forestry Project, Phase I. Govt. of Orissa (mimeographed).

Anon. (1987a). India's Forest, 1987. Survey and Utilisation Division, Ministry of Environment and Forests, Govt. of India, New Delhi.

Anon. (1988). Report of the Committee of Experts on Draft Outline of National Land-use Policy. National Land-use and Conservation Board. Ministry of Agriculture, New Delhi.

Anon. (1988a). India, 1987. Director of Publications Division, Ministry of Information and Broadcasting, Govt. of India, New Delhi.

Anon. (1988b). The State of Forest Report—1987. Ministry of Environment and Forests, Forest Survey of India, Dehra Dun.

Anon. (1988c). National Forest Policy, 1988. Ministry of Environment and Forest, New Delhi.

Anon. (1988d). Indian Agriculture in Brief. Govt. of India, New Delhi, 406 pp., 2nd ed.

Anon. (1989). Country Report. XIII Commonwealth Forestry Conference, Rotorua, New Zealand.

Anon. (1990). Evaluation Report—Farm Forestry. Gujarat Forest Department.

Arnold, J.E.M. (1983). Economic Considerations in Agroforestry Projects. *Agroforestry Systems*, 1:299–311.

Atkins, W.R.G. (1932). The measurement of daylight in relation to plant growth. *Empire For. Journ.*, 11: 42–52.

Bakshi, B.K. (1976). Forest Pathology—Principles and Practices in Forestry. Controller of publication, Govt. of India, Delhi, 400 pp.

Banerjee, A.K. (1972). Trial With *Agave* spp. in lateritic areas of West Bengal. *Indian Forester*, 98(7).

Banerjee, U. (1989). Social forestry in India. Paper in Seminar on Social Forestry for SAARC, F.R.I., Dehra Dun.

Bartle, J.R. (1977). Choosing trees for specific purposes. *Proc. Workshop Integrating Agriculture and Forestry*. Eunbury, Australia.

Barua, D.N. (1968). Light as a factor in metabolism of the tea plant (*Camellia sinensis*, L). In: *Physiology of Tree Crops* (Eds.: L.C. Luckwill and C.V. Cutting). Academic Press, London, N.Y., pp. 307–322.

Bassi, D., S. Sansavini and L. Guinchi. (1980). Tree efficiency and fruit quality of high density apple orchards. Ostbauweinbau 17, 236. Sudtiroler Beratungssing, Lana.

Bates, C.G. (1944). The windbreak as a farm asset. *Fmr's Bull*. 1405. US Deptt. Agri. 22 pp.

Bates, C.G. (1945). Shelterbelt influences. General description of studies made. *Jour. For.*, 43(2): 88–92.

Beason, C.F.C. (1941). Ecology and Control of Forest Insects of India and Neighbouring Countries. Controller of Publication, Govt. of India, New Delhi, 483 pp.

Bene, J.C., H.W. Beall and A. Cole. (1977). *Trees, Food and People*. IDRC, Ottawa.

Bennett, H. (1955). *Elements of Soil Conservation*. McGraw-Hill Book Comp., New York, 318 pp.

Bhattacharjee, A. (1988). "Eucalyptus: A distress story." *Hindustan Times*, September 25, New Delhi.

Bhimaya, C.P. (1960). Sand dune rehabilitation in Western Rajasthan. *Proc. IVth World For. Cong.*, Dehra Dun.

Bhimaya, C.P. and R.N. Kaul. (1965). Root system of four desert tree species. *Annals of Arid Zone*, 4(2): 185–194.

Bhimaya, C.P., D.K. Mishra and R.B. Das. (1958). Importance of shelterbelts in arid zone farming. *Proc. Farm Forestry Symp.* ICAR, New Delhi, pp. 65–72.

Biswas, T.C. and T.C. Bhuyan. (1983). On the identity of some food plants of Garo hills, Meghalaya. *Indian Jour. For.*, 6(3): 208–213.

Blencowe, J.W. (1969). Crop Diversification in Malaysia. Interpolated Society of Planters, Kuala Lumpur.

Bor, N.L. (1942). Ecology: Theory and Practice. Presidential address to the Botany Section. *Proc. 29th Indian Science Congr.*, Baroda, Part II, pp. 145–79.

Borthakur, R.N., S.P. Prasad Ghosh, A. Singh, R.P. Awasthi, R.N. Rai, A. Verma, H.N. Datta, J.N. Sachan and M.D. Singh. (1979). Agroforestry based farming system as an alternative to Jhuming. *Proc. of the Agroforestry Sem.*, ICAR, Imphal. pp. 109–32.

Bowen, G.D. (1984). Tree roots and the use of soil nutrients. In: *Nutrition of Plantation Forests* (Eds.: G.D. Bowen and E.K.S. Nambiar). Academic Press, London, 516 pp.

Brewbaker, J.L. and K.G. MacDicken. (1985). Why nitrogen fixing trees? In: *Reference Guide to Establishment and Management of Nitrogen-Fixing Trees.* (Eds.: K. MacDicken and P. Huxley). NFTA Waimanalo, 350 pp.

Buckingham, J. (1885). Papers Regarding the Sau Tree and its Remarkable Influence on the Tea Bush. Indian Tea Association, Calcutta.

Budelman, A. (1988). Leaf dry matter productivity of three selected perennial leguminous species in humid tropical Ivory Coast. *Agroforestry System*, 7(1): 42–62.

Bunting, A.H. (1976). Inaugural lecture. In: *Tree Physiology and Yield Improvement* (Eds.: M.G.R. Cannere and F.T. Last). Academic Press, London, New York, pp. 1–20.

Caborn, J.M. (1957). Shelterbelts and Microclimate. Forestry Commission Bulletin No. 2. Her Majesty's Stationery Office, 134 pp.

Caborn, J.M. (1965). *Shelterbelts and Windbreaks*. Faber and Faber Ltd., 24 London, 275 pp.

Chambers, R., N.C. Saxena and Tushaar Shah. (1989). *To the Hands of the Poor: Water and Trees*. Oxford and IBH, New Delhi.

Champion, H.G. and N.V. Brasnet. (1958). Choice of tree species. FAO Forestry Development Paper 13.

Champion, H.G. and S.K. Seth. (1968). General silviculture of India. Controller of Publications, Govt. of India, New Delhi, 511 pp.

Chandra, J.P. (1986). Poplar—a cash crop for North Indian Farmers. *Indian Forester*, 112(8): 698–710.

Chapman, G.W. and T.D. Allan. (1978). Establishment Techniques of Forest Plantations. Forest Resources Division, FAO Paper No. 9, Rome.

Chaturvedi, A.N. (1981). Poplar for Planting. Uttar Pradesh Forest Department Bull. No. 50, Lucknow, 27 pp.

Chaturvedi, A.N. (1983). Eucalyptus for Farming. U.P. Forest Bulletin No. 48, Lucknow 35 pp.

Chaturvedi, A.N. (1986). Aerial seeding in Yamuna Ravines. *Proc. National Sem. on Fuelwood and Fodder Production from Wastelands*. F.R.I., Dehra Dun.

Chaturvedi, A.N., S.C. Sharma and Ramji Shrivastava. (1984). Water consumption and biomass production of some forest trees. *Common. For. Rev.*, 63(3): 217–223.

Chaturvedi, J.K. (1983a). Afforestation of bauxite mined area in Central India. *Indian Forester*, 109(7): 458–465.

Chaudhari, K.A. (1939). Formation of growth rings in Indian trees, Part I. *Ind. For. Roc.* (NS) Silvi. 2(1).

Chaudhari, K.A. (1940). Formation of growth rings in Indian trees, Part II. *Ind. For. Rec.* (NS) Silvi. 2(2).

Chaudhari, Kamala. (1989). Social Forestry—roots of failure. *Indian Jour. of Public Adm.*, 35(3): 437–444.

Chhabra, R. and Virendra Kumar. (1989). Afforestation for Development of Salt-affected Soils. ICAR, New Delhi, 27 pp.

Chopra, Kanchan, G.K. Kadekodi and M.N. Murty. (1989). Peoples participation and common property resource. *Economic and Political Weekly*, Dec. 23–30.

Combe, J. (1982). Agroforestry techniques in tropical countries: Potential and limitations. *Agroforestry Systems*, 1: 13–28.

Cornelius, D.R., B.N. Bhat and R.L. Pathak. (1977). Windbreak plantation on sandy land in Northern Gujarat, *Indian Forester*, 103(4): 251–259.

Cornforth, J.S. (1970). Reafforestation and nutrient reserves in the humid tropics. *Jour. Appl. Ecology*, 7.

CRRI (1972). Annual Technical Report of Central Rice Research Institute. Cuttack, India, pp. 81–111.

Dabadghao, P.M. and K.A. Shankarnarayan. (1973). The Grass Cover of India. ICAR, New Delhi, 210 pp.

Dabral, B.G. and B.K. Subba Rao. (1968). Interception studies in chir and teak plantations in new forest. *Indian Forester*, 94(7): 541–551.

Dabral, B.G. and B.K. Subba Rao. (1969). Interception studies in sal (*Shorea robusta*) and Khair (*Acacia catechu*) plantation in new forest. *Indian Forester*, 95(6): 314–323.

Dabral, B.G. and A.S. Raturi. (1985). Water consumption by *Eucalyptus* hybrid. *Indian Forester*, 111(12): 1053–1069.

Dabral, B.G., B.K. Subba Rao and I.M. Qureshi. (1969). Some studies on air temperature and humidity inside *Pinus roxburghii* and *Dendrocalamus strictus* plantation at New Forest. *Indian Forester*, 95(8): 501–512.

Dabral, B.G., S.P. Pant and S.C. Pharasi. (1984). Micro-site characters vis-à-vis root behaviour in sal (*Shorea robusta*). *Indian Forester*, 110(10): 997–1013.

Dabral, B.G., S.P. Pant and S.C. Pharasi. (1987). Root habits of eucalyptus—Some observations. *Indian Forester*, 113(1): 11–32.

Dalvi, M.K. (1983). Social forestry—an overview of Indian experience. Conference on Forestry and Development in Asia. Asian Society Bangalore, India.

Das, D.C. (1981). Influence of current land use policies on some watershed problems, *Proc. Nat. Sem. on Watershed Management*. FRI. DehraDun, pp. 101–111.

Das, H.C., L.N. Hursh and K.A. Shankarnarayan. (1986). Technique for wasteland afforestation of arid areas in Rajasthan. In: *National Sem. on Fuelwood and Fodder Production*. F.R.I., Dehra Dun pp. 128–135

Das, M.C. and S. Mahapatra. (1983). Agroforestry in Orissa. In: *Proc. National Workshop on Agroforestry* (Eds.: R.S. Mathur and M.G. Gogate) Karnal, pp. 147–154.

Davidson, J. (1985). Setting aside the idea that eucalyptus are always bad. Working Paper No. 10. FAO Project BGD/79/017. pp 17.

De, R.N. (1932). Taungya in Garo Hills Division, Assam. *Indian Forester*, 58(2): 93–99.

Debell, D.S., C.D. Whitesell and T.H. Schubert. (1985). Mixed plantations of eucalypts and leguminous trees enhance biomass production. Research paper, PWW-195. USDA Forest Service, Washington, D.C.

Deb Roy, R. (1990). Forage and top feed production potential under silvipastural system. Paper in National Seminar on Forest Productivity. F.R.I., Dehra Dun.

Del Moral and C.H. Muller. (1970). The allelopathic effect of *Eucalyptus camaldulensis*. *Am. Midl. Nature*, 83:254–282.

Dent, T.V. (1948). Seed storage with particular reference to the storage of seed of Indian forest plants. *Ind. For. Rec.* (NS), Silvi. No. I.

Dhanda, R.S. (1983). Agroforestry in Punjab. In: *Proc. National Workshop on Agroforestry*. (Eds.: R.S. Mathur and M.G. Gogate). Karnal, pp. 178–185.

Dhillon, G.S., S.S. Grewal and A.S. Atwal. (1979). Developing agrisilviculture practices: effect of farm trees (*Eucalypts*) on adjoining crops. *Ind. Jour. Ecol.*, 6(1): 88–96.

Dhillon, M.S. Surjeet Singh, A.S. Atwal and G.S. Dhillon. (1984). Developing agrisilvicultural practices: Effect of *Dalbergia sissoo* and *Acacia nilotica* on the yield of adjoining crops. *Ind. Jour. Ecol.*, **11**(2): 249–253.

Durk, P. (1966). The influence of forests on the health of man. *VIth World Forestry Congress.* Madrid, Vol. III., pp. 3748–49.

Dwivedi, A.P. (1980). *Forestry in India.* Jugal Kishore & Co., Dehra Dun 455 pp.

Dwivedi, A.P. (1985). Price rise in some forest products in Madhya Pradesh. M.P. Forest Department Publication (mimeographed).

Dwivedi, A.P. (1986). Bamboo planting—generation of employment. Sem. on Plantation Strategy in Madhya Pradesh, Jabalpur.

Dwivedi, A.P. (1987). Effect of bund planting of eucalyptus on agricultural crops. Paper in Seminar on Eucalyptus. F.R.I., Dehra Dun.

Dwivedi, A.P. and K.K. Sharma (1989). Agroforestry: its potential. Paper in Seminar on Social Forestry and Agroforestry. F.R.I., Dehra Dun.

Dwivedi, A.P. (1989a). Food from forests. Paper in National Seminar on Minor Forest Produce. Institute of Deciduous Forests, Jabalpur.

Dwivedi, A.P. (1989b). Gregarious flowering in bamboo (*Dendrocalamus strictus*) in Shahdol (M.P.)—some management consideration. *Indian Forester,* **114**(a): 532–538

Dwivedi, A.P. (1990). *Silviculture—Principles and Practices.* Surya Publishers, Dehra Dun, 458 pp.

Dwivedi, A.P. and K.K. Sharma. (1990). Productivity under agroforestry systems. Paper in National Sem. on Forest Productivity F.R.I., Dehra Dun.

Ern, T. (1977). Study tour on forestry support for agriculture to the Peoples Republic of China. Preliminary report, 16 pp.

Evans, J. (1975). Two rotations of *Pinus patula* in Usutu forest, Swaziland. *Commonwealth For. Review,* **53**: 61–81.

Fairbairm, W.A. (1958). Methods of light intensity measurement instruments in the field. *Forestry,* **31**: 155–162.

FAO (1959). Tree Planting Practices in Temperate Asia—Burma, India and Pakistan. FAO Forestry Development Paper Rome, pp. 100–14.

FAO (1969). Report on FAO Study Tour on Shelterbelts and Windbreaks in the USSR (Parts I & II). FAO Report No. TA 2561, 145 pp.

FAO (1978). Forestry for Local Community Development. FAO Forestry Paper—7, Rome.

FAO (1980). Population Data Regarding Forestry Communities Practicing Shifting Cultivation. FAO/UNFPA Project. Rome.

FAO (1981). India and Srilanka—Agroforestry. Forestry for Local Community Development Programme. GCP/INT/347/SWE, FAO, Rome.

FAO (1985). Evaluation of the Gujarat Social Forestry Programme, Rome.

FAO (1988). Nitrogen-fixing Trees for Wastelands. RAPA (Regional Office for Asia and Pacific). FAO, Bangkok.

FAO (1989). Case Studies of Farm Forestry and Wasteland Development in Gujarat, India. FAO, Rome.

Faruqui, Q. (1972). Organic and mineral structure and productivity of plantations of sal and teak. Ph.D. Thesis, B.H.U.

Foley, Gerald and G. Barnard. (1984). Farm and Community Forestry. Earthscan—International Institute for Environment and Development. Technical Report No. 3, London, 236 pp.

Frank, A.B., D.G. Harris and W.O. Wills. (1976). Influence of windbreaks on crop performance and snow management in North Dacota. In: R.W. Tinus (Eds) *Shelterbelts on the Great Plains. Proc. Symp.* GPAC Publ. No. 78, Denver Co, 41–48 pp.

FRI (1985). Information on important tree species. F.R.I., Dehra Dun

FRI (1989). Annual Research Report, 1987–88. F.R.I., Dehra Dun, 103 pp.

FRI (1990). Annual Research Report, 1988–89. F.R.I., Dehra Dun, 95 pp.

FSI (1989). The State of Forest Report-1989. Govt. of India, Forest Survey of India. Ministry of Environment and Forests, Dehra Dun.

Ganguli, J.K. and R.N. Kaul. (1969). Wind erosion control. *I.C.A.R. Technical Bull.*, No. 20, 57 pp.

Gaonkar, P.K. (1989). Karnataka social forestry project—a status paper. In: Sem. on Social Forestry and Agroforestry. F.R.I., Dehra Dun.

Gatherum, G.E. (1961). Variation in measurements of light intensity under forest canopies. *For. Sc.*, 7: 144–145.

George, M. (1977). Organic productivity and nutrient cycling in *Eucalyptus* hybrid plantation. Ph.D. Thesis, Meerut University, 195 pp.

George, M. (1978). Interception, stem flow and throughfall in *Eucalyptus* hybrid plantations. *Indian Forester*, 104(11): 719–726.

Gholz, H.L. (1988). Agroforestry—Introduction. In: *Agroforestry: Realities, Possibilities, and Potential* (Ed.: H.L. Gholz). Martinus Nijhoff Publishers, Dor drecht, Netherlands, 227 pp.

Ghosh, R.C. (1961). Teak is introduced in laterite waste. *Proc. IXth Silvi. Conf.*, Dehra Dun, pp. 450–455.

Ghosh, R.C. (1974). The protective role of forestry to the land. Paper in IXth Commonwealth For. Conf., London.

Ghosh, R.C. (1977). Handbook of Afforestation Techniques. Controller of Publications, New Delhi, pp. 418.

Ghosh, R.C., O.N. Kaul and B.K. Subba Rao. (1978). Some aspects of water relations and nutrition in *Eucalyptus* plantation. *Indian Forester*, 104(7): 517–524.

Ghosh, R.C., B.K. Subba Rao and B.C. Ramola. (1980). Interception studies in sal (*Shorea robusta*) coppice forest. *Indian Forester*, 106(8): 513–525.

Gill, A.S., S.N. Tripathi and K.S. Gangwar. (1986). Performance of hybrid *Napier* (IGFRI) in association with *Leucaena* (K8) under various proportions for forage yield. *IGFRI Ann. Rep.*, 47 pp.

Griffith, A.L. and R.S. Gupta. (1948). Soils in relation to teak with special reference to laterization. *Ind. For. Bull.*, No. 141.

Gulati, N.K., D.C. Chaudhari and R.K. Suri. (1982). Nutritive value of some fodder trees. *Proc. National Seminar on Chemistry, Industry and Citizen.* Ind. Chem. Soc., Dehra Dun, pp. 82–92.

Gupta, T. and Deepinder Mohan. (1982). *Economics of Trees versus Annual Crops on Marginal Agricultural Lands.* Oxford IBH Publishing Co., New Delhi, 139 pp.

Gupta, P.N. (1982). Export potential of large cardamom. *Ind. Hort.*, 26: 21–25.

Gurmurti, K. and H.C.S. Bhandari. (1987). Comparative assessment of productivity in six fast-growing forest tree species. In: R.V. Singh's *Forestry and Food Security in India*, pp. 118.

Hardwood, R.R. and E.C. Price. (1976). Multiple cropping in tropical Asia. In: *Multiple Cropping*, pp. 11–40. ASA Special Pub. No. 27., Am. Soc. Agron. Madison, Wisconsin, USA.

Harold, R.W. and A.L. Warlito. (1985). Sloping agricultural land technology. A social forestry model in the Philippines. In: *Community Forestry—Lessons from Case Studies in Asia and Pacific Regions.* FAO, Bangkok.

Hegde, N.G. (1987). *Handbook of Wastelands Development.* BAIF, Kamdhenu Senapati Marg. Pune, 102 pp.

Heinick, D.R. (1966). Characteristics of McIntosh and Red Delicious apples as influenced by exposure to sunlight during the growing season. *Proc. Am. Soc. Hort. Sci.*, 89: 10–13.

Hesketh, J.D. (1963). Limitations to photosynthesis responsible for differences among species. *Crop. Sci.*, 3: 493–496.

Hill, M. (1906). A note on an enquiry by the Government of India into the relation between the forests and atmospheric and soil moisture in India. *For. Bull.*, No. 33, 41 pp.

Hoekstra, D.A. (1985). The use of economics in diagnosis and design of agroforestry systems. ICRAF, Nairobi, Working paper No. 29.

Hooda, R.S. (1983). Agroforestry and rural development in Haryana State. In: *Proc. National Workshop on Agroforestry* (Eds.: R.S. Mathur and M.G. Gogate). Karnal, pp. 55–64.

Howard, S.H. (1939). Effects of standards on field crops in taungya. *Indian Forester*, 65(3): 151–152.

Howard, S.H. (1944). Post-war Forest Policy of India. Govt. of India Press, New Delhi, 49 pp.

Hughes, C.E. and B.T. Styles. (1984). Exploration and seed collection of multipurpose dry zone trees in Central America. *International Tree Crops Jour.*, 3: 1–31.

Huxley, P.A. (1983). Comments on Agroforestry classification with special reference to plants. In: *Plant Research and Agroforestry* (Ed.: P.A. Huxley). ICRAF, Nairobi, pp. 161–171.

ICAR (1980). Handbook of Agriculture. ICAR, New Delhi, pp. 615.

ICAR (1985). Tree Plantation in Alkali Soils. Better Farming in Salt-affected Soils. CSSRI, Karnal, series 8.

ICRAF (1988). Newsletter and Agroforestry Review, No. 24.

IMRB (1989). Marketing of Farm Forestry Products in West Bengal. Indian Marketing Research Bureau, New Delhi.

Ingty, P.S. and N. Goswami. (1979). Shifting cultivation in north-eastern hills. *Proc. of the Agroforestry Sem.* ICAR, Imphal, pp. 138–47.

Iyengar, E.R.R. (1988). Personal communication from Central Salt and Marine Chemical Research Institute, Bhavnagar.

Jack, W.H. (1966). The afforestation of blanket bog areas in the British Islands. *Proc. VIth World For. Cong.* Madrid.

Jack Westoby (1968). Inaugural speech. Xth Commonwealth Forestry Conference, New Delhi.

Jackson, J.E. (1970). Aspects of light and climate within apple orchards. *J. Appl. Ecol.*, 7: 207–216.

Jackson, J.E. (1975). The seed bed, sowing the seed, tending the seedlings. In: Report on FAO/DANIDA. Training course on forest seed collection and handling, No. 2. FAO, Rome.

Jackson, J.E. (1980). Light interception and utilization by orchard systems. In: *Horticultural Review* (Ed.: J. Janick). AVI Publishing Co. Inc. Westport, Conn, U.S.A., pp. 208–267.

Jackson, J.E. and J. Landsberg. (1972). The value of research in orchard climatology. *Span*, 15: 63–65.

Jain, S. (1988). Case Studies in Farm Forestry in Gujarat. FAO, Rome.

Jambulingam, R. and E.C.M. Fernandes. (1986). Multipurpose trees and shrubs on farm lands in Tamil Nadu state (India). *Agroforestry Systems*, 4: 17–23.

Janardhan, K.V. and K.S. Murty. (1980). Effect of low light during vegetative stage on photosynthesis and growth attributes in rice. *Ind. Jour. Pl. Phy.*, 23: 156–162.

John, Beer (1988). Litter production and nutrient cycling in coffee (*Coffea arabica*) or Cacas (*Theobroina cacao*) plantations with shade. *Agroforestry Systems*, 7(2): 103–114.

Kang, B.T., G.F. Wilson and L. Sipkens. (1981). Alley cropping of maize (*Zea mays*) and leucaena (*Leucaena leucocephala*) in southern Nigeria. *Plant and Soil*, 63: 165–179.

Kang, B.T., H. Grimme and T.L. Lawson. (1985). Alley cropping sequentially cropped maize and cowpea with *Leucaena* on a sandy soil in southern Nigeria. *Plant and Soil*, 67: 267–277.

Kapur, S.K. (1989). Social Forestry in Punjab. Paper in Seminar on Social Forestry and Agroforestry. F.R.I., Dehra Dun.

Kaul, R.N. (1959). Shelterbelts to stop creep of the desert. *Indian Forester*, 85(3): 191–195.

Kaul, R.N. (1965). Afforestation of cold desert in India. *Indian Forester*, 91(1): 2–9.

Kaul, R.N. and B.N. Ganguli. (1963). Fodder potential of *Zizyphus* in the scrub grazing of arid zones. *Indian Forester*, 89(5): 623–630.

Khybri, K.L., P.K. Gupta and Sewa Ram. (1983). Studies on intercropping of field crops with fodder crops of subabul under rainfed condition. *Annual Report CSWCRTI*. Dehra Dun, 61 pp.

King, K.F.S. (1978). Agroforestry. Paper in the 50th Agricultural Conference, Amsterdam.

King, K.F.S. (1979). Some principles of Agroforestry. Keynote address. *Proc. of Agroforestry Seminar, Imphal.* ICAR, pp. 17–27.

King, K.F.S. (1984). *Agrisilviculture* (The Taungya System). University of Ibadan, Ibadan, 109 pp.

King, K.F.S. and M.T. Chandler. (1978). *The Wastelands.* ICRAF, Nairobi, 85 pp.

Kittredge, J. (1962). Forest Influences, FAO Forestry and Forestry Products Studies, No. 15, Rome. pp. 81–137.

Knight, R.K. (1935). The effect of shade on American cotton Empire. *Jour. of Expt. Agric.,* · 3: 31–40.

Kolar, M., R. Karschon, and J. Kaplan. (1966). Afforestation technique for difficult sites: Arid areas. *Proc. Sixth World For. Congress,* Madrid. Vol III, pp. 3675–81.

Kulkarni, M.M. (1979). People's participation in social forestry, fodder and pasture development under DPAP in Maharashtra. *Proc. Seminar on People's Participation in Social Forestry, Fodder and Pasture Development.* Indore (M.P.), pp. 175–180.

Kunkle, I.H. (1978). Forestry support for agriculture through watershed management: Windbreaks and other conservation measures, *Eighth World For. Cong.,* Jakarta.

Kushwah, B.L., E.V. Nelliat, V.T. Markose and A.F. Sunny. (1973). Rootir? pattern of coconut. *Indian Jour. Agron.,* 18: 71–74.

Lahiri, A.K. (1972). Intercropping trials with turmeric in North Bengal. *Indian forester,* 98(2): 109–115.

Lahiri, A.K. (1983). Agroforestry in West Bengal. Parts I and II. *Proc. Workshop on Agroforestry* (Eds.: R.S. Mathur and M.G. Gogate). Karnal, pp. 218–225.

Lahiri, A.K. (1987). Silvopisciculture as land management system for rural development in mangrove area. *Indian Agri.,* 31(4): 305–311.

Laurie, M.V. and A.L.Griffith. (1942). The problem of pure teak plantation. *Ind. For. Rec.* (NS) *Silvi,* 5.

Lewis, N.B. (1967). Regeneration of man-made forests. *Proc. of World Symp. on Man-made Forests and Their Industrial Importance.* FAO, Rome.

Lindbald, P. and R. Russo. (1986). $C_2H_2$ reduction by *Erythrina* in a Costa Rican coffee plantation. *Agroforestry Systems,* 4: 33–38.

Lines, R. and S.A. Neustein. (1966). Afforestation techniques for difficult sites—wetlands. *Proc. Sixth World For. Cong.,* Madrid, Vol. III, 3750–56.

Logan, K.T. and E.B. Peterson. (1964). A method of measuring and describing light patterns beneath the forest canopy. Ministry of Forestry, Catalogue No. Fo. 47–1073.

Lundgren, B. (1978). Soil conditions and nutrient cycling under natural and plantation forests in Tanzanian highlands. Report in *Forest Eco. and Forest Soils,* No. 31. Swedish University of Agricultural Science, Uppsala.

Lundgren, B.O. and J.B. Raintree. (1982). Sustained Agroforestry. In: *Agricultural Research for Development—Potentials and Challenges in Asia* (Ed. B. Nestel). IS NAR, pp. 37–49.

MacDicken, K.G. (1981). *Leucaena* as a fallow improvement crop: A first approximation. Paper in Agroforestry and Fuelwood Workshop, East-West Centre, Honolulu.

Malhotra, P.P., V.N. Tandon and P.P. Shankar. (1987). Distribution of nutrients and their return through litterfall in an age series of *Pinus patula* plantation in Nilgiris. *Indian Forester,* 112(5): 323–332.

Malik, H.C. (1958). Grazing by rotation is grazing the right way. *Indian Farming* (NS) 8(5): 25–29.

Mallik, P.S. (1989). Social forestry programme and its impact in Haryana. Paper in Seminar on Social Forestry and Agroforestry. FRI Dehra Dun.

Mann, H.S. and H.S. Daulay (1981). A review of forestry: agroforestry research in India with special reference to arid and semi-arid regions. *Indian Jour. Range Management* 2: (1 and 2): 87–93.

Mann, H.S., S.P. Malhotra and K.A. Shankarnarayan. (1977). Land and resource utilisation in the arid zone. In: *Desertification and Its Control.* ICAR, New Delhi, pp. 89–101.

Martin, O.M. (1944). Influence of forests on rainfall. *Indian Forester*, 70(10): 325–327.

Martins, P.J. and J.C. Nautiyal. (1987). Population supporting capacities—a case study in the central Himalayas. Paper at IUFRO Workshop on Agroforestry for Rural Needs, New Delhi.

Mathur, H.N. (1978). Environmental conservation and mining. *Ind. Jour. For.*, 1(4): 266–277.

Mathur, H.N. and P. Joshi. (1972). Annual Report. Central Soil and Water Cons. Research and Training Institute, Dehra Dun.

Mathur, H.N., Rambaboo, P. Joshi and Bakshish Singh. (1976). Effect of clearing and reforestation in run-off and peak rates in small watersheds. *Indian Forester*, 102(4): 219–226.

Mathur, R.B. and Bakshish Singh. (1973). Effect of different types of planting stock, nursery and plantation techniques on the growth and survival of *Eucalyptus* hybrid. *Proc. First For. Conf.* FRI, Dehra Dun, pp. 220–228.

Mathur, R.S. and K.K. Sharma. (1983). Poplars in India. *Indian Forester*, 109(9): 591–628.

Mathur, R.S., K.K. Sharma and M.Y. Ansari. (1984). Economics of *Eucalyptus* plantations under Agroforestry. *Indian Forester*, 110(2): 171–201.

Mathur, Suresh (1979). Socio-economic aspects of agroforestry. *Proc. of the Agroforestry Sem.*, ICAR, Imphal, pp. 87–108.

McQueen, I.M.P. (1977). Agroforestry in New Zealand. *Proc. Workshop Integrating Agriculture and Forestry*. Eunbury, Australia, pp. 103–12.

Mello, H.A. (1976). Management problems in man-made forests of short rotation in South America. *Proc. 16th IUFRO World Cong.*, Div.–3, pp.

Minckler, L.S. (1961). Measuring light in uneven-aged hardwood stands. Central States Forest Expt. Sta. Tech. Paper. 184, 9 pp.

Mishra, D.N. and S.C. Sharma. (1973). Afforestation of saline alkaline soils. *Proc. First For. Conf.* FRI, Dehra Dun, pp. 94–103.

Mishra, R. (1969). Primary production of Chakkia forest and IBP/PT study of organic productivity and nutrient cycling in monsoon forests, grasslands and croplands. IUCN (NS) No. 18.

Moneslise, S.P. (1951). Light distribution in citrus trees. *Bul. Res. Council of Israel*, 1: 36–53.

Muthana, K.D. (1980). Silviculture aspects of Khejri. In: *Khejri (Prosopis cineraria) in the Indian Desert* (Eds.: H.S. Mann and S.K. Saxena). CAZRI, Jodhpur, pp. 20–24.

Muthana, K.D. (1986). Afforestation of arid zone wastelands for fuelwood and fodder production. In: *National Sem. on Fuelwood and Fodder Production*. FRI, Dehra Dun, pp. 150–55.

Muthana, K.D. and K.A. Shankarnarayan. (1978). Scope of silvipastoral management in arid regions, Souvenier on Arid Zone Research, Silver Jubilee, CAZRI, Jodhpur.

Nageli, W. (1940). Light measurements in open and closed stands of timber. *Schweizerische Anatalt furdas forstliche versuchswesen*. Muttelungen Zurich, 21(2): 250–306.

Naik, S.K. and K.S. Murthy. (1978). Effect of varying light intensities on yield and growth parameters in rice. *Ind. J. Pl. Physiol.*, 23: 309–316.

Naik, S.K., K.V. Janardhan and K.S. Murty. (1978). Photosynthetic efficiency of rice is influenced by light intensities. *Ind. J. Pl. Physiol.*, 21: 48–52.

Nair, M.A. and C. Sreedharan. (1986). Agroforestry farming systems in the homesteads of Kerala, southern India. *Agroforestry Systems*, 4: 339–346.

Nair, P.K.R. (1979). Agroforestry Research: A retrospective and prospective appraisal. *Proc. Int. Conf. International Cooperation in Agroforestry*. ICRAF, Nairobi, pp. 275–296.

Nair, P.K.R. (1985). Classification of agroforestry systems. *Agroforestry Systems*, 3: 97–128.

Nair, P.K.R. (1987). Soil productivity under agroforestry. In: *Agroforestry: Realities, Possibilities and Potentials* (Ed.: H.L. Gholz). Martinus Nijhoff Publishers, The Netherlands, pp. 21–31.

Nair, P.K.R. (1989). Agroforestry defined. In: *Agroforestry Systems in the Tropics* (Ed.: P.K.R. Nair). Kluwer Academic Publishers, London, pp. 12–18.

Nambiar, E.K.S. (1983). Root development and configuration in intensively managed radiata pine plantation. *Plant Soil*, 71: 37–47.

Nautiyal, J.C. and P.S. Banor. (1985). Forestry in Himalayas—How to avert an environmental crisis. *Interdisciplinary Sci. Rev.*, 10(1): 27–41.

NCA (1972). Interim Report of National Commission of Agriculture-Production Forestry. Govt. of India, Ministry of Agriculture and Irrigation, New Delhi.

NCA (1976). Report of the National Commission on Agriculture, Part IX, Forestry. Govt. of India. Ministry of Agriculture and Irrigation, New Delhi.

NCA (1976a). Report of the National Commission on Agriculture, Part VII, Demand and Supply. GOI, Ministry of Agriculture and Irrigation, New Delhi.

NCA (1976b). Report of the National Commission on Agriculture, Part VII, Animal Husbandry. GOI, Ministry of Agriculture and Irrigation, New Delhi.

NCA (1976c). Report of the National Commission on Agriculture, Part VII, Policy and Strategy. GOI, Ministry of Agriculture and Irrigation, New Delhi.

NCAER (1988). Haryana Wood Balance Study. Report of National Council of Applied Economic Research, New Delhi.

Negi, J.D.S. and S.C. Sharma. (1984). Distribution of nutrients in an age series of *Eucalyptus globulus* plantation in Tamil Nadu. *Indian Forester*, 10(9): 944–953.

NFTA (1989). Why nitrogen-fixing trees? NFTA 89–03. NFT Association Publication, Waimanlo, USA.

Nicholson, J.W. (1960). The Influence of Forests on Climate and Water Supply in Kenya. Forest Department Pamphlet, No. 2.

NRSA (1984). Nationwide mapping of forest and non-forest areas using landsat FCC for period 1972–75 to 1980–82. Project Report, Vol I, Department of Space, Government of India, Hyderabad.

NWDB (1987). Description and classification of wastelands. National Wasteland Development Board. GOI, Ministry of Environment and Forests, New Delhi.

NWDB (1987a). Microplanning—a tool for social forestry implementation. National Wasteland Development Board. GOI, Ministry of Environment and Forests, New Delhi, 36 pp.

NWDB (1989). National mission on wastelands development. Government of India, Ministry of Environment and Forests, New Delhi, 11 pp.

NWDB (1989a). Report of the working group on wasteland development for formulation of 8th Five Year Plan. GOI, Ministry of Environment and Forests, New Delhi.

NWDB (1989b). Developing India's wastelands, GOI, Ministry of Environment and Forests, New Delhi, 82 pp.

O'Neil, R.V. and D.L. DeAngelis. (1981). *Dynamic Properties of Forest Ecosystem* (Ed.: D.E. Reichle). Cambridge University Press, New York, 449 pp.

Osmastan, F.C. (1968). *The Management of Forests*. George Allen and Unwin Ltd., London, 384 pp.

Ovington, J.D. ( 1959). The circulation of minerals in plantation of *Pinus sylvestris. Ann. Bot.* (NS) 23.

Ovington, J.D. (1965). *Woodlands.* The English University Press Ltd., London, 286 pp.

Ovington, J.D. (1968). Some factors affecting nutrient distribution within the ecosystem. In: *Functioning of Terrestrial Ecosystems at Primary Production Level* (Ed.: F.E. Eckardt). UNESCO, Paris, pp. 135–142.

Ovington, J.D. and H.A. Madgwick. (1955). A comparison of light in different woodlands. *Forestry*, 18: 141–146.

Pal, Mohinder, (1990). Biomass production in energy plantation using some fast-growing species. Paper in Sem. on Forest Productivity. F.R.I., Dehra Dun.

Pandey, D.C. (1961). A short note on the introduction of tree species in the usar lands of Uttar Pradesh. *Proc. IXth Silvi. Conf.* F.R.I., Dehra Dun, pp. 337–342.

Pant, M.M. (1984). *Forest Economics and Valuation.* Madhavi Publishers, Dehra Dun, 615 pp.

Parasnis, S.S. and M.B. Gogate. (1979). Productivity studies for manpower planning. National Seminar on Productivity of Forest Operations. National Productivity Council, New Delhi.

Pareek, O.P. and B.B. Vashishtha. (1978). Improvement of fruit and crop for increased productivity in arid zone. In: *Proc. of Int. Symp. on Arid Zone Research and Development*. CAZRI, Jodhpur.

Paroda, R.S. and K.D. Muthana. (1979). Agroforestry practices in arid zones. In: *Proc. of Agroforestry Seminar*. ICAR, Imphal, pp. 69—76.

Patel, V.J. (1988). A New Strategy to High Density Agroforestry. Jivrajbhai Patel Agroforestry Centre, Surendrabag, Gujarat, 57 pp.

Patil, B.D., R. Deb Roy and R.S. Pathak. (1979). Agroforestry: Research & Development with reference to the Indo-Gangetic plains. In: *Proc. of Agroforestry Seminar*. ICAR, Imphal, pp. 40—69.

Pathak, P.S. (1989). Management of subabul for optimising production. In: *Production of Fodder and Fuelwood Trees* (Eds.: N.G. Hegde and others). BAIF Publication, Pune, pp. 89—96.

Paul, D.K. (1972). A Handbook of Nursery Practice for *Pinus caribaea* and Other Conifers in West Malaysia. UNDP, Kaula Lumpur, 112 pp.

Pillai, V.K.K. (1983). Agroforestry relevance and scope. In: *Proc. National Workshop on agroforestry* (Eds.: R.S. Mathur and M.G. Gogate). Karnal, pp. 76—80.

Pradhan, I.P. (1973). Preliminary study of rainfall in interception through leaf litter. *Indian Forester*, 99(7): 440—445.

Prasad, R. (1965). Afforestation and soil conservation in South Bihar. *Indian Forester*, 91(1): 33—40.

Pryor, L.D.Z. (1978). Eucalyptus as exotics. Paper in International Training Course on Forest Tree Breeding. Australian Development Assistance Agency.

Puri, D.N. and P. Joshi. (1979). Economic utilisation of class VI and VII land for raising fuel (*Eucalyptus* hybrid) and fodder grass (*Chrysopogon fulvus*). Annual Report. Central Soil & Water Conservation Res. Training Institute, Dehra Dun.

Raghavachari, S. (1974). Population Projections, 1976—2001. In: *Population in India's Development*. Vikas Publishing House, New Delhi, 437 pp.

Raizada, H.C. (1989). Social forestry in U.P. Paper in Sem. on Social Forestry and Agroforestry. FRI, Dehra Dun.

Rakhmanov, R.A. (1966). Role of Forests in Water Conservation (Translated from Russian under Israel Programme for Scientific Translation, Jerusalem).

Ram Parkash and Drake Hocking. (1986). *Some Favourite Trees for Fuel and Fodder*. Society for Promotion Wasteland Development, New Delhi, 187 pp.

Ram Prasad (1988). Effectiveness of aerial seeding in reclamation of Chambal ravines in Madhya Pradesh. *Indian Forester*, 114(1): 1—18.

Ram Prasad and P.K. Shukla. (1986). Restoration of ecological balance to the bauxite mined areas of Madhya Pradesh. National Sem. on Fuelwood and Fodder Production from Wastelands. FRI, Dehra Dun.

Ram Prasad, A.K. Sah, A.S. Bhandari and O.B. Chaubey. (1984). Dry matter production by *Eucalyptus Camaldulensis* plantation in Jabalpur. *Indian Forester*, 110(9): 868—878.

Ranganathan, C.R. (1949). Protective functions of forests. In: *Proc. U.N. Conf. on Conservation and Utilization of Resources*, 5: 134—140.

Ranganathan, Shankar. (1979). Agroforestry: employment for millions. In: *Proc. Second Forestry Conf.* F.R.I., Dehra Dun, pp. 303—310.

Rao, C. Sarvotham (1983). Social forestry in Andhra Pradesh. In: *Proc. National Workshop on Agroforestry* (Eds.: R.S. Mathur and M.G. Gogate). Karnal, pp. 15—21.

Rao, D.V.M. and S.P. Singh. (1980). Studies on the effect of defoliation and shading on grain yield and its components in Bajra (*Pennisetum typhoides*). *Ind. J. Pl. Physiol.*, 23, 181—184.

Rao, M. and Sita Ram (1980). Influence of shelterbelts on annual crops. In: *Proc. Sec. For. Conf.* F.R.I., Dehra Dun, pp. 525—29.

Rao, S.N. and Reddy, P.C. (1984). Studies on the inhibitory effects of *Eucalyptus* hybrid leaf extract on germination of certain food crops. *Indian Forester*, 110(2).

Rawat, J.K. (1988). Modified volume table for Eucalyptus in Haryana, *Indian Forester*, 114(12): 837–843.

Rawat, R.B.S. (1990). A note on Tarai Central Forest Division, Haldwani (UP.). Personal Communication.

Ray, M.D. (1970). Preliminary observation on stem flow in *Alstonia scholaris* and *Shorea robusta* plantations at Arabari, West Bengal. *Indian Forester*, 96(7): 482–493.

Reddy, C.V.K. (1981). Agroforestry in coastal Andhra Pradesh. In: *Proc. Agrofor. Sem.*, ICAR, New Delhi, pp. 110–116.

Reddy Konda, C.V. (1979). Shelterbelts against storms and cyclones on the coast. *Indian Forester*, 105(10): 720–726.

Reddy, Sankara G.H., Y.Y. Rao and M.S. Rao. (1981). The effect of shelterbelt on the productivity of annual field crops. *Indian Forester*, 107(10): 624–629.

Rehr, S.S., P. Feeney and D.H. Janzen. (1973). Chemical defences in Central American non-ant acacias. *Jour. Animal Ecology*, 42: 405–416.

Rennie, P.J. (1955). The uptake of nutrients by mature forest growth. *Plant* and *Soil*, 1.

Richards, P.W. (1952). *The Tropical Rainforest*. The University Press, Cambridge, 450 pp.

Riley, L.F. (1980). Report on the Great Lakes. Forest Research Centre, Canada. No.6—X: 318.

Robert, R. (1977). When ambroisis beetles attack mahogany trees in Fiji. *Unasylva*, 117.

Royal Commission on Agriculture (1928). Royal Commission on Agriculture in India—Abridged Report. Govt. Central Press, Bombay, 755 pp.

Sale, P.J.M. (1976). Effect of shading at different times on the growth and yield of the potato. *Aust. J. Agric. Res.*, 27: 557–566.

Sanghal, P.M. (1983). Species compatibility considerations in agroforestry. The state of art in India. In: *Proc. National Seminar on Agroforestry*. Karnal, pp. 416–428.

Saukat Hussain, (1925). Suitable field crops for sal taungya in U.P. *Indian Forester*, 51(5): 192–198.

Saxena, N.C. (1989). Wasteland development, environmental protection for the poor. *Indian Jour. of Public Adm.*, V 35(3): 487–497.

Saxena, N.C. (1989a). Development of Degraded Village Lands in India. Regional Wood Energy Development Programme in Asia. Field document No. 5. FAO, Bangkok.

Saxena, N.C. (1990). Tree farming as a cash crop in North-West India: Recent experiences seen in a historical perspective. Paper in Seminar on Afforestation. Environment and Development. Gaziabad, U.P.

Sen, K.C., S.N. Ray and S.K. Ranjhans. (1978). *Nutritive Value of Indian Cattle Feeds and the Feeding of Animals.* ICAR, New Delhi.

Seth, S.K. (1960). Soil working technique in dry zone in relation to rainfall and soil types. *Indian Forester*, 98(4).

Shah, T. *et al.* (1980). Impact of increased dairy productivity on fairness, use of feed stuffs. *Eco. and Pol. Weekly*, xii(33): 1407–1412.

Shankarnarayan, K.A., L.N. Harsh and S. Katiuu. (1987). Agroforestry in the arid zones of India. *Agroforestry Systems*, 5: 67–88.

Shanmugasundaram, S. (1983). Agroforestry, its aspects and prospects in Tamil Nadu. In: *Proc. National Workshop on Agroforestry* (Eds.: R.S. Mathur and M.G. Gogate). Karnal, pp. 186–93.

Shanta, H.L. (1913). The effect of artificial shading on plant growth in Louisiana. USDA *Bureau of Pl. Industry Bull.*, 279.

Sharma, B.B. (1989). Rise and fall of Eucalyptus in Haryana. Paper in Seminar on Social Forestry and Agroforestry. F.R.I., Dehra Dun.

Sharma, D.G. (1984). Ravine reclamation through aerial seeding: an appraisal. *Jour. Trop. For.*, 1 (i): 1–15.

Sharma, K.K. (1987). Effect of trees on agricultural crops. Institutional Seminar, F.R.I., Dehra Dun.

Sharma, K.K. (1988). Personal communication—A note on Agroforestry. F.R.I., Dehra Dun.

Sharma, R.P. (1973). Afforestation of ravines in Uttar Pradesh *First For. Conf.*, F.R.I., Dehra Dun, pp. 107—10.

Sharma, Y.M.L. (1980). Environmental planning in mining with particular reference to Kudremukh Iron Ore Ltd. *Soc. For. Conf.*, F.R.I., Dehra Dun, pp. 281—86.

Shrivastav, A.K. and P. Narain (1980). Competition studies on tree crop association with *Eucalyptus tereticornis* under farm forestry at Kota. Annual Report, CSWCRTI, Dehra Dun, 119 pp.

Shrivastava, J.B.L. (1983). Practice of Agroforestry with special reference to Haryana. In: *Proc. National Workshop on Agroforestry* (Eds.: R.S. Mathur and M.G. Gogate), Karnal, pp. 81—85.

Shrivastava, T.N. and I.M. Qureshi. (1966). Afforestation of difficult sites. *Proc. Sixth World For. Conf.* Madrid, pp. 3769—3774.

Shyam Sunder. (1983). Social forestry including Agroforestry, Karnataka. *Proc. National Workshop on Agroforestry*, Karnal, Haryana (Eds.: R.S. Mathur and M.G. Gogate). F.R.I., Dehra Dun, pp. 100—109.

Singh, A.K., N.G. Totey and P.K. Khatri. (1988). Comparative performance of some important forest trees in Bhata soil of Raipur (M.P.). *Proc. Workshop on Technique of Wasteland Development through Agroforestry.* SFRI, Jabalpur (Mimeographed).

Singh, B. (1971). A comparative study on economics of various soil conservation-cum-grass land improvement practices for rejuvenation forage production in ravine lands. *Indian Forester*, 17(3): 387—391.

Singh B. and B. Verma. (1971). A comparative study on economics of various soil conservation-cum-grass land improvement practices in ravine land. *Indian Forester*, 97(4): 315—321.

Singh, Bakshish. (1972). Reclamation of ravine lands through afforestation. *Proc. Symp. on Man-made Forests*, Dehra Dun, pp. 73—78.

Singh, D. (1986a). Effect of low light intensity on growth yield of rainfed cotton. *Ind. J. Pl. Physiol.*, 29: 230—236.

Singh, Gurmel, C. Venkataramanan and Y. Shastry. (1981). Manual of Soil and Water Conservation Practices in India. Central Soil and Water Conservation Research and Training Institute, Dehra Dun, 434 pp.

Singh, H.S. (1988). Cost Benefit Analysis of Eucalyptus farming in U.P. with special reference to Kanpur. *Ind. Jour. of Agril. Eco.* (3): 56—60.

Singh, K.P. (1968). Litter production and nutrient turn-over in deciduous forests of Varanasi. In: *Proc. Symposium on Recent Advances in Tropical Ecology*, 2: 655—665.

Singh, R.P. (1984). Nutrient cycle in *Eucalyptus tereticornis* plantation. *Indian Forester*, 110(1): 76—85.

Singh, R.V. (1982). *Fodder Trees of India.* Oxford and IBH Publishing Co., New Delhi, 663 pp.

Singh, R.V. (1986). People's Participation in Social Forestry Programme in Himachal Pradesh. F.R.I., Dehra Dun, 28 pp.

Singh, R.V. (1988). Forestry and Food Security in India. F.R.I., Dehra Dun, 118 pp.

Singhal, R.M., E.R.C. Reynolds and S.P. Pant. (1987). Preliminary investigations on the spatial relations of roots of field crop and boundary tree. Paper presented in Workshop on Agroforestry for Rural Needs, New Delhi.

Singhal, R.M. and S.P. Pant. (1989). Tree crop interaction in certain agroforestry systems of Haryana. Sem. on Social Forestry and Agroforestry. F.R.I., Dehra Dun.

Sinha, R.L. (1975). The problem in silt in relation to irrigation and river projects and forests. In: *Proc. Fourth Irrigation and Pow. Seminar*, Hirakund, pp. 54—58.

Solanki, M.S. (1981). *Forests as a Source of Food*, Vols. I and II. FAO, Regional Office, Bangkok.

Stansel, J.W., C.N. Bollich and J.R. Thysell. (1965). The influence of light intensity and N fertility on rice yield and components of yield. *Rice J.*, **68**: 34–35.

Stebbing, F.P. (1952). Forests, catchment areas and water supplies. *Ind. For. Rec.* (NS) *Silvi*, 7(4).

Stenin, W.I., P.R. Slabaugh and A.P. Plummer. (1974). Harvesting, processing and storage of fruits and seeds. In: Seeds of Wood Plants in the U.S.A. Agriculture Handbook No. 450, Washington, D.C.

Stoler, A. (1978). Garden use and household economy in rural Jawa. *Bull. Indonesian Studies*, 14(2): 85–101.

Stewart, D. (1933). Financial prospect of departmental taungyas in Haldwani, U.P., *Indian Forester*, 59(3): 142–148.

Stewart, P.J. (1981). Forestry, agriculture and land husbandry. *Commonwealth Forestry Review*, 60(1): 29–34.

Stoeckeler, J.H. (1943). Windbreak and shelterbelt planting in the United States. In: *Proc. IIIrd For. Cong.* pp. 253–64.

Stoeckeler, J.H. (1965). The design of shelterbelts in relation to crop yields improvement. *World Crops*, 11(1): 27–32.

Storey, H.S. (1966). Forest ecological influence on climate soil, water resources and man. In: *Proc. VI World Forestry Congress*, Madrid, Vol. III, pp. 3774–3780.

Subhash Chand and S.N. Mahapatra. (1983). Production of sugar syrup from Mahua (*Madhuca latifolia*) flowers. *Research and Industry*, **28**: 29–31.

Tanaka, A., S.A. Navasero, C.V. Garcia, F.T. Parao and E. Ramirez. (1964). Growth habit of rice plant in the tropics and its effect on nitrogen response. *Tech. Bull.* 3. IRRI, Philippines, pp. 45–47.

Tejwani, K.G. (1987). Agroforestry practices and Research in India. In: *Agroforestry: Realities, Possibilities and Potentials* (Ed.: H.L. Gholz). Martinus Nijhoff Publishers, Dordrecht, the Netherlands, pp. 109–37.

Tejwani, K.G., V.V. Druvanarayan and T. Satyanarayan. (1960). Control of gully erosion in ravine land of Gujarat. *Jour. Sal. Water Cons.*, **8**: 65–74.

Tejwani, K.G., V. Shrivastav and M.S. Mistry. (1961). Gujarat can save its ravine land. *Indian Farming*, **11**: 20–21.

Thomas, J.F. (1977). Economics of agroforestry at the regional level. In: *Proc. Workshop Integrating Agriculture and Forestry*. Bunbury, Australia, pp. 78–87.

Tiwari, K.M. (1970). Interim results of intercropping of miscellaneous tree species with main crop of taungya plantation to increase the productivity, *Indian Forester*, 96(a): 142–148.

Tiwari, K.M. (1983). Hand Book on Social Forestry. Social Forestry Series 2. F.R.I., Dehra Dun, 43 pp.

Tiwari, K.M. and R.S. Mathur. (1983). Water conservation and nutrient uptake by Eucalyptus. *Indian Forester*, 109(12): 851–860.

Tripathi, K.P. (1983). Agroforestry in North Bihar. *Proc. National Workshop on Agroforestry* (Eds. R.S. Mathur and M.G. Gogate). Karnal, pp. 27–36.

Toumey, J.W. and C.F. Korstian. (1947). *Foundation of Silviculture upon and Ecological Basis.* John Wiley and Sons, New York, 421 pp.

Troup, R.S. (1926). Problem of forest ecology in India. In: *Aims and Methods in the Study of Vegetation* (Eds.: A.G. Transley and T.F. Chipp). London, pp. 283–313.

Turnbull, J.W. (1975). Assessment of seed crops and timing of seed collection. In: Report on Forest Seed Collection and Handling, Vol. 2, FAO, Rome.

USAID (1988). USAID/World Bank Evaluation of the Social Forestry Programme in Uttar Pradesh, Gujarat and Rajasthan.

Vaishnav, M.N. (1981). Position paper on ipil-ipil in Gujarat. In: *Proc. of Seminar on Ipil-ipil.* Gandhinagar, Gujarat, pp. 71–75.

Venkateswarlu, B. and T.E. Srinivasan. (1978). Influence of low light intensity on growth and productivity in relation to population pressure and varietal reaction in irrigated rice (*Oryza sativa L.*). *J. Pl. Physiol.*, **21**: 162–170

Vergara, N.T. (ed.). (1982). New Directions in Agroforestry. The Potential of Tropical Legume Trees. Environmental Policy Institute, East-West Centre, Honolulu, 32 pp.

Vergara, N.T. (1985). Agroforestry. A sustainable land use for fragile ecosystems in the humic tropics. In: *Agroforestry Realities, Possibilities and Potentials.* (Eds.: H.L. Gholz). Martinus Nijhoff Publishers, the Netherlands, pp. 7–21.

Vergara, N.T. (1985). Agroforestry systems. A primer. *Unasylva*, **37**: 22–28.

Verinumbe, I. (1983). Agroforestry, an independent system for improved food and wood production in an arid zone. Nigeria. *Nigerian Jour. of Forestry*, **13**: 1.

Verinumbe, I. (1987). Crop production on soil under some forest plantations in the Sahel. *Agroforestry Systems*, **5**(2): 185–188.

Verma, B., B. Singh, H.N. Saraf and K. Monnappa. (1969). Suitability and economics of grasses for reclamation and stabilisation of Mahi ravines in Gujarat. *Indian Forester*, **95**(1): 33–44.

Verma, C.M. (1975). Grassland improvement and management in arid and semi-arid regions of western Rajasthan. In: *Proc. National Symp. on Appraisal of Fodder Grasses and Their Seed Production.* Jhansi, pp. 150–54.

Verma, D.P.S. (1988). Fuel and fodder from village wood lots. A Gujarat (India) experience. *Agroforestry Systems*, **7**(1): 77–93.

Voelcker, J.A. (1897). *Report on the Improvement of Indian Agriculture.* Egre and Spottis Woode, London, 543 pp.

Vohra, B.B. (1986). Management of Natural Resources—Urgent Need for Fresh Thinking. Advisory Board on Energy, Govt. of India, New Delhi.

Warren, W.D.M. (1974). A study of climate and forests in the Ranchi Plateau. *Indian Forester*, **100**(4): 229–234.

Watt, G. and H.H. Mann. (1903). The Parts and Blights of the Tea Plant. Govt. of India, Calcutta, 2nd ed., pp. 140–145.

Wattal, P.N. and R.D. Asana. (1974). Responses of tall and dwarf varieties of wheat to light intensities. *Ind. J. Agric. Sc.*, **44**, 707–711.

Wellner, C.A. (1948). Light intensity related to stand density in mature stands of the Western White pine type. *Jour. For.*, **46**: 16–19.

Whyte, R.O. (1964). The Grassland and Fodder Resources in India. ICAR, New Delhi, 437 pp.

Wigston, D.L. (1985). Nitrogen demand and nitrogen fixation in woody plants—cautionary tale. *Klinkii*, **3**: 52–69.

Wilson, J. (1965). Soil erosion in Rameshwaram island. *Indian Forester*, **91**(1): 22–27.

Wright, W. (1959). The shade tree tradition in tea gardens of North India, I. The value of shade. Rep. Indian Tea Assoc. Sci. Dept. Tocklai Expt. Sta., Jorhat, pp. 75–98.

Yadav, J.S.P. (1980). Salt-affected soils and their afforestation. *Second For. Conf.* F.R.I., Dehra Dun, pp. 280–87.

Yadav, J.S.P. and K. Singh. (1970). Tolerance of certain forest species to varying degree of salinity and alkalinity. *Indian Forester*, **96**(8): 587–599.

Yoshida, S. (1975). Eco-physiology of rice. Paper presented at the Symposium on Tropical and Subtropical Crops Centro de Pesquisas do Cacan, Brazil, May 26–30.

Young, A. (1989). Ten hypotheses for soil-agroforestry research. *Agroforestry Today*, **1**(1): 13–16.

Zon, R. (1927). Forest and Water in the Light of Scientific Investigation. Final Report, National Waterways Commission, 1912.

# Appendix

Technical Information on

| Sl. No. | Species | Common name | Climatic zone | Fruit/seed collection season | Number of seeds per kg | Seed storage temperature | Seed longevity | Pre-sowing seed treatment |
|---|---|---|---|---|---|---|---|---|
| 1 | 2 | 3 | 4 | 5 | 6 | 7 | 8 | 9 |
| 1 | *Abies pindrow* | Silver fir | MT | Sep.–Nov. | 27,200 | LT | SL | Not required |
| 2 | *Abies spectabilis* | Himalayan fir | WT | Oct.–Nov. | 16,700 | LT | SL | -do- |
| 3 | *Acacia auriculiformis* | Australian wattle | Dtr, Mtr | Dec.–Feb. | 40,000 | RT | MLL | Scarification, Hot water |
| 4 | *Acacia catechu* | Khair | Dtr, Mtr | Jan.–March | 40,000 | RT (Requires insecticide treatment before storage) | MLL | -do- |
| 5 | *Acacia dealbata* | Silver wattle | T | June | 86,500 | RT | VLL | -do- |
| 6 | *Acacia decurrens* | Green wattle | T | Oct.–Nov. | 80,357 | RT | VLL | -do- |
| 7 | *Acacia farnesiana* | Pissi babul | Dtr | July–onwards | 9,700–12,100 | RT | LL | Scarification, Hot water |
| 8 | *Acacia mearnsii* | Black wattle | T | November | 95,000 | RT | LL | -do- |
| 9 | *Acacia melanoxylon* | Australian black wood | T | July–Nov. | 66,000–77,000 | RT | LL | -do- |
| 10 | *Acacia modesta* | Phulai | DSt, Dtr | Oct.–Dec. | 21,179–35,298 | RT | LL | -do- |
| 11 | *Acacia nilotica* | Babul | Mtr, Dtr | April–June | 7,000–11,000 | RT | VLL | -do- |
| 12 | *Acacia senegal* | Khor | Dtr | Spring | 7,165 | RT | LL | -do- |
| 13 | *Acacia tortilis* | Israeli kikkar | Dtr | Nov.–Feb. | 12,000 | RT | LL | -do- |
| 14 | *Acer caesium* | Maple | MT, T | July–Oct. | 12,142 (fruits) | LT | SL | Not required |

Important Tree Species

| Sowing season | Germination per cent | Normal germination period in nursery (days) | Optimum spacing in nursery (cm) | Age of normal planting stock (months) | Planting season | Method of planting | Uses |
|---|---|---|---|---|---|---|---|
| 10 | 11 | 12 | 13 | 14 | 15 | 16 | 17 |
| Oct.–Nov. | 10–65 | – | 20 x 10 | 36–48 | July | EP, DS | Timber |
| February | Good | 42 | 20 x 10 | 4 | June–July | EP | Timber |
| April | 50–67 | 30 | 10 x 10 | 2–3 | July–Aug. | DS, EP, ST | Timber, Fuel, Sand dune fixation, Ornamental |
| March–April | 60–80 | 30 | 8 x 8 | 2–3 | July | DS, EP, ST | Fodder, Timber, Fuel, Gum, *Katha*, Soil Conservation |
| October | 68 | 30 | 10 x 10 | 6–9 | July | DS, EP | Fuel, Gum, Tannin, Fodder |
| October | 50–90 | 30 | 10 x 10 | 9–12 | June–July | DS, EP, RS | Fuel, Tannin, Fodder |
| May | 60–70 | 30 | 10 x 10 | – | July | DS, EP | Fuel, Tannin, Fodder, Gum, Soil Conservation |
| Sept.–Oct. | 57 | 30 | 10 x 10 | 6–9 | June–July | DS, EP | Tannin, Fuel, Soil Conservation, Fodder |
| Sept.–Oct. | 60–70 | 30 | 10 x 10 | -do- | -do- | DS, EP | Fuel, Timber, Tannin, Fodder |
| Feb.–March | 60–90 | 30 | 10 x 10 | -do- | -do- | DS, EP, BC | Fuel, Timber, Gum, Tannin, Soil Conservation, Fodder |
| March | 88 | 30 | 15 x 15 | 4 | -do- | DS, EP | Fuel, Timber, Tannin, Gum, Fodder |
| June–July | 60–70 | 30 | 15 x 15 | 12 | July | DS, EP | Fuel, Fodder, Tannin, Soil Conservation, Gum |
| Feb.–March | 40–60 | 60 | 15 x 15 | 9 | July | EP | Fodder, Fuel |
| Feb.–March | High | 14–28 | 20 x 20 | 12 | Winter | EP, ST | Fuel, Timber |

Contd.

| 1 | 2 | 3 | 4 | 5 | 6 | 7 | 8 | 9 |
|---|---|---|---|---|---|---|---|---|
| 15 | *Acer campbellii* | Maple | MT, WT, T | Nov.–Dec. | 18,00,000 | LT | SL | -do- |
| 16 | *Acer oblongum* | Maple | T, St | Jan.–April | 20,120–40,480 (fruits) | LT | MLL | -do- |
| 17 | *Acrocarpus fraxini-folius* | Mundani | Mtr, St | April–June | 32,000 | — | MLL | Hot water, $H_2SO_4$ |
| 18 | *Adina cordifolia* | Haldu | Mtr, Wtr | Jan.–Mar. | 11,000,000 | — | SL–MLL | Not requi-red |
| 19 | *Aegle marmelos* | Bel | Wtr, Mtr | March–May | 5,300 | — | SL | -do- |
| 20 | *Aesculus indica* | Horse chestnut | T, St | Sept.–Oct. | 36 | — | VSL | Stratifica-tion |
| 21 | *Ailanthus altissima* | Tree of Haven | Mtr | November | 2,100 | LT | — | Stratifica-tion |
| 22 | *Ailanthus excelsa* | Maharukh | Dtr, Mtr | May–June | 9,500 | — | SL | Not requi-red |
| 23 | *Ailanthus integri-folia* | Gogul | Mtr, Wtr | Jan.–May | 1,235–1,750 | — | SL | -do- |
| 24 | *Albizia amara* | Lallei | Dtr | Feb.–April | 14,472 | RT | VLL | Scarifica-tion, Hot water |
| 25 | *Albizia chinensis* | Sirin | Mtr, St | Dec.–March | 32,143 | RT | LL | -do- |
| 26 | *Albizia falcataria* | Sengon | Mtr | May–June | 42,240 | RT | LL | -do- |
| 27 | *Albizia lebbeck* | Siris | Dtr, Mtr | Jan.–March | 8,000–13,000 | RT | VLL | -do- |
| 28 | *Albizia odora-tissima* | Kala Siris | Mtr, Wtr | Jan.–Feb. | 15,600 | LT | VLL | Scarifica-tion, Hot water |
| 29 | *Albizia procera* | Safed Siris | Dst, Mtr, WSt, Wtr | Jan.–April | 21,786 | LT | VLL | -do- |
| 30 | *Alnus nepal-ensis* | Alder (Utis) | St, T | Dec.–Jan. | 570,000 | LT | SL | Stratifica-tion |
| 31 | *Alnus nitida* | Alder (Sarol) | MSt, MT | Nov.–Dec. | 885,987 | LT | SL | -do- |
| 32 | *Amoora wallichii* | Amari | Mtr, Wtr | May–July | 132–155 | — | SL | Not requi-red |

| 10 | 11 | 12 | 13 | 14 | 15 | 16 | 17 |
|---|---|---|---|---|---|---|---|
| March–April | 75 | 14 | 25 x 20 | 3 | June–July | DS, EP | Fuel, Timber |
| March | 50 | — | — | 2 | July | DS, EP | Agricultural implements, Timber |
| May–June | 75 | 14–21 | 25 x 20 | 1 | June–July | DS, EP, ST | Fuel, Box wood, Planks, Boards |
| May | 50 | 30 | 15 x 10 20 x 10 | 3 | August | DS, EP, ST | Bobbins, Timber |
| Soon after collection | 56 | 21 | 20 x 10 | 12–24 | July | DS, EP | Fuel, Fruit, Gum, Bark and Fruit (medicinal) |
| -do- | 80–90 | — | 15 x 15 | Few months | July | EP, ST | Timber, Fodder, Fruit, Ornamental |
| Mar.–April | 80–90 | 12–13 | — | 12 | Winter/Monsoon | EP, ST, BC | Ornamental |
| Soon after collection | 70–90 | 45 | 20 x 10 | 4–5 | July–Aug. | DS, EP, ST, BC | Timber, Packing cases, Boards, Bark (medicinal), Fishing floats |
| May–June | 60 | 120 | 8 x 8 | — | July/Winter | DS, EP, ST | Plywood, Ornamental |
| — | — | — | — | 12 | July | DS, EP | Fodder, Timber, Tool handles, Agricultural implements |
| Soon after collection | 50 | 7 | — | — | — | DS, EP, ST | Timber, Fodder, Fuel |
| — | 70 | 10 onwards | — | — | — | DS, EP, ST | Fuel, Match-wood, Pulp |
| Feb.–July | 60–94 | 60 | 15 x 15 | 2–3 | July | DS, EP, ST | Timber, Fodder, Fuel, Medicinal |
| April | 30–45 | 30 | 15 x 15 | 2–3 | August | DS, EP, ST | Timber, Fodder |
| May | 80–90 | 21 | 10 x 10 | 1 | July | DS, EP, ST | Fuel, Fodder, Timber |
| March | 70 | 28–42 | 15 x 10 | 3 | July | DS, EP | Timber, Fuel, Soil Conservation |
| February | — | — | — | 24–36 | Jan.–Feb., July | DS, EP | -do- |
| June–July | 90 | 10–20 | — | 12–24 | June–July | DS, EP | Plywood, Fuel |

Contd.

| 1 | 2 | 3 | 4 | 5 | 6 | 7 | 8 | 9 |
|---|---|---|---|---|---|---|---|---|
| 33 | *Anacardium occidentale* | Cashew-nut | Dtr, Coastal | Mar.–June | 145–165 | — | MLL | Soak in water |
| 34 | *Anogeissus latifolia* | Axlewood (Bakli) | DSt, MSt, Mtr | Mar.–May | 135,800 | — | — | Soak in water |
| 35 | *Anogeissus pendula* | Kardhai | Dtr | Dec.–March | 96,400 (fruits) | — | — | Not required |
| 36 | *Anthocephalus chinensis* | Kadam | Mtr | Jan.–Feb Aug.–Oct. | 932,142– 2,721,400 | — | MLL | -do- |
| 37 | *Artocarpus chaplasha* | Chaplash | Mtr | June–Aug. | 2,000 | — | VSL | Not required |
| 38 | *Artocarps hirsutus* | Aini | Mtr, Wtr | May–July | 1,412 | — | SL | -do- |
| 39 | *Artocarpus heterophyllus* | Kathal | Mtr, Wtr | May–July | 43–50 | — | SL | -do- |
| 40 | *Azadirachta indica* | Neem | Dtr, Mtr | June–Aug. | 3,330 | RT | VSL | -do- |
| 41 | *Bambusa arundinacea* | Bans | Mtr | Feb.–July | 75,000– 105,000 | LT | MLL | Soak in water |
| 42 | *Barringtonia acutangula* | Hijal | MSt. | Aug.–Sept. | 1,412 (fruits) | — | SL | Not required |
| 43 | *Bauhinia purpurea* | Khairwal | Mtr | Jan.–May | 4,000– 5,000 | — | MLL | -do- |
| 44 | *Bauhinia racemosa* | Jhinjhora | St, Tr | Dec.–Mar. | 7,000– 9,000 | — | LL | Soak in water |
| 45 | *Bauhinia variegata* | Kachnar | Mtr, St, Tr | May–June | 2,800– 3,520 | — | MLL | Not required |
| 46 | *Betula alnoides* | Bhojpatra | T | Jan.–Mar. | 13,875,000 | RT | MLL | Stratification |
| 47 | *Betula cylindrostachys* | Birch | St, T | Jan.–Mar. | 14,000,000 | — | SL | Stratification |
| 48 | *Bischofia javanica* | Uriam | Mtr | Dec.–Mar. | 58,400– 1,04,000 | — | SL | — |
| 49 | *Bombax ceiba* | Semul | Mtr | Mar.–May | 21,430– 38,500 | — | MLL | Not required |
| 50 | *Butea monosperma* | Dhak | Dtr | May–July | 9,850– 14,790 | — | MLL | -do- |

| 10 | 11 | 12 | 13 | 14 | 15 | 16 | 17 |
|---|---|---|---|---|---|---|---|
| May–June | 90 | 50 | 15 x 15 | 12 | June | DS, EP, AL | Fuel, Fruits |
| April–May | Low | 15 | 15 x 15 | — | July | EP | Fuel, Fodder |
| June | Very low | 15 | 15 x 15 | 1 | August | EP | Fuel, Fodder |
| February | Fair | 21 | 20 x 20 | 4–5 | June–July | ST, EP | Timber, Plywood, Ornamental |
| July–Aug. | 80 | 14 | 8 x 8 | 1 or 12 | July–Aug. | DS, EP, ST | Timber, Fruit, Fodder |
| — | 60 | 42 | — | — | — | DS, EP | -do- |
| June–July | 75 | 21 | 20 x 20 | 12 | July | DS, EP | -do- |
| July | 100 | 21 | 15 x 15 | 12 | July | DS, EP, RS | Bark, leaf and seed (medicinal), Insecticidal, Fertiliser, Fodder, Timber |
| April | 50 | 14 | 15 x 15 | 24–36 | July | EP, R,O | In place of timber for house construction, Ladders, Boats and Rafts |
| — | 90–100 | 15–20 | 8 x 8 (fruits) | — | July–Oct. | DS, EP | Fuel |
| March–Apr. | 80–100 | 30 | 15 x 15 | — | — | DS, ST, EP | Gum, Fodder, Fuel |
| April–May | 60–95 | 21 | 15 x 15 | — | — | DS, ST, EP | Tannin, Fibre, Gum |
| May | 95 | 30 | 15 x 15 | 2–3 | June–July | DS, ST, EP | Cherry gum, Bark (dye and medicinal), Fodder, Flower buds eaten |
| Sept. | 10 | 56 | 10 x 10 | 7–8 | June–July | EP | Plywood, Furniture, Tool handles |
| Soon after collection | 10 | 6 | 10 x 10 | — | July | EP | Essential oils, Fuel, Timber, Charcoal |
| -do- | 70–90 | 14–20 | 15 x 15 | — | July | DS, EP, RC, BC | Timber, Fuel |
| May | 14–75 | 25 | 20 x 20 | 12 | Jan.–Feb. | DS, EP, ST, BC | Match, Plywood |
| May | 75–100 | 15 | 15 x 15 | — | July–Aug. | DS, ST, EP | Reclamation of saline soil, Host tree for lac-insect, Ruby-red gum, Cheap board wood, Well curbs, water scoops |

Contd.

| 1 | 2 | 3 | 4 | 5 | 6 | 7 | 8 | 9 |
|---|---|---|---|---|---|---|---|---|
| 51 | *Cassia fistula* | Amaltas | Mtr | Mar.–April | 6000–7090 | RT | VLL | Scarification, Hot water |
| 52 | *Cassia javanica* | Java cassia (Mazeli) | Mtr | Dec.–April | 5460–6400 | RT | VLL | –do– |
| 53 | *Cassia siamea* | Kassod | Mtr, Dtr | Mar.–April | 37,040 | RT | VLL | –do– |
| 54 | *Casuarina equisetifolia* | Beef wood (Sura) | Mtr Coastal | June–Dec. | 760,000 | LT | — | Soak in Water for 24 hrs. |
| 55 | *Cedrus deodara* | Deodar | T | Oct.–Nov. | 7000–9000 | LT | SL | Stratification |
| 56 | *Celtis australis* | Kharik | T,St | Oct.–Dec. | — | — | LL | –do– |
| 57 | *Chloroxylon swietenia* | Satinwood (Bhera) | Mtr | May–Aug. | 35,000–60,000 | — | SL–MLL | Not required |
| 58 | *Chukrasia velutina* | Chickrassi | Mtr | Jan.–Feb. | 100,000 | RT | MLL | –do– |
| 59 | *Cinnamomum camphora* | Camphor (Kapur) | Mtr,St | Oct.–Nov. | 5,000–11,000 | — | SL | Soak in luke-warm water for 24 hrs. |
| 60 | *Cinnamomum cecidodaphne* | Rohu (Gundroi) | St, T | Aug.–Nov. | 2000–3000 | — | VSL | — |
| 61 | *Cryptomeria japonica* | Suji (Dhupi) | T,St | July–Dec. | 338,000 | LT | MLL | Not required |
| 62 | *Cupressus torulosa* | Himalayan cypress | T,St | April–July | 201,786 | RT | MLL | Stratification |
| 63 | *Dalbergia latifolia* | Rosewood | Mtr | Jan.–Mar. | 18,500–40,000 | — | SL–MLL | Not required |
| 64 | *Dalbergia sissoo* | Shisham | Mtr | Nov.–Mar. | 53,000 | RT | MLL | –do– |
| 65 | *Dendrocalamus strictus* | Narbans | Dtr, Mtr | April–June | 32,000 | LT | MLL | Soak in water |
| 66 | *Diospyros melanoxyln* | Tendu | Dtr | April–June | 880–1410 | — | MLL | Alternate wetting and drying |
| 67 | *Diospyros peregrina* | Kala tendu | Mtr | May–Aug. | 786–895 | — | SL | — |
| 68 | *Dipterocarpus macrocarpus* | Hollong | Mtr | Feb.–Mar. | 120 (fruits) | — | VSL | Not required |

| 10 | 11 | 12 | 13 | 14 | 15 | 16 | 17 |
|---|---|---|---|---|---|---|---|
| Mar.–April | 22–60 | 1–2 | 15 × 10 | – | May–June | DS, ST, EP | Medicinal (Bark, Fruit and Leaves), Timber, Tannin (Root and Bark) |
| – | 60 | 20 | 15 × 10 | – | – | DS, ST, EP | Fuel, Ornamental |
| June–Aug. | 98 | 28 | 15 × 15 | 2 | July | DS, ST, EP | Fuel, Timber |
| Jan.–Feb. | 70 | 3–4 | 15 × 15 | 9 | June | EP | Timber, Fuel, Ornamental |
| Nov.–Dec. | 80–90 | 4 | 15 × 20 | 24–28 | October, March | DS, EP | Timber |
| Oct.–Nov. | 60–70 | 3–4 | 15 × 15 | 15 | Sept.–Oct. | DS, EP | Timber, Fuel, Fodder, Sports goods, Fruit— edible and medicinal, Utensils |
| Soon after collection | 70 | 2 | 15 × 15 | 12–24 | July | DS, EP, ST | Furniture, Building timber, Agricultural implements, Sleepers, Bobbins. |
| Jan.–Feb. | 70–90 | 7–10 | 30 × 20 | 4–5 | June–July | DS, EP, ST | Timber, Furniture, Decorative plywood |
| March | 47 | 200–360 | 20 × 20 | – | June–July | DS, ST, EP | Camphor oil, Medicinal |
| Soon after collection | 5–25 | 2 | 10 × 10 | – | July–Aug. | EP, ST | Cabinet making, Planks, Construction work, Essential oil |
| February | 80 | 21–28 | 10 × 10 | 12–24 | June or Dec.–Jan. | EP | Timber, Pulpwood |
| June | 50–60 | 90 | 10 × 10 | 12–24 | July | DS, EP | Timber, Furniture, Leaves (oil) |
| Mar.–April | 45–80 | 21 | 10 × 10 | 12 | Mar.– April | DS, EP, ST, RC, BC | Timber, Furniture, Cabinet |
| March | 90–100 | 15 | 10 × 10 | 3–4 | July–Aug. | DS, EP, ST, RC, BC | Timber, Furniture, Sleeper, Plywood, Fuel, Fodder |
| May–June | 25–80 | 10–30 | 15 × 15 | 1–2 | – | DS, EP, R,O | Cottage industry, Paper, Mathematical instrument |
| Mar.–April | 50–60 | 90 | 15 × 15 | – | – | DS, ST, EP | Timber, Fuel, Bidi industry |
| July | 12 | 12–21 days onward | 10 × 10 | – | July | DS, ST, EP | Gum, Tannin—dyeing, tanning and medicine. Seed-medicinal oil, Constructional wood |
| Soon after collection | 25–80 | 13 days onward | 15 × 15 | – | May–June | DS, EP | Timber, Plywood, Teachest, Sleepers |

Contd.

| 1 | 2 | 3 | 4 | 5 | 6 | 7 | 8 | 9 |
|---|---|---|---|---|---|---|---|---|
| 69 | *Dipterocarpus turbinatus* | Common gurjan | Mtr | May–June | 154 (fruits) | – | VSL | –do– |
| 70 | *Emblica officinalis* | Amla | Mtr,Dtr | Nov.–Feb. | 68,000– 89,000 | – | SL | Hot water |
| 71 | *Eucalyptus citriodora* | Lemon scented spotted gum | St, Mtr | April–June | 155,000 | LT | SL | Not required |
| 72 | *Eucalyptus globulus* | Blue gum | T | Mar.–May | 230,000 | LT | MLL | –do– |
| 73 | *Eucalyptus tereticornis* | Mysore gum | Dtr, Mtr | Sept.–Oct. | 367,400 | LT | – | –do– |
| 74 | *Exbucklandia populnea* | Pipli | Str, T | Dec.–Jan. | 270,000 | – | SL | – |
| 75 | *Fraxinus excelsa* | European Ash | T | – | – | LT | – | Stratification |
| 76 | *Fraxinus floribunda* | Angu | Str, T | Sept.–Oct. | 9,000 (fruits) | – | MLL | – |
| 77 | *Garuga pinnata* | Kharpat | Mtr | July–Oct. | 4,000– 5,000 | – | MLL | – |
| 78 | *Gmelina arborea* | Gamhar | Mtr | May–June | 2,500– 2,600 | RT, | MLL | Soak in water |
| 79 | *Grevillea robusta* | Silver oak | Str | June | 100,000 | LT, | LL | Stratification |
| 80 | *Hardwickia binata* | Anjan | Dtr | April–May | 3900 (fruits) | – | MLL | Not required |
| 81 | *Hevea brasiliensis* | Pararubber tree | Mtr | July–Sept. | – | – | SL | –do– |
| 82 | *Holoptelea integrifolia* | Kanju | Dtr, Mtr | April–May | 27,000 | – | SL | –do– |
| 83 | *Hymenodictyon excelsum* | Bartu | Tr | Dec.–Feb. | 142,000– 172,000 | RT, | MLL | Not required |
| 84 | *Juglans regia* | Walnut | T | Sept.–Oct. | 50–100 | – | SL | Stratification |
| 85 | *Kydia calycina* | Pula | Mtr | Dec.–Mar. | 32,000 | – | MLL | Not required |
| 86 | *Lagerstroemia parviflora* | Lendia | Mtr | Dec.–May | 28,000 | – | MLL | –do– |

| 10 | 11 | 12 | 13 | 14 | 15 | 16 | 17 |
|---|---|---|---|---|---|---|---|
| Soon after collection | 82 | 7–28 | 15 × 15 | – | – | DS | Timber, Railway sleepers |
| March | 40 | 30 | 10 × 10 | – | July | BC, EP, BS | Fruit, Hairdyes, Fodder, Tannin, Timber, Fuel |
| Sept.–Oct. | 100 | 60 | 10 × 10 | – | July | EP | Fuel, Oil-medicinal |
| Mar.–April | 30–80 | 10 | 10 × 10 | 12 | Aug.–Oct. | EP | Timber, Paper, Essential oil |
| Sept.–Oct. | 90 | 5–15 | 10 × 10 | 4–8 | July–Aug. | EP | Timber, Paper, Essential oil, Fuel, Plywood |
| March | 75 | 24 | 10 × 10 | 12–24 | July–Aug. | EP,ST | Soil Conservation, Timber, Furniture, Plywood, Ornamental |
| Nov.–Jan. | – | 48 | 15 × 15 | 24 | July–Aug., Feb.–March | EP | Timber, Furniture, Axe handles, Gum, Carriages, Boat building, Sports goods |
| Sept.–Oct., Mar.–April | 75 | 180–300 | 15 × 15 | 24 | – | EP | – |
| July | 44 | – | 15 × 15 | 2 | Aug.–Sept. | DS, EP, ST | Timber, Furniture, Bark–tannin, Fodder |
| Mar.–April | 13–85 | 10–15 | 15 × 20 | 1–2 | June | DS, EP, ST | Timber, Printing block, Musical instrument, Cart axles, Fuel, Medicinal |
| Nov.–Dec. | 60–80 | 21 | 15 × 15 | 6–7 | July | DS, EP | Gum, Resin, Timber, Fuel, Ornamental |
| Apr.–May | 79 | 28 | 10 × 10 | – | July | DS, EP | Timber, Posts and beams, Fibres for ropes, Oleo resins |
| Soon after collection | – | 7–21 | 15 × 15 | – | July | DS, EP, BS | Rubber |
| –do– | 60 | 10 | 10 × 10 | – | July | DS, EP, ST | Fuel, Charcoal, Brush back, Light furniture, Cabinets, Plywood, Fodder |
| Apr.–May | 30–35 | 60–120 | 15 × 15 | 2 | April–May | DS, EP, ST | Match wood |
| November | 70–80 | 35–50 | 20 × 20 | 18–36 | Nov.–Dec. | DS, EP | Timber, Gunstocks, Fruits |
| Soon after collection | 13–16 | 35–45 | 10 × 10 | 2–3 or 12 | July | DS, EP, ST, TS | Timber, Match box, Plywood, Fibre for coarse ropes, Light packing cases |
| Feb.–Mar. | Very poor | 28 | 10 × 10 | – | June | EP, ST | House posts, beams and rafters, Frames for doors and |

Contd.

| 1 | 2 | 3 | 4 | 5 | 6 | 7 | 8 | 9 |
|---|---|---|---|---|---|---|---|---|
| 87 | *Lagerstroemia speciosa* | Jarul | Mtr | Jan.–Feb. | 120,000 | — | LL | Soak in water |
| 88 | *Lophopetalum fimbriatum* | Narikeli, Rumu | Mtr | Aug.–Sept. | 8000 | — | VSL | — |
| 89 | *Machilus edulis* | Dudri | St | Nov.–Jan. | 29 | — | SL | — |
| 90 | *Machilus gammieana* | Lali | St | Nov.–Dec. | 282 | — | SL | — |
| 91 | *Madhuca longifolia* | Mahua | Dtr | June–Aug. | 450 | — | SL | Not required |
| 92 | *Mangifera indica* | Mango | Dtr, Mtr, Wtr | April–July | 10–237 | LT | VSL | Soak in water |
| 93 | *Melia azedarach* | Bakain | Dtr, Mtr | Jan.–Feb. | 750–917 | RT (Sealed Tin) | LL | Not required |
| 94 | *Mesua ferrea* | Nahor | MSt, Mtr, Wtr, WSt | July–Sept. | 260 | — | SL-MLL | Soak in water |
| 95 | *Michelia champaca* | Champa | Mtr, Wtr | Aug.–Sept. | 15,000 | LT | VSL | Stratification |
| 96 | *Michelia doltsopa* | Safed champ | WT | Nov. | 8,470–10,580 | LT | SL | — |
| 97 | *Michelia lanuginosa* | Gogay champ | St | Nov.–Dec. | 10,000 | LT | VSL | — |
| 98 | *Moringa oleifera* | Sanjna | Mtr | April–June | 8000–9000 | — | SL | Soak in water |
| 99 | *Morus alba* | Mulberry (Tul) | Mtr | May–June | 428,000–465,000 | — | VLL | Not required |
| 100 | *Morus indica* | Tutri | St | Mar.–April | 512,500 | — | VSL-SL | Stratification |
| 101 | *Morus laevigata* | Bola | St | April–June | 443,000 | — | SL | — |
| 102 | *Olea ferruginea* | Indian olive (Kan) | T | Sept.–Dec. | 9000 | RT | MLL–LL | Acid scarification |
| 103 | *Ougeinia oogeinensis* | Sandan | Dtr, Mtr | May–June | 28,000–33,000 | — | — | Soak in water |
| 104 | *Phoebe attenuata* | Bansum | Mtr | July | 3000 | — | SL | — |
| 105 | *Phoebe hainesiana* | Angania | Mtr | Sept.–Oct. | 900 | — | VSL–SL | — |
| 106 | *Picea smithiana* | Himalayan spruce | T | Oct.–Nov. | 63,928 | LT | MLL | — |

| 10 | 11 | 12 | 13 | 14 | 15 | 16 | 17 |
|---|---|---|---|---|---|---|---|
| | | | | | | | Windows, Carts and Boats, Sleepers |
| Feb.–Mar. | 90 | 10–30 | 15 × 15 | — | June | DS, EP, ST | Timber, Medicinal, Fodder, Furniture poles, Sleepers |
| Soon after collection | 40 | 42 | 10 × 10 | 15 | May | EP | Tools, Boxes, Teachests, Aeroplane frames |
| February | 50 | 180 | 10 × 10 | 12 | June | DS, EP, ST | Timber |
| February | 90 | 180 | 10 × 10 | — | July | DS, EP, ST | –do– |
| July–Aug. | High | 15 | 15 × 15 | 12 | July | DS, EP, ST | Timber, Fuel, Fruits and flowers edible, Oil |
| June–July | High | 30 | 20 × 20 | 12 | July | EP, RC | Edible fruits, Fatty oils, Plywood, Shoe heels |
| Feb.–May | 70–80 | 20–40 | 15 × 15 | 12 | July | DS, EP, ST, BC, RC | Box planks, Fuel, Paper pulp, Medicinal |
| Soon after collection | 90 | 60 | 10 × 10 | 9 or 12 | July | DS, EP | Timber, Fuel, Fatty oil, Resin |
| -do- | 70–80 | 60 | 15 × 15 | 12 | June–July, Dec.–Jan. | EP | Decorative timber, Linear, Fuel, Ornamental, Planks, Furniture, House building |
| -do- | 25–40 | 28–75 | 10 × 10 | 24 | April–May | EP | -do- |
| Soon after collection | 60 | 180 | 10 × 10 | 12 | June | EP | Timber, Fuel, Planks, Charcoal |
| — | 66 | 20–30 | 15 × 15 | 6 | July–Aug. | DS, BC | Edible fruits, Medicinal, Fodder |
| June | 70 | 45 | 15 × 15 | — | July | DS, EP, ST, BC | Edible fruits, Timber, Sports goods, Fodder |
| May | 40 | 90 | 15 × 15 | 2 | July | EP, BC, ST | Timber, Cultivation of silkworms |
| Soon after collection | 50–80 | 35 | 15 × 15 | 2 | Aug. or Feb. | EP, ST, BC | Timber, Furniture, Planks, Sports goods, Fodder |
| Sept.– Oct. | 45–65 | 28 | 20 × 20 | 18 | July–Aug. | DS, EP, RS, BC | Fodder, Fatty oil, Tool handles, Turnery articles, Walking sticks |
| May– June | 75 | 30 | 15 × 15 | — | July–Aug. | DS, EP, ST, RC | Cart carriage, Building construction, Fodder, Agricultural implements |
| Soon after collection | 65 | 120 | 15 × 15 | — | June | DS, EP | Timber, Furniture, Plywood, Fuel, Bobbins |
| — | 80 | 60–150 | 15 × 15 | 8 | June–July, Feb.-March | EP | Plywood, Planks, Fuel, Ceiling and Partition boards |
| June– July | 22–65 | 21 | 15 × 15 | 36 | June | EP | Planks, Packing cases, General fittings and joinery |

Contd.

| 1 | 2 | 3 | 4 | 5 | 6 | 7 | 8 | 9 |
|---|---|---|---|---|---|---|---|---|
| 107 | *Pinus elliottii* | Slash pine | St, Tr | Oct.–Nov. | 28,204–30,000 | LT | SL | Not required |
| 108 | *Pinus kesiya* | Khasia pine | St | Feb.–Mar. | 49,000–58,000 | LT | MLL | Not required |
| 109 | *Pinus patula* | Patula pine | St, T | – | 88,123–132,275 | LT | MLL | Soak in water |
| 110 | *Pinus roxburghii* | Chir | St | Feb.–April | 8800–12,300 | LT | MLL | Not required |
| 111 | *Pinus wallchiana* | Kail | MT | Sept.–Nov. | 19,240 | LT | VLL | -do- |
| 112 | *Pithecellobium dulce* | Jangli jalebi | Tr | April | 6700 | – | – | Soak in water |
| 113 | *Pongamia pinnata* | Karanj | Dtr | Mar.–May | 800–1500 | – | SL–MLL | Soak in water |
| 114 | *Prosopis chilensis* | Kikkar | Dtr | May–June | 12,500 | RT | LL–VLL | Scarification, soak in water |
| 115 | *Prosopis cineraria* | Jand | Dtr | June–Aug. | 25,000 | RT | LL | -do- |
| 116 | *Prunus nepalensis* | Saiong | T | Oct.–Nov. | 1350 | LT | SL | Stratification |
| 117 | *Pterocarpus dalbergioides* | Padauk | Mtr, Wtr | Jan.–Mar. | 1450 (fruits) | – | LL | Soak in water |
| 118 | *Pterocarpus marsupium* | Bijasal | Mtr | Dec.–April | 1590–1940 (fruits) | – | MLL | Soak in camphor water |
| 119 | *Pterocarpus santalinus* | Red sanders | Dtr | Feb.–April | 1000 (fruits) | – | MLL | Soak in water |
| 120 | *Pterocymbium tinctorium* | – | Tr, Islands region | Mar.–May | 4940 | – | VSL | Not required |
| 121 | *Pterygota alata* | Buddha narikal | Mtr | Feb.–April | 1060–1660 | – | – | – |

| 10 | 11 | 12 | 13 | 14 | 15 | 16 | 17 |
|---|---|---|---|---|---|---|---|
| Sept.–Oct. | 70–92 | 30 | 10 × 10 | 21 | July–Aug. | EP | Timber, Resin, Pulpwood |
| February | 87–100 | 20 | 10 × 10 | 4–6 | July | DS, EP | Timber, Resin, Pulpwood |
| Feb.–Mar. | 30–60 | 21–50 | 10 × 10 | 12 or 24 | Aug.–Sept. | EP | -do- |
| Soon after collection | 80–90 | 15–21 | 15 × 15 | 4 | August | DS, EP | -do- |
| Mar.–April | Poor | 50 | 15 × 15 | 24 | July–Aug. | EP | -do- |
| Feb.–Mar. | 45 | 35 | 15 × 15 | 5–9 | -do- | DS, BC, EP | Packing cases, Fuel, Fodder, Edible fruits, Fatty oil |
| July–Aug. | 80 | 30 | 15 × 20 | 10–12 | June–July | DS, ST, EP | Seeds—Fatty oil, Furniture, Turnery articles, Fuel, Fodder, Medicinal |
| March | 85–95 | 14 | 10 × 10 | 4 | July | DS, ST, EP | Fuel, Fodder, Small timber, Soil Conservation |
| June | 65 | 14 | 10 × 10 | 12 | July | DS, EP | Fodder, House construction, Fuel and Charcoal, Soil conservation |
| Mar.–April | 85 | 42 | 15 × 15 | — | June, Feb. | DS, EP | Planks, Fuel |
| May | 35–40 | 90 | 10 × 10 | 4 | June–July | EP | Billiard tables, Furniture, Musical instruments, Decorative timber, Ornamental |
| June | 40–90 | 56 | 10 × 10 | 4 | June–July | DS, ST, EP | Gum, Kino-medicinal, Railway carriages, Firewood, Fodder, Constructional purposes |
| May–June | 10–80 | 10 days onwards | 20 × 20 | 12 | July | EP, ST | Carving, Dye, Medicinal, Fuel, Fodder, Musical instruments |
| Soon after collection | 90 | 21–28 | 10 × 10 | 4 | July | DS, ST, EP | Match wood, Packing cases, Plywood, Fuel, Light furniture |
| — | 80–90 | 25 | — | 3–4 | July | EP | Ornamental, Timber |

Contd.

| 1 | 2 | 3 | 4 | 5 | 6 | 7 | 8 | 9 |
|---|---|---|---|---|---|---|---|---|
| 122 | *Quercus lineata* | Indian oak | T | Nov.–Dec. | 170 | LT | SL | Stratification |
| 123 | *Robinia pseudacacia* | Black locust | T | Sept.–Oct. | 30,000–50,000 | LT | — | Scarification |
| 124 | *Santalum album* | Chandan | Dtr | May–Aug. Oct.–Dec. | 7000–10,000 | — | MLL –LL | Soak in water |
| 125 | *Schima wallichii* | Needle wood | Mtr, Wtr, Tr | Feb.–Mar. | 300,000 | — | VSL | — |
| 126 | *Schleichera oleosa* | Kusum | Tr | June–July | 1410–2190 | — | VSL | Not required |
| 127 | *Shorea assamica* | Makai | Mtr, Wtr | Jan.–April | 2000 | LT | VSL | –do– |
| 128 | *Shorea robusta* | Sal | Mtr | May–July | 575–1000 (fruits) | LT | VSL | –do– |
| 129 | *Sterculia villosa* | Udal | Dtr | May–June | 5600–6000 | — | SL | Not required |
| 130 | *Sterculia urens* | Karaya | Dtr | April | 4410–6360 | — | SL | –do– |
| 131 | *Stereospermum suaveolens* | Paral | MSt | Dec.–Jan., April | 27,356 | — | SL | Soak in water |
| 132 | *Swietenia macrophylla* | Mahogany | MSt | Dec.–Mar. | 2150 | — | MLL | — |
| 133 | *Syzygium cumini* | Jaman | Mtr | June | 1200 | — | VSL | Not required |
| 134 | *Tamarindus indica* | Imli | Dtr | Mar.–April | 1800 | RT | LL | Soak in hot water |
| 135 | *Tectona grandis* | Teak | Dtr, Mtr | Nov.–Jan. | 1850–3100 (fruits) | RT | MLL –LL | Alternate wetting and drying |
| 136 | *Terminalia alata* | Laurel | Tr | Feb.–May | 13,000 | — | MLL | Soak in water |

| 10 | 11 | 12 | 13 | 14 | 15 | 16 | 17 |
|---|---|---|---|---|---|---|---|
| Feb.–Mar. | 40 | — | 15 x 15 | 24–60 | June | DS, EP | Fuel, Bark and Leaves—tannin |
| April–May | 80–90 | 30 | 10 x 10 | 12–24 | July | DS, EP | Agricultural implements, Tool handles, Sports goods, Boat ribs, Brackets, Sleepers, Light construction, Furniture |
| Feb.–Mar. | 75 | 90 | 20 x 20 | — | July–Aug. | DS, EP | Sandal oil, Carving work, Medicinal, Perfumery, Fodder, Ornamental |
| Soon after collection | Poor | 21 days onwards | 10 x 10 | 4 | June–July | EP | Plywood |
| –do– | 58 | 10–90 | 15 x 15 | — | July | DS, ST, RC, RS | Timber, Turnery articles, Tool handles, Constructional purposes, Sleepers, Beams |
| –do– | 83 | 20–68 | 15 x 15 | 12 | July–Aug. | DS, EP | Timber, Constructional purposes, Plywood, Paper pulp in mixture |
| –do– | 75–90 | 10–28 | 20 x 20 | 12 | July–Aug. | DS, EP, ST | Beams, Planks, Bridges, Gun—carriages, Carts, Railway sleepers |
| May–June | 50–70 | — | 20 x 20 | — | July–Aug. | DS, ST, EP | Fibres for ropes and coarse bags |
| Soon after collection | 88 | 15–20 | 20 x 20 | 4 | July | EP | Gum karaya, Timber, Toys, Match boxes and splints, Fuel, Seed—edible |
| Mar.–May | 45 | 90 | 10 x 10 | — | — | EP, DS, RS | Timber, Carriages, Wagons, Furniture, Fuel, Charcoal, Medicinal |
| Soon after collection | 20–40 | 90 | 10 x 10 | 4–9 | July | DS, EP, ST | Jetty piles, Furniture, Plywood |
| –do– | 90 | 30 | 15 x 15 | 12 | July | DS, EP, ST | Fruit—edible and medicinal, Timber, Plywood, Fuel, Fodder, Match wood |
| April | 66 | 15–20 | 15 x 15 | 24 | July | DS, EP | Fruit pulp—sauces and chutney, Medicinal, Fodder, Timber, Fuel, Charcoal |
| April | 10–60 | 15 days onwards | 15 x 15 | — | July–Aug. | ST, EP | Timber, Railway carriages, Wagons, Cabinets, Furniture, Ship building |
| April | 35–70 | 25 | 15 x 15 | — | July | DS, ST, EP | Telegraphic and electric poles, Constructional purposes, Agricultural implements, Tool handles, Railway wagons, Pulp, Fodder |

Contd.

| 1 | 2 | 3 | 4 | 5 | 6 | 7 | 8 | 9 |
|---|---|---|---|---|---|---|---|---|
| 137 | *Terminalia arjuna* | Arjun | Tr | Feb.–May | 775 | — | MLL | Soak in water |
| 138 | *Terminalia bellirica* | Bahera | Mtr | Nov.–Feb. | 423 | RT | MLL | Alternate wetting and drying |
| 139 | *Terminalia chebula* | Harra | Mtr, Tr | Jan.–Mar., Dec.–May | 141–220 (fruits) | RT | MLL | -do- |
| 140 | *Terminalia crenulata* | Laurel | Mtr, Dtr, WSt | Mar.–April | 500 (fruits) | RT | SL–MLL | — |
| 141 | *Terminalia myriocarpa* | Hollock | Mtr | Jan.–Feb. | 500,000 (fruits) | RT | MLL | — |
| 142 | *Toona ciliata* | Toon | Tr | April–June | 550,000 | LT | SL | Not required |
| 143 | *Trewia nudiflora* | Gutel | Tr | July, Sept.–Dec. | 5360–8200 | — | SL–MLL | Soak in water |
| 144 | *Tsuga dumosa* | Hemlock | T | Nov.–Dec. | 400,000 | — | VSL | — |
| 145 | *Xylia dolabriformis* | Jambu | Mtr | Dec.–Jan. | 4300 | — | LL | — |
| 146 | *Xylia xylocarpa* | Irul | Tr | Jan.–Mar. | 3300–4000 | — | MLL | Not required |
| 147 | *Ziziphus mauritiana* | Ber | Dtr, Mtr | Feb.–Mar. | 1240–1760 | — | LL | Soak in water |

| 10 | 11 | 12 | 13 | 14 | 15 | 16 | 17 |
|---|---|---|---|---|---|---|---|
| April–May | 50–60 | 50 | 15 × 15 | 9 | June–July | DS, ST, EP | Timber, Mine props, Plywood, Bark-tannin and oxalic acid, Fodder, Leaves for tasar silkworm |
| Mar.–April | 86–100 | 30–60 | 15 × 15 | 12 | June–July | DS, ST, EP | Timber, Pulp, Plywood, Fruit, Medicinal, Fodder |
| June–July | 60 | 90 | 15 × 15 | 12 | July | DS, EP | Fruits, Medicinal, Timber |
| Soon after collection | 70 | 15–35 | 8 × 8 | 2 | July | EP, DS | Timber, Fodder, Bark—oxalic acid |
| -do- | 20–30 | 28 | 10 × 10 | 12 | July | DS, EP | Timber, Plywood, Jute mill rollers, Fuel, Fodder |
| Soon after collection | 60–80 | 10–15 | 15 × 15 | 12 | July | DS, EP, ST | Timber, Plywood, Fodder, Ornamental |
| Feb.–Mar., May–June | 90–100 | 60 | 15 × 15 | — | June–July | DS, EP, ST | Timber, Plywood, Fuel, Medicinal |
| February | 80–90 | 45 | 10 × 10 | 14–15 | July | EP | Ornamental, Timber, Bobbins and Reels |
| June–July | 80 | 50 | 15 × 15 | 18 | July–Aug. | DS, EP, ST | Railway sleepers, Telegraph posts, Boat building, Agricultural implements, Carts, Tool handles |
| June–July | 80 | 12 | 15 × 15 | — | — | DS, ST, EP | Sleepers, Construction of boats, Ship building, Beams, Posts, Agricultural implements |
| April–May | 31–95 | 16–87 | 15 × 15 | 2–3 | July | DS, BC, EP | Fruits, Fuel, Tannin, Medicinal, Fodder |

# Abbreviations Used

## Abbreviations for Climatic Zone

DSt   : Dry-Subtropical
Dtr   : Dry-tropical
MSt   : Moist-Subtropical
St   : Subtropical
MT   : Moist-Temperate
WT   : Wet-Temperate
T   : Temperate
Mtr   : Moist-tropical
Tr   : Tropical
Wtr   : Wet-tropical

## Abbreviations of Seed Viability Period

VSL   : Very Short-Lived (< one month)
SL   : Short-Lived (1 to 6 months)
MLL   : Moderate Long-Lived
       (6 to 12 months)
LL   : Long-Lived (1 to 2 years)
VLL   : Very Long-Lived (> 2 years)

## Abbreviations for Seed Storage Method

RT   : Room Temperature
LT   : Low Temperature
      (Store in Refrigerator)

## Abbreviations for Planting Method

EP   : Entire Planting
DS   : Direct Sowing
BC   : Branch Cutting
ST   : Stump Planting
RS   : Root Suckers
AL   : Air Layering
BS   : Budded Stump
R   : Rhizome
O   : Offset
TS   : Tree Stumping
BG   : Budded Grafting
RC   : Root Cutting

# INDEX